Multidimensional Integral Representations

Alexander M. Kytmanov • Simona G. Myslivets

Multidimensional Integral Representations

Problems of Analytic Continuation

 Springer

Alexander M. Kytmanov
Institute of Mathematics and Computer
  Science
Siberian Federal University
Krasnoyarsk, Russia

Simona G. Myslivets
Institute of Mathematics and Computer
  Science
Siberian Federal University
Krasnoyarsk, Russia

Multidimensional Integral Representations
ISBN 978-3-319-37400-0   ISBN 978-3-319-21659-1   (eBook)
DOI 10.1007/978-3-319-21659-1

Mathematics Subject Classification (2010): 32A25, 32A26, 32A40, 32A50

Springer Cham Heidelberg New York Dordrecht London
© Springer International Publishing Switzerland 2015
Softcover reprint of the hardcover 1st edition 2015

Printed on acid-free paper

Springer International Publishing AG Switzerland is part of Springer Science+Business Media
(www.springer.com)

Foreword

This monograph is devoted to integral representations for holomorphic functions in several complex variables such as: Bochner–Martinelli, Cauchy–Fantappiè, Koppelman, etc. and their applications to analytic continuation functions with a one-dimensional property of holomorphic extension. This book also contains multidimensional boundary analogues of the Morera theorem.

Tel-Aviv, Israel
June 2015

Lev Aizenberg

Preface

The Bochner–Martinelli integral representation for holomorphic functions of several complex variables appeared in the works of Martinelli (1938) and Bochner (1943). It was the first essentially multidimensional representation with integration taking place over the whole boundary of the domain. This integral representation has a universal kernel (not depending on the form of the domain), like the Cauchy kernel in $\mathbb{C}^1$. However, in $\mathbb{C}^n$ when $n > 1$, the Bochner–Martinelli kernel is harmonic, but not holomorphic. For a long time, this circumstance hindered the wide application of the Bochner–Martinelli integral in multidimensional complex analysis.

Interest in the Bochner–Martinelli representation grew in the 1970s in connection with the increased attention to integral methods in multidimensional complex analysis. Moreover, it turned out that the very general Cauchy–Fantappiè representation suggested by Leray is easily obtained from the Bochner–Martinelli representation (Khenkin). Koppelman's representation for exterior differential forms, which has the Bochner–Martinelli representation as a special case, emerged at the same time.

The Cauchy–Fantappiè and Koppelman representations were extensively used in multidimensional complex analysis: yielding good integral representations for holomorphic functions, explicit solution of the $\bar{\partial}$-equation and estimates of this solution, uniform approximation of holomorphic functions on compact sets, etc.

In the early 1970s, it was shown that, notwithstanding the non-holomorphicity of the kernel, the Bochner–Martinelli representation holds only for holomorphic functions. In 1975, Harvey and Lawson obtained a result for odd-dimensional manifolds on spanning by complex chains; the Bochner–Martinelli formula lies at its foundation. In the 1980s and 1990s, the Bochner–Martinelli formula was successfully exploited in the theory of function of several complex variables: in multidimensional residues, in complex (algebraic) geometry, in questions of rigidity of holomorphic mappings, in finding analogues of Carleman's formula, etc.

The school of multidimensional complex analysis promoted by L.A. Aizenberg and A.P. Yuzhakov in Krasnoyarsk in the 1960s last century was involved in the development of the theory of integral representations and residues and their applications. A series of monographs on integral representations and residues by L.A. Aizenberg, Sh.A. Dautov, A.P. Yuzhakov, A.K Tsih, A.M. Kytmanov, and

N.N. Tarkhanov were published in the 1980s and 1990s. Over the 20 years since then, new results have been obtained and new areas of research explored.

Our monograph summarizes the results obtained by the authors in recent years, including in particular the studies on different families of complex lines and curves sufficient for analytic continuation of functions from the boundary of a bounded domain, multidimensional boundary analogues of the Morera theorem.

In a sense, this monograph is a sequel to an earlier book of one of the authors [45]. In any case, the first two chapters of our book are almost entirely taken from [45].

The results of the monograph were delivered as part of specialized courses at the Institute of Mathematics and Computer Science of the Siberian Federal University between 1995 and 2015.

Chapters are numbered throughout the monograph, sections are numbered throughout the chapters. All statements, comments, formulas, and examples are tied to the number of the respective section.

Krasnoyarsk, Russia   Alexander Kytmanov
June 2015   Simona Myslivets

Acknowledgements

The authors used the financial support of RFBR, grant 14-01-00544, and grant 14.Y26.31.0006 of the Russian Government to support research schools under the supervision of leading scientists in the Siberian Federal University.

Acknowledgements

Contents

Chapter 1
Multidimensional Integral Representations

Abstract The theory of integral representations is important in multidimensional complex analysis. It continues to develop rapidly and is finding new applications in multidimensional complex analysis, as well as in other areas of mathematics [see, for example, monographs Aizenberg and Yuzhakov (Integral Representations and Residues in Multidimensional Complex Analysis. AMS, Providence, 1983), Khenkin (Several Complex Variables I. Encyclopedia of Mathematical Sciences, vol. 7, pp. 19–116. Springer, New York, 1990), Krantz (Function Theory of Several Complex Variables, 2nd edn. Wadsworth & Brooks/Cole, Pacific Grove, 1992), Kytmanov (The Bochner–Martilnelli Integral and Its Applications. Birkhäuser Verlag, Basel, 1995), Rudin (Function Theory in the Unit Ball of $\mathbb{C}^n$. Springer, New York, 1980), Shabat (Introduction to Complex Analysis. Part 2: Functions of Several Complex Variables. AMS, Providence, 1992), Vladimirov (Methods of the Theory of Functions of Many Complex Variables. MIT Press, Cambridge, 1966)]. This chapter provides those integral representations, which are then used in other chapters. Of course, we do not have space to mention all integral formulas known at this time. We leave out of the scope of this book the formulas of integration by manifolds of smaller dimension (such as the multiple Cauchy formula). The theory of multidimensional residues will be used just a little in the final chapters. We will only dwell on the formulas where integration is performed over the entire boundary of domain. The presentation is designed to show the logic of proceeding from the classical Bochner–Green formula to the Khenkin–Ramirez formula that has found a number of important applications in multidimensional complex analysis.

1.1 The Bochner–Green Integral Representation

We consider an n-dimensional complex space $\mathbb{C}^n$ with the variables $z = (z_1, \ldots, z_n)$. If z and w are points in $\mathbb{C}^n$, then we write

$$\langle z, w \rangle = z_1 w_1 + \cdots + z_n w_n, \qquad |z| = \sqrt{\langle z, \bar{z} \rangle},$$

© Springer International Publishing Switzerland 2015

A.M. Kytmanov, S.G. Myslivets, *Multidimensional Integral Representations*,
DOI 10.1007/978-3-319-21659-1_1

1

where $\bar{z} = (\bar{z}_1, \ldots, \bar{z}_n)$. The topology in $\mathbb{C}^n$ is given by the metric $(z, w) \mapsto |z - w|$. If $z \in \mathbb{C}^n$, then

$$\operatorname{Re} z = (\operatorname{Re} z_1, \ldots, \operatorname{Re} z_n) \in \mathbb{R}^n, \qquad \operatorname{Im} z = (\operatorname{Im} z_1, \ldots, \operatorname{Im} z_n) \in \mathbb{R}^n.$$

We write $\operatorname{Re} z_j = x_j$ and $\operatorname{Im} z_j = y_j$, i.e., $z_j = x_j + iy_j$ for $j = 1, \ldots, n$. Thus $\mathbb{C}^n \simeq \mathbb{R}^{2n}$. Orientation of $\mathbb{C}^n$ is determined by the coordinate order $(x_1, \ldots, x_n, y_1, \ldots, y_n)$. Accordingly, the volume form dv is given by

$$dv = dx_1 \wedge \cdots \wedge dx_n \wedge dy_1 \wedge \cdots \wedge dy_n = dx \wedge dy$$

$$= \left(\frac{i}{2}\right)^n dz \wedge d\bar{z} = \left(-\frac{i}{2}\right)^n d\bar{z} \wedge dz.$$

As usual, a function f on an open set $U \subset \mathbb{C}^n$ belongs to the space $\mathscr{C}^k(U)$, i.e., $f \in \mathscr{C}^k(U)$, if f is k times continuously differentiable in U as $0 \leq k \leq \infty$, and $\mathscr{C}^0(U) = \mathscr{C}(U)$. If M is a closed set in $\mathbb{C}^n$, then $f \in \mathscr{C}^k(M)$, when f extends to some neighborhood U of M as a function of class $\mathscr{C}^k(U)$. We will also consider the space $\mathscr{C}^r(U)$ or $\mathscr{C}^r(M)$ when $r \geq 0$ is not necessarily an integer. A function $f \in \mathscr{C}^r(U)$ if it lies in the class $\mathscr{C}^{[r]}(U)$, where $[r]$ is the integer part of r, and all its derivatives of order $[r]$ satisfy the Hölder condition on U with the exponent $(r - [r])$.

The space $\mathscr{O}(U)$ consists of those functions f that are holomorphic on the open set U. When M is a closed set, $\mathscr{O}(M)$ consists of those functions f, that are holomorphic in some neighborhood of M (a different neighborhood for each function). A function f belongs to the space $\mathscr{A}(U)$, if f is holomorphic in U and continuous on the closure $\overline{U}$, i.e., $f \in \mathscr{O}(U) \cap \mathscr{C}(\overline{U})$.

We will also consider the Sobolev space $\mathscr{W}^s(U) = \mathscr{W}_2^s(U)$, $s \in \mathbb{N}$. This space consists of the measurable functions $f \in \mathscr{L}^2(U)$ such that all generalized derivatives $\partial^\alpha f$ of order up to s lie in the Lebesgue space $\mathscr{L}^2(U)$.

As usual, we will denote $\mathscr{D}(U)$ the space of infinitely differentiable functions with compact support on the open set U with the inductive limit topology, and $\mathscr{E}(U) = \mathscr{C}^\infty(U)$ will denote the space of infinitely differentiable functions on U with the topology of uniform convergence of the functions and all their derivatives on compact subsets of U.

A domain D in $\mathbb{C}^n$ has a boundary ∂D of class $\mathscr{C}^k$ (we write $\partial D \in \mathscr{C}^k$), if

$$D = \{z \in \mathbb{C}^n : \rho(z) < 0\},$$

where ρ is the real-valued function of class $\mathscr{C}^k$ in some neighborhood of the closure of D, and the differential $d\rho \neq 0$ on ∂D. If $k = 1$, then we say that D is a domain with a smooth boundary. We will call the function ρ a *defining function* for the domain D. The orientation of the boundary ∂D is induced by the orientation of D.

A domain D with *a piecewise-smooth boundary* ∂D will be understood as a smooth polyhedron, that is, a domain of the form

$$D = \{z \in \mathbb{C}^n : \rho_j(z) < 0, \ j = 1, \ldots, m\},$$

where the real-valued functions ρ_j are class $\mathscr{C}^1$ in some neighborhood of the closure $\overline{D}$, and for every set of distinct indices $j_1, \ldots, j_s$ we have $d\rho_{j_1} \wedge \cdots \wedge d\rho_{j_s} \neq 0$ on the set $\{z \in \mathbb{C}^n : \rho_{j_1}(z) = \cdots = \rho_{j_s}(z) = 0\}$. It is well known that Stokes' formula holds for such domains D and surfaces ∂D.

We denote a ball of radius $\varepsilon > 0$ with the center at the point $z \in \mathbb{C}^n$ by

$$B(z, \varepsilon) = \{\zeta \in \mathbb{C}^n : |\zeta - z| < \varepsilon\},$$

and its boundary by $S(z, \varepsilon)$, i.e., $S(z, \varepsilon) = \partial B(z, \varepsilon)$.

Consider the exterior differential form (*the Bochner–Martinelli kernel*) $U(\zeta, z)$ of type $(n, n - 1)$ given by

$$U(\zeta, z) = \frac{(n - 1)!}{(2\pi i)^n} \sum_{k=1}^{n} (-1)^{k-1} \frac{\bar{\zeta}_k - \bar{z}_k}{|\zeta - z|^{2n}} d\bar{\zeta}[k] \wedge d\zeta, \qquad (1.1.1)$$

where $d\bar{\zeta}[k] = d\bar{\zeta}_1 \wedge \cdots \wedge d\bar{\zeta}_{k-1} \wedge d\bar{\zeta}_{k+1} \wedge \cdots \wedge d\bar{\zeta}_n$, $d\zeta = d\zeta_1 \wedge \ldots \wedge d\zeta_n$. When $n = 1$, the form $U(\zeta, z)$ reduces to the Cauchy kernel $\dfrac{1}{2\pi i} \dfrac{d\zeta}{\zeta - z}$. It is clear that the form $U(\zeta, z)$ has the coefficients that are harmonic in $\mathbb{C}^n \setminus \{z\}$, and it is closed with respect to ζ, i.e., $d_\zeta U(\zeta, z) = 0$.

Let $g(\zeta, z)$ be a fundamental solution to the Laplace equation:

$$g(\zeta, z) = \begin{cases} -\dfrac{(n-2)!}{(2\pi i)^n} \dfrac{1}{|\zeta - z|^{2n-2}}, & n > 1, \\[2mm] \dfrac{1}{2\pi i} \ln |\zeta - z|^2, & n = 1. \end{cases} \qquad (1.1.2)$$

Then

$$U(\zeta, z) = \sum_{k=1}^{n} (-1)^{k-1} \frac{\partial g}{\partial \zeta_k} d\bar{\zeta}[k] \wedge d\zeta = (-1)^{n-1} \partial_\zeta g \wedge \sum_{k=1}^{n} d\bar{\zeta}[k] \wedge d\zeta[k], \qquad (1.1.3)$$

where the operator ∂ is given by

$$\partial = \sum_{k=1}^{n} d\zeta_k \frac{\partial}{\partial \zeta_k}.$$

We will write the Laplace operator Δ in the following form:

$$\Delta = \sum_{k=1}^{n} \frac{\partial^2}{\partial \zeta_k \partial \bar{\zeta}_k} = \frac{1}{4} \sum_{k=1}^{n} \left(\frac{\partial^2}{\partial x_k^2} + \frac{\partial^2}{\partial y_k^2} \right) = \frac{1}{4} \Delta^R.$$

If $\zeta_k = x_k + i y_k$, then

$$\frac{\partial}{\partial \zeta_k} = \frac{1}{2} \left(\frac{\partial}{\partial x_k} - i \frac{\partial}{\partial y_k} \right), \qquad \frac{\partial}{\partial \bar{\zeta}_k} = \frac{1}{2} \left(\frac{\partial}{\partial x_k} + i \frac{\partial}{\partial y_k} \right).$$

When $f \in \mathscr{C}^1(U)$, we define the differential form μ_f via

$$\mu_f = \sum_{k=1}^{n} (-1)^{n+k-1} \frac{\partial f}{\partial \bar{\zeta}_k} d\zeta[k] \wedge d\bar{\zeta}.$$

Theorem 1.1.1 (Green's Formula in a Complex Form) *Let D be a bounded domain in $\mathbb{C}^n$ with a piecewise-smooth boundary, and let $f \in \mathscr{C}^2(\overline{D})$. Then*

$$\int_{\partial D} f(\zeta) U(\zeta, z) - \int_{\partial D} g(\zeta, z) \mu_f(\zeta) + \int_D g(\zeta, z) \Delta f(\zeta) \, d\bar{\zeta} \wedge d\zeta$$

$$= \begin{cases} f(z), & z \in D, \\ 0, & z \notin \overline{D}, \end{cases} \qquad (1.1.4)$$

where the integral in (1.1.4) converges absolutely.

Proof Since

$$d_\zeta \big(f(\zeta) U(\zeta, z) - g(\zeta, z) \mu_f(\zeta) \big) + g(\zeta, z) \Delta f \, d\bar{\zeta} \wedge d\zeta = 0, \qquad (1.1.5)$$

Stokes's formula implies that (1.1.4) holds for $z \notin \overline{D}$. If $z \in D$, then from (1.1.5) and Stokes' formula we obtain that

$$\int_{\partial D} f(\zeta) U(\zeta, z) - \int_{\partial D} g(\zeta, z) \mu_f(\zeta) + \int_{D \backslash B(z, \varepsilon)} g(\zeta, z) \Delta f(\zeta) \, d\bar{\zeta} \wedge d\zeta$$

$$= \int_{S(z, \varepsilon)} f(\zeta) U(\zeta, z) - \int_{S(z, \varepsilon)} g(\zeta, z) \mu_f(\zeta)$$

for sufficiently small positive ε. When $n > 1$

$$\left| \int_{S(z,\varepsilon)} g(\zeta, z) \mu_f(\zeta) \right| \leq \frac{(n-2)!}{(2\pi)^n \varepsilon^{2n-2}} \int_{S(z,\varepsilon)} |\mu_f| \leq C\varepsilon,$$

i.e.,

$$\lim_{\varepsilon \to +0} \int_{S(z,\varepsilon)} g(\zeta, z) \mu_f(\zeta) = 0.$$

(The argument for $n = 1$ is analogues.) However,

$$\int_{S(z,\varepsilon)} f(\zeta) U(\zeta, z) = \frac{(n-1)!}{(2\pi i)^n \varepsilon^{2n}} \int_{S(z,\varepsilon)} f(\zeta) \sum_{k=1}^{n} (-1)^{k-1} (\bar{\zeta}_k - \bar{z}_k) d\bar{\zeta}[k] \wedge d\zeta$$

$$= \frac{(n-1)!}{(2\pi i)^n \varepsilon^{2n}} \int_{B(z,\varepsilon)} \left[nf(\zeta) + \sum_{k=1}^{n} \frac{\partial f}{\partial \bar{\zeta}_k} (\bar{\zeta}_k - \bar{z}_k) \right] d\bar{\zeta} \wedge d\zeta.$$

Since

$$\lim_{\varepsilon \to +0} \frac{1}{\varepsilon^{2n}} \int_{B(z,\varepsilon)} \sum_{k=1}^{n} \left(\frac{\partial f}{\partial \bar{\zeta}_k} (\bar{\zeta}_k - \bar{z}_k) \right) d\bar{\zeta} \wedge d\zeta = 0,$$

we have,

$$\lim_{\varepsilon \to +0} \int_{S(z,\varepsilon)} f(\zeta) U(\zeta, z)$$

$$= \lim_{\varepsilon \to +0} \frac{n!}{(2\pi i)^n \varepsilon^{2n}} \int_{B(z,\varepsilon)} f(\zeta) d\bar{\zeta} \wedge d\zeta = \lim_{\varepsilon \to +0} \frac{n!}{\pi^n \varepsilon^{2n}} \int_{B(z,\varepsilon)} f(\zeta) dv = f(z)$$

(by the mean-value theorem). $\qquad \square$

1.2 The Bochner–Martinelli Integral Representation

Let us formulate some consequences of the Bochner–Green formula (1.1.1) for various classes of functions f.

Corollary 1.2.1 (Bochner [16]) *Let D be a bounded domain in $\mathbb{C}^n$ with a piecewise-smooth boundary, and let f be a harmonic function in D of class $\mathscr{C}^1(\overline{D})$.*

Then

$$\int_{\partial D} f(\zeta)U(\zeta, z) - \int_{\partial D} g(\zeta, z)\mu_f(\zeta) = \begin{cases} f(z), & z \in D, \\ 0, & z \notin \overline{D}. \end{cases} \qquad (1.2.1)$$

Corollary 1.2.2 (Koppelman [35]) *Let D be a bounded domain in $\mathbb{C}^n$ with a piecewise-smooth boundary, and let f be a function in $\mathscr{C}^1(\overline{D})$. Then*

$$\int_{\partial D} f(\zeta)U(\zeta, z) - \int_D \bar{\partial} f(\zeta) \wedge U(\zeta, z) = \begin{cases} f(z), & z \in D, \\ 0, & z \notin \overline{D}, \end{cases} \qquad (1.2.2)$$

where

$$\bar{\partial} = \sum_{k=1}^n d\bar{\zeta}_k \frac{\partial}{\partial \bar{\zeta}_k},$$

and the integral in (1.2.2) converges absolutely.

Formula (1.2.2) is the Bochner–Martinelli formula for smooth functions.

Proof Suppose first that $f \in \mathscr{C}^2(\overline{D})$. We transform the integral

$$\int_D \bar{\partial} f(\zeta) \wedge U(\zeta, z) = \int_D \sum_{k=1}^n \frac{\partial f}{\partial \bar{\zeta}_k} \frac{\partial g}{\partial \zeta_k} d\bar{\zeta} \wedge d\zeta = \int_D \partial_\zeta g \wedge \mu_f$$

$$= \int_D d_\zeta(g\mu_f) - \int_D g\Delta f \, d\bar{\zeta} \wedge d\zeta = \int_{\partial D} g\mu_f - \int_D g\Delta f \, d\bar{\zeta} \wedge d\zeta$$

(here we have applied Stokes's formula, since all the integrals converge absolutely). Then for $z \in D$, formula (1.1.4) implies that

$$\int_D \bar{\partial} f(\zeta) \wedge U(\zeta, z) = \int_{\partial D} f(\zeta)U(\zeta, z) - f(z).$$

Now if $f \in \mathscr{C}^1(\overline{D})$, we obtain (1.2.2) by approximating f (in the metric of $\mathscr{C}^1(\overline{D})$) by functions of class $\mathscr{C}^2(\overline{D})$. ☐

Corollary 1.2.3 (Bochner [16] and Martinelli [62]) *If D is a bounded domain in $\mathbb{C}^n$ with a piecewise-smooth boundary, and f is a holomorphic function in D of class $\mathscr{C}(\overline{D})$, then*

$$\int_{\partial D} f(\zeta)U(\zeta, z) = \begin{cases} f(z), & z \in D, \\ 0, & z \notin \overline{D}. \end{cases} \qquad (1.2.3)$$

Formula (1.2.3) was obtained by Martinelli, and later separately by Bochner by different methods. It is the first integral representation for holomorphic functions in $\mathbb{C}^n$ where the integration is carried out over the whole boundary of the domain. By now this formula has become classical and found its place in many textbooks on multidimensional complex analysis (see, for example, [9, 39, 45, 69, 71, 73, 81]).

Formula (1.2.3) reduces to Cauchy's formula when $n = 1$, but unlike Cauchy's formula, the kernel in (1.2.3) is not holomorphic in z and ζ when $n > 1$. By splitting the kernel $U(\zeta, z)$ into real and imaginary parts, it is easy to show that

$$\int_{\partial D} f(\zeta) U(\zeta, z)$$

is the sum of the double-layer potential and the tangential derivative of a single-layer potential. Namely, Martinelli observed that if two continuous vector fields $v(\zeta)$ and $s(\zeta)$ are chosen on the boundary of a bounded domain, provided $v = is$ (v is the field of outer unit normals to ∂D), then the restriction of the kernel $U(\zeta, z)$ on ∂D coincides with

$$\left(\frac{\partial}{\partial v} + i \frac{\partial}{\partial s} \right) g(\zeta, z) d\sigma.$$

Consequently, the Bochner–Martinelli integral inherits some of the properties of the Cauchy integral and some of the properties of the double-layer potential. It differs from the Cauchy integral in not being a holomorphic function, and it differs from the double-layer potential in having a somewhat worse boundary behavior. At the same time, it establishes a relation between harmonic and holomorphic functions in $\mathbb{C}^n$ when $n > 1$.

Formula (1.2.2) implies the jump theorem for the Bochner–Martinelli integral. Let D be a bounded domain in $\mathbb{C}^n$ with piecewise-smooth boundary, and let f be a function in $\mathscr{C}^1(\overline{D})$. We consider the Bochner–Martinelli integral

$$Mf(z) = \int_{\partial D} f(\zeta) U(\zeta, z), \quad z \notin \partial D. \tag{1.2.4}$$

We will write $M^+f(z)$ for $z \in D$ and $M^-f(z)$ for $z \notin \overline{D}$. The function $Mf(z)$ is a harmonic function for $z \notin \partial D$ and $Mf(z) = O(|z|^{1-2n})$ as $|z| \to \infty$.

Corollary 1.2.4 *Under these conditions the function M^+f has a continuous extension on $\overline{D}$, the function M^-f has a continuous extension on $\mathbb{C}^n \setminus D$ and*

$$M^+f(z) - M^-f(z) = f(z), \quad z \in \partial D. \tag{1.2.5}$$

Formula (1.2.5) is the simplest jump formula for the Bochner–Martinelli integral. There exist many jump theorems for different classes of functions: Hölder functions [60], continuous functions [19, 30], integrable functions [42, 43], distributions [18], hyperfunctions [54].

Later on we will need formula (1.2.3) for the Hardy spaces $\mathcal{H}^p(D)$, so we now recall some definitions [34, 76]. Let D be a bounded domain and suppose that ∂D is a connected Lyapunov surface, i.e., $\partial D \in \mathscr{C}^{1+\alpha}$, $\alpha > 0$. It is known that the Green function (for the Laplace equation) $G(\zeta, z)$ has a good boundary behavior in such domains: for fixed $z \in D$, the function $G(\zeta, z) \in \mathscr{C}^{1+\alpha}(\overline{D})$.

We say that a holomorphic function f belongs to $\mathcal{H}^p(D)$ (where $p > 0$), if

$$\sup_{\varepsilon>0} \int_{\partial D} |f(\zeta - \varepsilon \nu(\zeta))|^p \, d\sigma < \infty,$$

were $d\sigma$ is the surface area element on ∂D and $\nu(\zeta)$ is the outer unit normal vector to the surface ∂D. A holomorphic function f belongs to $\mathcal{H}^\infty(D)$, if $\sup_D |f(z)| < \infty$.

The class $\mathcal{H}^p(D)$ may also be defined in the following way. Let domain $D = \{z \in \mathbb{C}^n : \rho(z) < 0\}$ for the defining function ρ, and let $D_\varepsilon = \{z \in D : \rho(z) < -\varepsilon\}$ for $\varepsilon > 0$. A holomorphic function $f \in \mathcal{H}^p(D)$, if

$$\sup_{\varepsilon>0} \int_{\partial D_\varepsilon} |f(\zeta)|^p \, d\sigma_\varepsilon < \infty.$$

As is shown in [76], this definition does not depend on the choice of the smooth defining function ρ.

Corollary 1.2.5 *If $p \geq 1$ then formula (1.2.3) holds for the function $f \in \mathcal{H}^p(D)$.*

Proof If $f \in \mathcal{H}^p(D)$ for $p \geq 1$, then f has a normal boundary values almost everywhere on ∂D (see [34, 76]) producing a function of class $\mathscr{L}^p(\partial D)$ (we denote these boundary values again by f). Moreover, the function f can be reconstructed in D from its boundary values by's

$$f(z) = \int_{\partial D} f(\zeta) P(\zeta, z) \, d\sigma,$$

(where $P(\zeta, z)$ is the Poisson kernel for D). Since the Green function $G(\zeta, z) = g(\zeta, z) + h(\zeta, z)$, where for fixed $z \in D$ the function $h(\zeta, z)$ is harmonic in D of class $\mathscr{C}^{1+\alpha}(\overline{D})$, we have

$$P(\zeta, z) \, d\sigma = U(\zeta, z)\big|_{\partial D} + \sum_{k=1}^{n} (-1)^{k-1} \frac{\partial h}{\partial \zeta_k} \, d\bar{\zeta}[k] \wedge d\zeta \big|_{\partial D}.$$

Since the differential form

$$\sum_{k=1}^{n} (-1)^{k-1} \frac{\partial h}{\partial \zeta_k} \, d\bar{\zeta}[k] \wedge d\zeta$$

is closed, we have

$$\int_{\partial D} f(\zeta) \sum_{k=1}^{n} (-1)^{k-1} \frac{\partial h}{\partial \zeta_k} \, d\bar{\zeta}[k] \wedge d\zeta$$

$$= \int_D f(\zeta) d \left(\sum_{k=1}^{n} (-1)^{k-1} \frac{\partial h}{\partial \zeta_k} \, d\bar{\zeta}[k] \wedge d\zeta \right) = 0.$$

Consequently, formula (1.2.3) holds for $f \in \mathscr{H}^p(D)$. $\qquad\square$

1.3 The Cauchy–Fantappiè Integral Representation

1.3.1 The Leray (Cauchy–Fantappiè) Integral Representation

We start by noting that the Cauchy–Fantappiè formula, which was obtained by Leray [58, 59], can be derived from the Bochner–Martinelli formula (1.2.3).

Let D be a bounded domain with a piecewise-smooth boundary, and suppose that for a fixed point $z \in D$ there is defined on ∂D a continuously differentiable vector-valued function

$$\eta(\zeta) = \big(\eta_1(\zeta), \ldots, \eta_n(\zeta) \big)$$

such that

$$\sum_{k=1}^{n} (\zeta_k - z_k) \eta_k(\zeta) = 1, \quad \zeta \in \partial D.$$

Theorem 1.3.1 (Leray) *Every function $f \in \mathscr{A}(D)$ satisfies the equation*

$$f(z) = \frac{(n-1)!}{(2\pi i)^n} \int_{\partial D} f(\zeta) \omega'(\eta) \wedge d\zeta, \quad z \in D, \tag{1.3.1}$$

where

$$\omega'(\eta) = \sum_{k=1}^{n} (-1)^{k-1} \eta_k \, d\eta[k]. \tag{1.3.2}$$

Proof Khenkin's proof goes as follows. In the space $\mathbb{C}^{2n}$ of the variables

$$(\eta, \zeta) = (\eta_1, \ldots, \eta_n, \zeta_1, \ldots, \zeta_n)$$

consider the analytic hypersurface

$$L_z = \left\{(\eta, \zeta) \in \mathbb{C}^{2n} : \sum_{k=1}^{n} (\zeta_k - z_k)\eta_k(\zeta) = 1\right\},$$

on which the form $\omega'(\eta) \wedge d\zeta$ is closed. The two cycles

$$\Gamma_1 = \{(\eta, \zeta) : \zeta \in \partial D, \eta_j = (\bar{\zeta}_j - \bar{z}_j)|\zeta - z|^{-2}, j = 1, \ldots, n\}$$

and

$$\Gamma_2 = \{(\eta, \zeta) : \zeta \in \partial D, \eta_j = \eta_j(\zeta), j = 1, \ldots, n\}$$

in L_z are homotopic in L_z, the homotopy being given by the formula

$$\tilde{\eta}_j = t\frac{\bar{\zeta}_j - \bar{z}_j}{|\zeta - z|^2} + (1 - t)\eta_j(\zeta), \quad 0 \le t \le 1.$$

That is, these are homologous cycles. Consequently,

$$\int_{\Gamma_1} f(\zeta)\omega'(\eta) \wedge d\zeta = \int_{\Gamma_2} f(\zeta)\omega'(\eta) \wedge d\zeta$$

when f is a holomorphic function. But

$$\omega'\left(\frac{\bar{\zeta}_1 - \bar{z}_1}{|\zeta - z|^2}, \ldots, \frac{\bar{\zeta}_n - \bar{z}_n}{|\zeta - z|^2}\right) = \frac{(2\pi i)^n}{(n-1)!} U(\zeta, z).$$

Hence (1.3.1) follows.   □

Differential form (1.3.2) is called the *Leray form*. The Cauchy–Fantappiè representation has turned out to be very useful, and it has many applications in multidimensional complex analysis.

1.3.2 The Khenkin–Ramirez Integral Representation

The bounded domain D is called *a strongly pseudo-convex domain* if there exists a neighborhood $U \supset \overline{D}$ and a real-valued function $\rho \in \mathscr{C}^2(U)$ such that

$$D = \{z \in U : \rho(z) < 0\},$$

where grad $\rho \neq 0$ on ∂D and the function ρ is *a strongly plurisubharmonic function* in U, i.e., the inequality

$$\sum_{j,k=1}^{n} \frac{\partial^2 \rho(z)}{\partial z_j \partial \bar{z}_k} w_j \bar{w}_k > 0$$

holds for $z \in U$ and all $w \in \mathbb{C}^n$, $w \neq 0$.

Strongly pseudo-convex domains play an important role in multidimensional complex analysis. Any domain of holomorphy can be approximated from inside by these domains [9, Sect. 25].

Here are some auxiliary notations and statements we will need in the future. We denote

$$D_\delta = \{z \in U : \rho(z) < \delta\}, \quad V_\delta = \{z \in U : |\rho(z)| < \delta\},$$

$$U_{\varepsilon,\delta} = \{(\zeta, z) : \zeta \in V_\delta, z \in D_\delta, |\zeta - z| < \varepsilon\},$$

where $\varepsilon > 0$, $\delta > 0$. Let $\mathscr{C}^1(V_\delta, \mathscr{H})$ be the space of functions of class $\mathscr{C}^1$ in V_δ with values in space $\mathscr{H}$.

Lemma 1.3.1 ([9]) *For every strongly pseudo-convex domain D there exist positive constants ε, δ and functions $F \in \mathscr{C}^1(U_{\varepsilon,\delta})$, $G \in \mathscr{C}^1(U_{\varepsilon,\delta})$, $\Phi \in \mathscr{C}^1(V_\delta, \mathscr{O}(D_\delta))$ such that:*

1. $\Phi = FG$ on $U_{\varepsilon,\delta}$; $F(z,z) = 0$; $|G| > \delta$ on $U_{\varepsilon,\delta}$; $|\Phi| > \delta$ outside $U_{\varepsilon,\delta}$;
2. We have the inequality

$$2 \operatorname{Re} F(\zeta, z) \geq \rho(\zeta) - \rho(z) + \gamma |\zeta - z|^2, \quad \gamma > 0;$$

on $U_{\varepsilon,\delta}$
3. $\left| d_\zeta F(\zeta, z) \right|_{\zeta=z} = \left| d_z F(\zeta, z) \right|_{\zeta=z} = \partial \rho.$

Proof See the monograph Aizenberg and Yuzhakov [9, Lemma 10.1].

Lemma 1.3.2 ([9]) *Let $D = \{z \in U : \rho(z) < 0\}$ be a strongly pseudo-convex domain. For every point $\tilde{\zeta} \in \partial D$ there is a biholomorphic map of the neighborhood $\tilde{U}_{\tilde{\zeta}}$ of the point $\tilde{\zeta}$ to a neighborhood W of zero in space $\mathbb{C}^n_w$, such that the mapping domain $D \cap \tilde{U}_{\tilde{\zeta}}$ is biholomorphically equivalent to the convex domain in $\mathbb{C}^n_w$, while the inverse mapping turns the strongly plurisubharmonic function ρ into a strongly convex function.*

Proof Parallel translation can be arranged so that the point $\tilde{\zeta}$ is zero, then in some neighborhood of zero the Taylor expansion is valid

$$\rho(z) = 2 \operatorname{Re} \sum_{j=1}^{n} \frac{\partial \rho(0)}{\partial z_j} z_j + \frac{1}{2} \sum_{j,k=1}^{n} \frac{\partial^2 \rho(0)}{\partial z_j \partial \bar{z}_k} z_j \bar{z}_k + \operatorname{Re} \sum_{j,k=1}^{n} \frac{\partial^2 \rho(0)}{\partial z_j \partial z_k} z_j z_k + o(|z|^2)$$

as $z \to 0$.

By the implicit function theorem the local coordinates in a smaller neighborhood of zero can be assumed to be given by the function

$$w_1(z) = 2 \sum_{j=1}^{n} \frac{\partial \rho(0)}{\partial z_j} z_j + \sum_{j,k=1}^{n} \frac{\partial^2 \rho(0)}{\partial z_j \partial z_k} z_j z_k$$

and any linear functions $w_2, \ldots, w_n$ being the coordinates of the complex hyperplane

$$\left\{ z \in \mathbb{C}^n : \frac{\partial \rho(0)}{\partial z_1} z_1 + \ldots + \frac{\partial \rho(0)}{\partial z_n} z_n = 0 \right\}.$$

Strong pseudoconvexity ρ at 0 now means that

$$\sum_{j,k=1}^{n} a_{jk} \xi_j \bar{\xi}_k \geq \gamma |\xi|^2,$$

where $\gamma > 0$ and $a_{jk} = \dfrac{\partial^2 \rho(0)}{\partial w_j \partial \bar{w}_k}$.

In the new local coordinates of w the function ρ is of the form

$$\rho(w) = \operatorname{Re} w_1 + \sum_{j,k=1}^{n} a_{jk} w_j \bar{w}_k + o(|w|^2). \qquad (1.3.3)$$

Function (1.3.3), as is easily seen, is strongly convex in the coordinates of w. If $\varepsilon > 0$ is sufficiently small, then $W \cap D$ is a convex domain in the coordinates of w, where

$$W = \{ w \in \mathbb{C}^n : |w| < \varepsilon \}.$$

$\square$

Remark 1.3.1 If we first perform a unitary transformation and make a shift so that the plane $\operatorname{Re} z_1 = 0$ is now a tangent plane at the point 0, and then repeat the whole procedure described in the previous lemma, we will find that locally the domain D is given by the function

$$\rho(w) = \operatorname{Re} w_1 + \sum_{j,k=2}^{n} a_{jk} w_j \bar{w}_k + o(|w|^2).$$

Applying unitary transformation and stretching to the last equation, we can get a function ρ that will have the form [15, Chap. 6, Sect. 4]

$$\rho(w) = \operatorname{Re} w_1 + \sum_{k=2}^{n} |w_k|^2 + o(|w|^2). \qquad (1.3.4)$$

Lemma 1.3.3 ([9]) *Under the conditions of Lemma 1.3.1*

$$\Phi(\zeta, z) = \sum_{k=1}^{n} P_k(\zeta, z)(\zeta_k - z_k), \tag{1.3.5}$$

where $P_k(\zeta, z) \in \mathscr{C}^1(V_\delta, \mathscr{O}(D_\delta))$, $k = 1, \ldots, n$, $\zeta \in V_\delta$, $z \in D_\delta$.

Proof We denote

$$C(\zeta, z, w) = \Phi(\zeta, z) - \Phi(\zeta, w),$$

then

$$C(\zeta, z, w) \in \mathscr{C}^1(V_\delta, \mathscr{O}(D_\delta \times D_\delta)).$$

For a fixed $\zeta \in V_\delta$ the function C belongs to the ideal J of holomorphic functions equal to zero on the set

$$\{(z, w) : z \in D_\delta, \ w \in D_\delta, \ z = w\}.$$

By the Hefer theorem (see, for example, [9, Sect. 25]) the ideal J has generators

$$w_1 - z_1, \ldots, w_n - z_n$$

and decomposition of $C(\zeta, z, w)$ for these generators can be made continuously differentiable to the parameter ζ (see [9, Theorem 25.2']), i.e., there are functions

$$Q_k(\zeta, z, w) \in \mathscr{C}^1(V_\delta, \mathscr{O}(D_\delta \times D_\delta)), \quad k = 1, \ldots, n,$$

such that

$$C(\zeta, z, w) = \sum_{k=1}^{n} Q_k(\zeta, z, w)(w_k - z_k).$$

Note that $F(\zeta, \zeta) = 0$, hence $\Phi(\zeta, \zeta) = 0$, therefore

$$\Phi(\zeta, z) = C(\zeta, z, \zeta) = \sum_{k=1}^{n} P_k(\zeta, z)(\zeta_k - z_k),$$

where $P_k(\zeta, z) = Q_k(\zeta, z, \zeta)$, $k = 1, \ldots, n$. So formula (1.3.5) holds. $\qquad\square$

Consider the Leray form (1.3.2)

$$\omega'(\eta) = \sum_{j=1}^{n} (-1)^{j-1} \eta_j d\eta[j]$$

for a given smooth function $\eta = \eta(\zeta, z, \lambda)$ with values in $\mathbb{C}^n$, where

$$(\zeta, z, \lambda) \in \mathbb{C}^n \times \mathbb{C}^n \times \mathbb{R}.$$

Then we can write

$$\left(\omega'(\eta)\right) \wedge d\zeta \wedge dz = \left(\sum_{q=0}^{n} \omega'_q(\eta)\right) \wedge d\zeta \wedge dz,$$

where $\omega'_q(\eta)$ is the differential form of type $(n - q - 1)$ by $d\bar{\zeta}$ and $d\lambda$ and type q by $d\bar{z}$. In particular, the form ω'_0 is a form of type 0 by $d\bar{z}$ and type $(n - 1)$ by $d\bar{\zeta}$ and $d\lambda$. In what follows we will assume that $\omega'_{-1} = 0$. We note the obvious identity satisfied by the form ω'_q [33, Sect. 8.2]:

$$d_\lambda \omega'_q + \bar{\partial}_\zeta \omega'_q + \bar{\partial}_z \omega'_{q-1} = 0. \tag{1.3.6}$$

From Theorem 1.3.1 and Lemma 1.3.3 we get the Khenkin–Ramirez integral formula (see [33, Sect. 4.2]).

Theorem 1.3.2 (Khenkin, Ramirez) *For any function $f \in \mathscr{A}(D)$ the following integral representation is true*

$$f(z) = \frac{(n-1)!}{(2\pi i)^n} \int_{\partial D} f(\zeta) \frac{\omega'_0(P(\zeta, z)) \wedge d\zeta}{[\Phi(\zeta, z)]^n}, \quad z \in D, \tag{1.3.7}$$

where the vector function $P(\zeta, z) = \bigl(P_1(\zeta, z), \ldots, P_n(\zeta, z)\bigr)$.

Formula (1.3.7) is one of the most successful realizations of the general Cauchy–Fantappiè formula in multidimensional complex analysis (see [33, 71]).

1.3.3   The Cauchy–Szegö (Hua Loken) Integral Representation

Let B be a unit ball in $\mathbb{C}^n$, i.e.,

$$B = \{z \in \mathbb{C}^n : |z| < 1\}$$

then its boundary S has the form

$$S = \{z \in \mathbb{C}^n : |z| = 1\}.$$

We define the Cauchy–Szegö kernel $K(\zeta, z)$ for the ball by the formula

$$K(\zeta, z) = \frac{(n-1)!}{(2\pi i)^n} \frac{1}{(1 - \langle \bar{\zeta}, z \rangle)^n}.$$

It can also be written as a matrix product: if we assume z to be a column vector, then $\langle \bar{\zeta}, z \rangle = \bar{\zeta}^T \cdot z$, where the superscript T denotes the transpose of the matrix. We define the differential form $\sigma(\zeta)$ by the formula

$$\sigma(\zeta) = \sum_{k=1}^{n} (-1)^{k-1} \bar{\zeta}_k \, d\bar{\zeta}[k] \wedge d\zeta,$$

where $d\zeta = d\zeta_1 \wedge \ldots \wedge d\zeta_n$, and $d\bar{\zeta}[k]$ is obtained from $d\bar{\zeta}$ by removing the differential $d\bar{\zeta}_k$. On the boundary of the ball the restriction of the form $\sigma(\zeta)$ coincides up to a constant with the Lebesgue boundary measure for S.

Theorem 1.3.3 (Hua Loken [32])   *If a function* $f \in \mathscr{A}(B)$, *then*

$$f(z) = \int_S f(\zeta) \, K(\zeta, z) \, \sigma(\zeta), \quad z \in B. \tag{1.3.8}$$

Moreover, the integral operator defined by (1.3.8), yields an orthogonal projection of the Hilbert space $\mathscr{L}^2(S)$ onto the subspace of functions allowing holomorphic extension from S to B (i.e., $\mathscr{H}^2(B)$).

Proof is immediately obtained from formula (1.3.7), if we put $\rho(\zeta) = 1 - |\zeta|^2$.   □

1.3.4   The Andreotti–Norguet Integral Representation

Another generalization of formula (1.2.3) is the Andreotti–Norguet formula (a different method of proof can be found in [63]). Suppose D is a bounded domain with a piecewise-smooth boundary, $\alpha = (\alpha_1, \ldots, \alpha_n)$ is a multi-index, f is a function holomorphic in D and continuous on $\overline{D}$, and

$$\partial^\alpha f = \frac{\partial^{\|\alpha\|} f}{\partial z_1^{\alpha_1} \ldots \partial z_n^{\alpha_n}},$$

where $\|\alpha\| = \alpha_1 + \cdots + \alpha_n$. Consider the following differential form:

$$U_\alpha(\zeta, z) = \frac{(n-1)!\alpha_1! \ldots \alpha_n!}{(2\pi i)^n} \sum_{k=1}^n \frac{(-1)^{k-1}(\bar{\zeta}_k - \bar{z}_k) \, d\bar{\zeta}^{\alpha+I}[k] \wedge d\zeta}{(|\zeta_1 - z_1|^{2(\alpha_1+1)} + \cdots + |\zeta_n - z_n|^{2(\alpha_n+1)})^n},$$

where

$$d\bar{\zeta}^{\alpha+I}[k] = d\bar{\zeta}_1^{\alpha_1+1} \wedge \cdots \wedge d\bar{\zeta}_{k-1}^{\alpha_{k-1}+1} \wedge d\bar{\zeta}_{k+1}^{\alpha_{k+1}+1} \wedge \cdots \wedge d\bar{\zeta}_n^{\alpha_n+1}.$$

Theorem 1.3.4 (Andreotti, Norguet) *The formula*

$$\partial^\alpha f(z) = \int_{\partial D} f(\zeta) U_\alpha(\zeta, z). \tag{1.3.9}$$

holds for every point $z \in D$ and every multi-index α.

Proof (given in [9, p. 60]) goes as follows. First verify that U_α is a closed form, so integration over ∂D can be replaced by integration over the set

$$\{\zeta \in \mathbb{C}^n : |\zeta_1 - z_1|^{2\alpha_1+2} + \cdots + |\zeta_n - z_n|^{2\alpha_n+2} = \varepsilon^2\}.$$

Expand the function f in powers $(\zeta - z)$ in a neighborhood of z, and integrate the series termwise against the form $U_\alpha(\zeta, z)$. We obtain $\partial^\alpha f(z)$ as the result of direct calculation. When $\alpha = (0, \ldots, 0)$, formula (1.3.9) reduces to (1.2.3).   □

We note that (1.3.9) can be generalized in the spirit of the Cauchy–Fantappiè formula [9, p. 61]. Analogues of the Bochner–Martinelli formula have also been considered in quaternionic analysis [80] and in Clifford analysis [75].

1.4   The Logarithmic Residue Formula

Let D be a bounded domain in $\mathbb{C}^n$ with a piecewise-smooth boundary ∂D, and let $w = \psi(z) = (\psi_1, \ldots, \psi_n)$ be a holomorphic map from $\overline{D}$ into $\mathbb{C}^n$ with a finite number of zeros E_ψ in D and no zeros on ∂D.

Recall the (dynamic) definition of the *multiplicity* of zero of a map ψ [9, Sect. 2]. We denote a ball of radius $R > 0$ with the center at the point $z \in \mathbb{C}^n$ by $B(z, R) = \{\zeta : |\zeta - z| < R\}$ and its boundary by $S(z, R) = \partial B(z, R)$. Let a be a zero of the map ψ and $B(a, R)$ have no other zeros in $B(a, R)$. Then there is a ball $B(0, r)$ such, that for almost all $\zeta \in B(0, r)$ the map $w = \psi - \zeta$ has the same number of zeros in $B(a, R)$. This number is called the multiplicity of zero a and is denoted by μ_a .

Consider the differential form

$$U(w) = \frac{(n-1)!}{(2\pi i)^n} \sum_{k=1}^{n} (-1)^{k-1} \frac{\bar{w}_k \, d\bar{w}[k] \wedge dw}{|w|^{2n}},$$ (1.4.1)

where $dw = dw_1 \wedge \ldots \wedge dw_n$, and $d\bar{w}[k]$ is obtained from the form $d\bar{w}$ by eliminating the differential dw_k, i.e., $U(w) = U(w, 0)$ is the Bochner–Martinelli kernel at zero.

Theorem 1.4.1 ([45]) *Let $f \in \mathscr{C}^1(\overline{D})$, then the formula*

$$\int_{\partial D_\zeta} f(\zeta) \, U(\psi(\zeta)) - \int_{D_\zeta} \bar{\partial} f \wedge U(\psi(\zeta)) = \sum_{a \in E_\psi} \mu_a f(a).$$ (1.4.2)

holds. (The integral over D converges absolutely.)

Proof Let $a \in E_\psi$. Let us prove that the integral

$$\int_D \bar{\partial} f \wedge U(\psi(\zeta))$$

converges absolutely. It is enough to show that the integral

$$\int_K d\bar{\zeta}_j \wedge U(\psi(\zeta))$$

converges absolutely, where K is some compact, containing the point a and not containing any other points of E_ψ, $j = 1, \ldots, n$. Consider compacts of the form:

$$B_\psi(r) = \{z \in \mathbb{C}^n : |\psi| \leq r\}, \quad S_\psi(r) = \{z \in \mathbb{C}^n : |\psi| = r\}$$

and $B_\psi(r) \subset K$. The surface $S_\psi(r)$ is smooth and compact for almost all r, $0 \leq r \leq r_0$ by Sard's theorem. Then

$$\int_{B_\psi(r_0)} \left| d\bar{\zeta}_j \wedge U(\psi(\zeta)) \right| = \int_0^{r_0} \int_{S_\psi(r)} \left| d\bar{\zeta}_j \wedge U(\psi) \right|$$

$$\leq C \int_0^{r_0} dr \int_{S_\psi(r)} |U(\psi)| = C_1 \int_0^{r_0} \frac{dr}{r^{2n}} \int_{S_\psi(r)} \left| \sum_{k=1}^{n} (-1)^{k-1} \bar{\psi}_k \, d\bar{\psi}[k] \wedge d\psi \right|$$

by Fubini's theorem. Restriction of the form

$$\sigma_\psi = \sum_{k=1}^{n} (-1)^{k-1} \bar{\psi}_k \, d\bar{\psi}[k] \wedge d\psi$$

to $S_\psi(r)$ (up to a constant) is a positive measure, since restriction of the form
$\sigma = \sum_{k=1}^{n} (-1)^{k-1} \bar{w}_k \, d\bar{w}[k] \wedge dw$ to the sphere $S = \{w : |w| = 1\}$ is the Lebesgue
measure on S up to a constant. And σ_ψ is obtained from σ by replacing the variables
$w \to \psi(\zeta)$. Then the integral

$$\int_{S_\psi(r)} \left| \sum_{k=1}^{n} (-1)^{k-1} \bar{\psi}_k \, d\bar{\psi}[k] \wedge d\psi \right| = C_2 \int_{S_\psi(r)} \sigma_\psi$$

$$= C_2 n \int_{B_\psi(r)} d\bar{\psi} \wedge d\psi = C_2 n \mu \int_{B(r)} d\bar{w} \wedge dw = C_3 \, r^{2n},$$

where $B(r) = \{w : |w| < r\}$.

Further proof is standard. Considering an auxiliary domain

$$D_\varepsilon = \{\zeta \in D : |\psi(\zeta)| > \varepsilon\},$$

yields

$$\int_{\partial D_\zeta} f \, U(\psi(\zeta)) - \int_{D_\varepsilon} \bar{\partial} f \wedge U(\psi(\zeta)) = \int_{S_\psi(\varepsilon)} f(\zeta) \, U(\psi(\zeta)).$$

If the zero of the map ψ is simple, i.e., ψ is biholomorphic in the neighborhood of
a, then choosing ε small enough and making the replacement of variables $w = \psi(\zeta)$,
we obtain that

$$\int_{S_\psi(\varepsilon)} f(\zeta) \, U(\psi(\zeta)) = \int_{S(0,\varepsilon)} f(\psi^{-1}(w)) \, U(w) = \int_{S(0,\varepsilon)} f(\psi^{-1}(w)) \, U(w, 0).$$

The last integral tends to $f(\psi_1(0)) = f(a)$ at $\varepsilon \to 0$ (see the proof of Bochner–Green
formula (1.1.4)). Thus formula (1.4.2) is proved for a map with simple zeros.

If a is a multiple zero of the map ψ, then, by considering the map $\psi_\rho = \psi - \rho$ in
a neighborhood K of the point a, we obtain that the map ψ_ρ has μ_a simple zeros in
K for almost all smaller-module ρ (this is a (dynamic) definition of the multiplicity
of zero (see [9, Sect. 2])). Applying (1.4.2) in K to map ψ_ρ, using closedness of
the form $U(\psi)$ and passing over to the limit as $|\rho| \to 0$ we obtain the required
assertion.   □

Let D be a bounded domain in $\mathbb{C}^n$, $n > 1$, with a connected piecewise-smooth
boundary, and let $\psi = (\psi_1, \ldots, \psi_n)$ be a map consisting of holomorphic functions
ψ_j, defined in some neighborhood

$$K_D = \{w : w = \zeta - z, \ z, \zeta \in \overline{D}\}$$

and having a unique zero as the origin of multiplicity μ.

Corollary 1.4.1 *If $f \in \mathscr{C}^1(\overline{D})$, then*

$$\int_{\partial D_\zeta} f(\zeta)\, U(\psi(\zeta - z)) - \int_{D_\zeta} \bar{\partial} f \wedge U(\psi(\zeta - z)) = \begin{cases} \mu f(z),\ z \in D, \\ \qquad\quad 0,\ z \notin \overline{D}. \end{cases} \qquad (1.4.3)$$

(The integral over D converges absolutely, and z is fixed.)

Corollary 1.4.2 *The following formula holds*

$$\mu f(z) = \int_{\partial D_\zeta} f(\zeta)\, U(\psi(\zeta - z)), \quad z \in D, \qquad (1.4.4)$$

where the function f is holomorphic in D and continuous on $\overline{D}$ (i.e., $f \in \mathscr{A}(D)$). The vector z in the form $U(\psi(\zeta - z))$ is fixed.

Formula (1.4.4) is a special case of the multidimensional logarithmic residue for the map ψ (see [9, Chap. 1]).

Corollary 1.4.3 *If $f \in \mathscr{C}^1(\partial D)$, and*

$$\int_{\partial D_\zeta} f(\zeta)\, U(\psi(\zeta - z)) = \begin{cases} M_\psi^+ f(z),\ z \in D, \\ M_\psi^- f(z),\ z \notin \overline{D}, \end{cases} \qquad (1.4.5)$$

then the functions $M_\psi^\pm f(z)$ are continuous up to the boundary of the domain and

$$M_\psi^+ f(z) - M_\psi^- f(z) = \mu f(z), \quad z \in \partial D.$$

Proof directly follows from (1.4.2) and the continuity of the integral on the domain D in (1.4.2). □

Remark 1.4.1 As shown by the proof of Theorem 1.4.1, integrals of the form

$$\int_D \frac{\bar{\psi}_j}{|\psi|^{2n}} \frac{\partial \bar{\psi}_{s_1}}{\partial \bar{\zeta}_{j_1}} \cdots \frac{\partial \bar{\psi}_{s_{n-1}}}{\partial \bar{\zeta}_{j_{n-1}}} \frac{\partial \psi_{i_1}}{\partial \zeta_1} \cdots \frac{\partial \psi_{i_n}}{\partial \zeta_n}\, d\bar{\zeta} \wedge d\zeta$$

absolutely converge.

Chapter 2
Properties of the Bochner–Martinelli Integral and the Logarithmic Residue Formula

Abstract In this chapter, we will consider the boundary behavior of the Bochner–Martinelli integral. Most of the statements have been collected in the book (Kytmanov, The Bochner–Martilnelli Integral and Its Applications. Birkhäuser Verlag, Basel, 1995). Some of these results can be obtained from the general theory of integral operators. But we seek to provide independent and more elementary proofs thereof. Since many of them will be used in the subsequent chapters, we decided to reproduce these in the book. The last section of this chapter contains the results of possible connection of the holomorphic continuation of functions with the homogeneous $\bar{\partial}$-Neumann problem, emphasizing the relationship between the harmonic and complex analysis in $\mathbb{C}^n$.

2.1 Boundary Behavior of the Bochner–Martinelli Integral

2.1.1 The Sokhotskiĭ–Plemelj Formula

Let D be a bounded domain with a piecewise-smooth boundary, and let f be an integrable function on ∂D ($f \in \mathscr{L}^1(\partial D)$). We consider the Bochner–Martinelli (type) integral (1.2.4):

$$Mf(z) = \int_{\partial D} f(\zeta)U(\zeta,z), \quad z \notin \partial D.$$

We recall that this is a function which is harmonic in both D and $\mathbb{C}^n \setminus \overline{D}$, moreover $M(z) = O(|z|^{1-2n})$ for $|z| \to \infty$. Like in Sect. 1.2, we will write $M^+f(z)$ for integral (1.2.4) when $z \in D$, and $M^-f(z)$ when $z \notin \overline{D}$. When $z \in \partial D$, integral (1.2.4) generally speaking does not exist as an improper integral, since the integrand has the singularity $|\zeta - z|^{1-2n}$. Therefore, for $z \in \partial D$, we will consider the Cauchy *principal value* of the Bochner–Martinelli integral:

$$\text{p.v.} \int_{\partial D} f(\zeta)U(\zeta,z) = \lim_{\varepsilon \to +0} \int_{\partial D \setminus B(z,\varepsilon)} f(\zeta)U(\zeta,z), \quad z \in \partial D.$$

© Springer International Publishing Switzerland 2015

A.M. Kytmanov, S.G. Myslivets, *Multidimensional Integral Representations*,
DOI 10.1007/978-3-319-21659-1_2

Below we will frequently omit the sign of the principal value p.v., that is, we will always assume an integral of the form (1.2.4) to be understood in terms of the principal value when $z \in \partial D$.

In this section we are interested in analogues of the Sokhotskiĭ–Plemelj formula for the Bochner–Martinelli integral, that is, in the relation between the boundary values of the functions $M^{\pm}(z)$ and the singular integral. First we will consider a simple case when density f satisfies the Hölder condition with the exponent $\alpha > 0$, i.e.,

$$|f(\zeta) - f(\eta)| \leq C|\zeta - \eta|^{\alpha} \tag{2.1.1}$$

for $\zeta, \eta \in \partial D$ and $C = \text{const}$. Generally speaking, these formulas can be deduced from the properties of potentials, but we will provide a direct proof.

We need to compute the restriction to ∂D of the differential forms $d\bar{\zeta}[k] \wedge d\zeta$ and $d\zeta[k] \wedge d\bar{\zeta}$ in terms of the Lebesgue surface measure $d\sigma$. Suppose $D = \{z \in \mathbb{C}^n : \rho(z) < 0\}$, where $\rho \in \mathscr{C}^1(\mathbb{C}^n)$ and $d\rho \neq 0$ on ∂D.

Lemma 2.1.1 *Restriction of the form $d\bar{\zeta}[k] \wedge d\zeta$ to the boundary ∂D is equal to*

$$2^{n-1}i^n(-1)^{k-1}\frac{\partial \rho}{\partial \bar{\zeta}_k}\frac{d\sigma}{|\,\text{grad}\,\rho|},$$

and restriction of the form $d\zeta[k] \wedge d\bar{\zeta}$ to ∂D is equal to

$$2^{n-1}i^n(-1)^{n+k-1}\frac{\partial \rho}{\partial \zeta_k}\frac{d\sigma}{|\,\text{grad}\,\rho|},$$

where

$$\text{grad}\,\rho = \left(\frac{\partial \rho}{\partial \zeta_1}, \ldots, \frac{\partial \rho}{\partial \zeta_n}\right).$$

Proof It is well known that restriction of the forms $dx[k] \wedge dy$ and $dx \wedge dy[k]$ to the boundary ∂D are equal to

$$dx[k] \wedge dy\Big|_{\partial D} = (-1)^k \gamma_k d\sigma,$$

$$dx \wedge dy[k]\Big|_{\partial D} = (-1)^{n+k-1}\gamma_{k+n}d\sigma, \tag{2.1.2}$$

where γ_k are the direction cosines of the normal vector to ∂D: namely

$$\gamma_k = \frac{\partial \rho}{\partial x_k} \frac{1}{\sqrt{\sum\limits_{j=1}^{n}\left[\left(\frac{\partial \rho}{\partial x_j}\right)^2 + \left(\frac{\partial \rho}{\partial y_j}\right)^2\right]}},$$

$$\gamma_{k+n} = \frac{\partial \rho}{\partial y_k} \frac{1}{\sqrt{\sum\limits_{j=1}^{n}\left[\left(\frac{\partial \rho}{\partial x_j}\right)^2 + \left(\frac{\partial \rho}{\partial y_j}\right)^2\right]}}.$$

We obtain the assertion of the lemma by using (2.1.2) and the formulas

$$\frac{\partial \rho}{\partial z_k} = \frac{1}{2}\left(\frac{\partial \rho}{\partial x_k} - i\frac{\partial \rho}{\partial y_k}\right), \qquad \frac{\partial \rho}{\partial \bar{z}_k} = \overline{\frac{\partial \rho}{\partial z_k}},$$

$$|\operatorname{grad} \rho| = \frac{1}{2}\sqrt{\sum\limits_{j=1}^{n}\left[\left(\frac{\partial \rho}{\partial x_j}\right)^2 + \left(\frac{\partial \rho}{\partial y_j}\right)^2\right]}, \qquad dz_k \wedge d\bar{z}_k = -2i\,dx_k \wedge dy_k.$$

$\square$

From Lemma 2.1.1, we have

$$U(\zeta, z)\big|_{\partial D} = \frac{(n-1)!}{2\pi^n}\sum\limits_{k=1}^{n}\frac{\bar{\zeta}_k - \bar{z}_k}{|\zeta - z|^{2n}}\frac{\partial \rho}{\partial \bar{\zeta}_k}\frac{d\sigma(\zeta)}{|\operatorname{grad}\rho(\zeta)|} = F(\zeta, z)d\sigma(\zeta). \qquad (2.1.3)$$

When $z \in \partial D$, we use $\tau(z)$ for the expression

$$\tau(z) = \lim_{\varepsilon \to +0}\frac{\operatorname{vol}\{S(z, \varepsilon) \cap D\}}{\operatorname{vol} S(z, \varepsilon)}.$$

In other words, $\tau(z)$ is a solid angle of the tangent cone to the surface ∂D at z. Since we consider a domain D with a piecewise-smooth boundary, the quantity $\tau(z)$ is defined and different from zero.

Lemma 2.1.2

$$\text{p.v.}\int_{\partial D}U(\zeta, z) = \tau(z)$$

for $z \in \partial D$.

Proof By definition

$$\text{p.v.} \int_{\partial D} U(\zeta, z) = \lim_{\varepsilon \to +0} \int_{\partial D \setminus B(z,\varepsilon)} U(\zeta, z).$$

But

$$\int_{\partial D \setminus B(z,\varepsilon)} U(\zeta, z) = \int_{\partial(D \setminus B(z,\varepsilon))} U(\zeta, z) + \int_{S^+(z,\varepsilon)} U(\zeta, z),$$

where $S^+(z, \varepsilon)$ is the part of the sphere $S(z, \varepsilon)$ lying in D, i.e., $S^+(z, \varepsilon) = D \cap S(z, \varepsilon)$. The sign of the second term has been changed because the orientation of $S(z, \varepsilon)$ (induced by the orientation of the ball $B(z, \varepsilon)$) is opposite to that of ∂D. Since $z \notin D \setminus B(z, \varepsilon)$ while the form $U(\zeta, z)$ is closed, the integral

$$\int_{\partial(D \setminus B(z,\varepsilon))} U(\zeta, z) = 0,$$

so

$$\int_{\partial D \setminus B(z,\varepsilon)} U(\zeta, z) = \int_{S^+(z,\varepsilon)} U(\zeta, z)$$

$$= \frac{(n-1)!}{(2\pi i)^n \varepsilon^{2n}} \int_{S^+(z,\varepsilon)} \sum_{k=1}^{n} (-1)^{k-1} (\bar{\zeta}_k - \bar{z}_k) d\bar{\zeta}[k] \wedge d\zeta.$$

From Lemma 2.1.1 it follows that the restriction of the forms equals

$$\sum_{k=1}^{n} (-1)^{k-1} (\bar{\zeta}_k - \bar{z}_k) d\bar{\zeta}[k] \wedge d\zeta \Big|_{S(z,\varepsilon)} = \varepsilon^{2n} 2^{n-1} i^n d\sigma,$$

where $d\sigma$ is the area element on the sphere.

Thus

$$\int_{\partial D \setminus B(z,\varepsilon)} U(\zeta, z) = \frac{\text{vol } S^+(z, \varepsilon)}{\text{vol } S(z, \varepsilon)} \to \tau(z)$$

as $\varepsilon \to +0$. $\square$

We extend $f(z)$ to a neighborhood $V(\partial D)$ as a function satisfying the Hölder condition on $V(\partial D)$ with the same exponent α, and we again denote it by $f(z)$. Consider the integral

$$\Phi(z) = \int_{\partial D} (f(\zeta) - f(z)) U(\zeta, z). \tag{2.1.4}$$

If $z \notin \partial D$, then integral (2.1.4) has no singularity, whereas if $z \in \partial D$, then

$$|f(\zeta) - f(z)| \, |U(\zeta, z)| \leq C |\zeta - z|^{\alpha+1-2n} d\sigma(\zeta),$$

so the integral $\Phi(z)$ is absolutely convergent.

Lemma 2.1.3 *If the function f satisfies the Hölder condition in $V(\partial D)$ with the exponent α, where $0 < \alpha < 1$, then the function $\Phi(z)$ satisfies the Hölder condition in $V(\partial D)$ with the same exponent α.*

Proof Let z^1 and z^2 be points in $V(\partial D)$ with $|z^1 - z^2| = \delta$, where δ is sufficiently small. Consider a ball $B(z^1, 2\delta) \subset V(\partial D)$, and set $\sigma_\delta = \partial D \cap B(z^1, 2\delta)$. Then

$$\left| \int_{\sigma_\delta} (f(\zeta) - f(z^j)) U(\zeta, z^j) \right| \leq C_1 \int_{\sigma_\delta} |\zeta - z^j|^{1+\alpha-2n} d\sigma \leq C_2 \delta^\alpha$$

for $j = 1, 2$. When σ_δ is a smooth surface, it is easy to obtain this inequality by replacing z^j by their projections onto σ_δ and using the integral over σ_δ instead of the integral over a $(2n-1)$-dimensional sphere of radius δ, and passing over to polar coordinates in this sphere. If σ_δ is piecewise smooth, we estimate the integral over each smooth piece of σ_δ that way.

We consider the difference of integrals (2.1.4) over $\partial D \setminus \sigma_\delta$ at the points z^1 and z^2, which equals

$$\int_{\partial D \setminus \sigma_\delta} (f(\zeta) - f(z^2))(U(\zeta, z^2) - U(\zeta, z^1))$$

$$+ (f(z^1) - f(z^2)) \int_{\partial D \setminus \sigma_\delta} U(\zeta, z^1). \qquad (2.1.5)$$

We have already dealt with the second integral in Lemma 2.1.2 (except that there $z^1 \in \partial D$), from which we obtain

$$\left| \int_{\partial D \setminus \sigma_\delta} U(\zeta, z^1) \right| \leq 1.$$

Consequently,

$$|f(z^1) - f(z^2)| \left| \int_{\partial D \setminus \sigma_\delta} U(\zeta, z^1) \right| \leq C_3 \delta^\alpha.$$

Now we estimate the first term in (2.1.5). If $\zeta \in \partial D \setminus \sigma_\delta$, then

$$
\left| \frac{\bar{\zeta}_j - \bar{z}_j^1}{|\zeta - z^1|^{2n}} - \frac{\bar{\zeta}_j - \bar{z}_j^2}{|\zeta - z^2|^{2n}} \right| \leq \frac{|z_j^2 - z_j^1|}{|\zeta - z^1|^{2n}} + |\bar{\zeta}_j - \bar{z}_j^2| \left| \frac{1}{|\zeta - z^1|^{2n}} - \frac{1}{|\zeta - z^2|^{2n}} \right|
$$

$$
\leq \frac{|\bar{\zeta}_j - \bar{z}_j^2||z^2 - z^1|}{|\zeta - z^1||\zeta - z^2|} \sum_{s=0}^{2n-1} \frac{|\zeta - z^2|^{s+1-2n}}{|\zeta - z^1|^{s}} + \frac{|z^2 - z^1|}{|\zeta - z^1|^{2n}} \leq C_4 \delta |\zeta - z^1|^{-2n}
$$

since $|\zeta - z^1| \leq 2|\zeta - z^2|$. Thus

$$
|\Phi(z^1) - \Phi(z^2)| \leq C_5 \delta^\alpha + C_6 \delta \int_{\partial D \setminus \sigma_\delta} |\zeta - z^1|^{\alpha - 2n} d\sigma.
$$

If σ_δ is a smooth surface, then by replacing the point z^1 by its projection onto σ_δ we obtain

$$
\int_{\partial D \setminus \sigma_\delta} |\zeta - z^1|^{\alpha - 2n} d\sigma \leq C_7 \delta^{\alpha - 1}.
$$

$\square$

Remark 2.1.1 As in the case of the Cauchy-type integral, when $\alpha = 1$ the function $\Phi(z)$ will satisfy the condition

$$
|\Phi(z^1) - \Phi(z^2)| \leq C|z^1 - z^2| \, |\ln |z^1 - z^2||,
$$

since

$$
\int_{\partial D \setminus \sigma_\delta} |\zeta - z^1|^{1 - 2n} d\sigma \leq C_8 |\ln \delta|.
$$

Theorem 2.1.1 *Let D be a bounded domain with a piecewise-smooth boundary ∂D and let $f \in \mathscr{C}^\alpha(\partial D)$, where $0 < \alpha < 1$. Then the Bochner–Martinelli integral M^+f extends continuously to $\overline{D}$ as a function of class $\mathscr{C}^\alpha(\overline{D})$, while M^-f extends continuously to $\mathbb{C}^n \setminus D$ as a function of class $\mathscr{C}^\alpha(\mathbb{C}^n \setminus D)$. Moreover, the Sokhotskiĭ–Plemelj formulas are valid for $z \in \partial D$:*

$$
M^+f(z) = (1 - \tau(z))f(z) + \text{p.v.} \int_{\partial D} f(\zeta)U(\zeta, z),
$$

$$
M^-f(z) = -\tau(z) + \text{p.v.} \int_{\partial D} f(\zeta)U(\zeta, z).
$$

(2.1.6)

Proof The first part of the theorem follows from Lemma 2.1.3. We consider the integral

$$\text{p.v.} \int_{\partial D} f(\zeta)U(\zeta, z) = \int_{\partial D} (f(\zeta) - f(z))U(\zeta, z) + \tau(z)f(z)$$

(by Lemma 2.1.2). Since $\Phi(z)$ is continuous in $V(\partial D)$ (by Lemma 2.1.3),

$$\int_{\partial D} (f(\zeta) - f(z))U(\zeta, z) = M^+f(z) - f(z),$$

that is,

$$M^+f(z) = (1 - \tau(z))f(z) + \text{p.v.} \int_{\partial D} f(\zeta)U(\zeta, z).$$

On the other hand,

$$\int_{\partial D} (f(\zeta) - f(z))U(\zeta, z) = M^-f(z).$$

$\square$

Remark 2.1.2 If we introduce the norm

$$\|f\|_{\mathscr{C}^\alpha} = \sup_{\partial D} |f| + \sup_{\zeta, \eta \in \partial D} \frac{|f(\zeta) - f(\eta)|}{|\zeta - \eta|^\alpha} \quad .$$

in the space $\mathscr{C}^\alpha(\partial D)$ of functions f satisfying the Hölder condition with the exponent α, then Lemmas 2.1.2 and 2.1.3 show that the Bochner–Martinelli integral and the Bochner–Martinelli singular integral define bounded operators in this space for $0 < \alpha < 1$ (when $\partial D \in \mathscr{C}^1$).

Lemma 2.1.3 is contained in the paper by Chirka [18]. Various versions of Theorem 2.1.1 have been quoted on numerous occasions. Look and Zhong [60] proved (2.1.6) for domains with a boundary of class $\mathscr{C}^2$. Later these formulas were obtained by Harvey and Lawson [30] for domains with a smooth boundary.

Corollary 2.1.1 *If* $\partial D \in \mathscr{C}^1$, *then for* $z \in \partial D$ *formula (2.1.6) takes the form*

$$M^+f(z) = \frac{1}{2}f(z) + \text{p.v.} \int_{\partial D} f(\zeta)U(\zeta, z),$$

$$M^-f(z) = -\frac{1}{2}f(z) + \text{p.v.} \int_{\partial D} f(\zeta)U(\zeta, z),$$

(2.1.7)

and therefore the Bochner–Martinelli singular integral also satisfies the Hölder condition with the exponent α *on* ∂D.

Corollary 2.1.2 *If ∂D is piecewise smooth, then*

$$M^+ f(z) - M^- f(z) = f(z)$$

for $z \in \partial D$.

2.1.2 Analogue of Privalov's Theorem

In this subsection, we consider a bounded domain D with a boundary of class $\mathscr{C}^1$ and functions f that are integrable on ∂D (i.e., $f \in \mathscr{L}^1(\partial D)$). Let $z^0 \in \partial D$. Consider a right circular cone V_{z^0} with the vertex at z^0 and the axis that coincides with the normal to ∂D at z^0, the angle β between the axis and the generator of the cone being less than $\dfrac{\pi}{2}$. Let $z \in D \cap V_{z^0}$. Suppose that z^0 is *a Lebesgue point* for f, i.e.,

$$\lim_{\varepsilon \to +0} \varepsilon^{1-2n} \int_{\partial D \cap B(z^0, \varepsilon)} \left| f(\zeta) - f(z^0) \right| d\sigma = 0.$$

Theorem 2.1.2 ([45]) *If $z \in D \cap V_{z^0}$, then*

$$\lim_{\substack{z \to z^0 \\ z \in V_{z^0}}} \left[\int_{\partial D} \left(f(\zeta) - f(z^0) \right) U(\zeta, z) - \int_{\partial D \setminus B(z^0, |z - z^0|)} \left(f(\zeta) - f(z^0) \right) U(\zeta, z^0) \right] = 0.$$

This theorem is an analogue of Privalov's theorem for an integral of Cauchy type.

Proof We make a unitary transformation of $\mathbb{C}^n$ and a translation so that z^0 goes to 0 and the tangent plane to ∂D at z^0 goes to the plane $T = \{ w \in \mathbb{C}^n : \operatorname{Im} w_n = 0 \}$. The surface ∂D in a neighborhood of the origin will then be given by the equations

$$\zeta_1 = w_1, \ldots, \zeta_{n-1} = w_{n-1}, \ \zeta_n = u_n + i\varphi(w),$$

where $w = (w_1, \ldots, w_{n-1}, u_n) \in T$; the function $\varphi \in \mathscr{C}^1(W)$, where W is a neighborhood of the origin in the plane T; and $\varphi(w) = o(|w|)$ as $w \to 0$. We denote the projection of z onto the $\operatorname{Im} w_n$ axis by $\tilde{z}$. Then

$$|z - \tilde{z}| \le |\tilde{z}| \tan \beta, \quad |z| \le |\tilde{z}| \frac{1}{\cos \beta}.$$

Fix $\varepsilon_0 > 0$, and choose a $(2n-1)$-dimensional ball B' in the plane T with the center at 0 and radius ε such that

1. $B' \subset W$;
2. $|w - \tilde{z}| \le C|\zeta(w) - z|$ for $w \in B'$, where C is a constant independent of w and z.

Condition (2) is ensured by the relations

$$|w - \zeta(w)| = |\varphi(w)| = o(|w|), \quad |w| \to 0;$$

$$|w| \le |w - \tilde{z}|, \quad |\tilde{z}| \le |w - \tilde{z}|,$$

$$|w - \tilde{z}| \le |w - \zeta(w)| + |\zeta(w) - z| + |z - \tilde{z}| \le |\varphi(w)| + |\zeta(w) - z| + |\tilde{z}| \tan \beta$$

$$\le |\varphi(w)| + |\zeta(w) - z| + \tan \beta (|w - \zeta(w)| + |\zeta(w) - z|)$$

$$= (1 + \tan \beta)(|\varphi(w)| + |\zeta(w) - z|) \le C|\zeta(w) - z|.$$

We note that the ball B' and the constant C may be taken to be independent of the point $z^0 = 0$. If 0 is a Lebesgue point for the function $f(\zeta)$, then 0 is also a Lebesgue point for the function $f(\zeta(w))$. It is clear that the form of the kernel $U(\zeta, z)$ is not affected by the translation.

Lemma 2.1.4 *The kernel $U(\zeta, z)$ is invariant with respect to unitary transformations.*

Proof Suppose the unitary transformation has the form $\zeta = A\zeta'$, where A is the unitary matrix $A = \|a_{jk}\|_{j,k=1}^n$. Then the distance $|\zeta - z|$ will not change, $d\zeta = \det A \, d\zeta' = e^{i\psi} d\zeta'$, and

$$\sum_{k=1}^n (-1)^{k-1} (\zeta_k - z_k) \, d\zeta[k] = \sum_{k=1}^n (-1)^{k-1} \sum_{j=1}^n a_{jk}(\zeta_j' - z_j') \sum_{p=1}^n A_{pk} \, d\zeta'[p],$$

where A_{pk} is the minor of the matrix A corresponding to the element a_{pk}, so

$$\sum_{k=1}^n (-1)^{k-1} a_{jk} A_{pk} = \begin{cases} 0, & j \ne p, \\ (-1)^{p-1} \det A, & j = p. \end{cases}$$

$\square$

We now continue with the proof of Theorem 2.1.2. Let $|z| = \varepsilon$. Transform the difference of the integrals as follows

$$\int_{\partial D} (f(\zeta) - f(0)) U(\zeta, z) - \int_{\partial D \backslash B(0,\varepsilon)} (f(\zeta) - f(0)) U(\zeta, 0)$$

$$= \int_{\partial D \backslash B(0,\varepsilon)} (f(\zeta) - f(0))(U(\zeta, z) - U(\zeta, 0)) + \int_{\partial D \cap B(0,\varepsilon)} (f(\zeta) - f(0)) U(\zeta, z).$$

Now

$$\frac{1}{|\zeta - z|^{2n-1}} \leq \frac{C_1}{\varepsilon^{2n-1}},$$

since

$$C|\zeta(w) - z| \geq |w - \tilde{z}| \geq |\tilde{z}| \geq \varepsilon \cos\beta,$$

so

$$\left| \int_{\partial D \cap B(0,\varepsilon)} (f(\zeta) - f(0))U(\zeta,z) \right| \leq \frac{C_2}{\varepsilon^{2n-1}} \int_{\partial D \cap B(0,\varepsilon)} |f(\zeta) - f(0)|\, d\sigma \longrightarrow 0$$

as $\varepsilon \to 0$. Now consider the difference

$$\frac{\bar{\zeta}_j - \bar{z}_j}{|\zeta - z|^{2n}} - \frac{\bar{\zeta}_j}{|\zeta|^{2n}} = \bar{\zeta}_j \left(\frac{1}{|\zeta - z|^{2n}} - \frac{1}{|\zeta|^{2n}} \right) - \frac{\bar{z}_j}{|\zeta - z|^{2n}}.$$

We have

$$\frac{|z_j|}{|\zeta - z|^{2n}} \leq C_3 \frac{|\tilde{z}|}{|w - \tilde{z}|^{2n}} = C_3 \frac{|\tilde{z}|}{(|w|^2 + |\tilde{z}|^2)^n}.$$

However

$$|\bar{\zeta}_j| \left| \frac{1}{|\zeta - z|^{2n}} - \frac{1}{|\zeta|^{2n}} \right| = \frac{|\zeta_j| \, ||\zeta| - |\zeta - z||}{|\zeta| \, |\zeta - z|} \sum_{s=0}^{2n-1} \frac{1}{|\zeta|^s |\zeta - z|^{2n-s-1}}$$

$$\leq |z| \sum_{s=0}^{2n-1} \frac{1}{|\zeta|^s |\zeta - z|^{2n-s}}.$$

We have

$$|\zeta(w)| \geq C_4 |w| \geq C_5 |w - \tilde{z}|,$$

since $\zeta \notin B(0,\varepsilon)$, and the fraction $\dfrac{|w|}{|w - \tilde{z}|}$ is bounded from below by a positive constant because this fraction equals the cosine of the angle between the vectors w and $w - \tilde{z}$, and that angle cannot be greater than $\dfrac{\pi}{4}$. Thus

$$|U(\zeta,z) - U(\zeta,0)| \leq \frac{C_6 |\tilde{z}|\, d\sigma}{(|w|^2 + |\tilde{z}|^2)^n}.$$

Since $d\sigma \leq C_7\, dS$, where dS is the area element of the plane T, we obtain

$$\int_{B(0,\varepsilon_0)\cap\partial D\setminus B(0,\varepsilon)} |f(\zeta) - f(0)|\, |U(\zeta, z) - U(\zeta, 0)|$$

$$\leq C_8 \int_{B(0,\varepsilon_0)\cap T\setminus B(0,\varepsilon)} |f(\zeta(w)) - f(0)|\, \frac{|\bar{z}|\, dS}{(|w|^2 + |\bar{z}|^2)^n}$$

$$\leq C_8 \int_{B(0,\varepsilon_0)\cap T} |f(\zeta(w)) - f(0)|\, \frac{|\bar{z}|\, dS}{(|w|^2 + |\bar{z}|^2)^n}. \qquad (2.1.8)$$

If $\varepsilon \to +0$, then $|\bar{z}| \to 0$, while the expression $\dfrac{|\bar{z}|}{(|w|^2 + |\bar{z}|^2)^n}$ is the Poisson kernel for the half-space. Since 0 is the Lebesgue point of $f(\zeta(w))$, it is well known that this integral converges to zero as $\varepsilon \to +0$ (see, for example, [77]).   □

Theorem 2.1.2 shows that the existence of the Bochner–Martinelli singular integral at z^0 is equivalent to the existence of the limit of $M^+(z)$ as $z \to z^0$ along nontangential paths. Therefore, if the singular integral exists, so $\lim\limits_{z\to z^0} M^+(z)$ exists, and so the Sokhotskiĭ–Plemelj formula (2.1.7) holds.

2.2 Jump Theorems for the Bochner–Martinelli Integral

We saw in Sect. 2.1 that the Sokhotskiĭ–Plemelj formula implies a jump theorem (see Corollary 2.1.2). As a rule, the jump theorem is more readily proved than the Sokhotskiĭ–Plemelj formula, and moreover the difference $M^+f - M^-f$ may have a limit on ∂D even when the functions M^+f and M^-f themselves do not. Therefore jump theorems hold for a wider class of functions than do Sokhotskiĭ–Plemelj formulas.

2.2.1 Integrable and Continuous Functions

First we study the case when D is a bounded domain with a boundary of class $\mathscr{C}^1$, and $f \in \mathscr{L}^1(\partial D)$. Let us consider a right circular cone V_{z^0} with the vertex at $z^0 \in \partial D$ whose axis coincides with the normal to ∂D at z^0, the angle β between the axis and the generator being less than $\dfrac{\pi}{2}$. We take two points $z^+ \in V_{z^0} \cap D$ and $z^- \in V_{z^0} \cap (\mathbb{C}^n \setminus \overline{D})$ such that $a|z^+ - z^0| \leq |z^- - z^0| \leq b|z^+ - z^0|$, where a and b are constants not depending on $z^\pm$, and $0 < a \leq b < \infty$.

Theorem 2.2.1 ([45]) *If z^0 is a Lebesgue point of the function $f \in \mathscr{L}^1(\partial D)$, then*

$$\lim_{z^{\pm} \to z^0} \left(Mf(z^+) - Mf(z^-) \right) = f(z^0) \tag{2.2.1}$$

(where Mf is defined by (1.2.4)). If $f \in \mathscr{C}(\partial D)$, then limit (2.2.1) exists for all points $z^0 \in \partial D$, and it is attained uniformly if the angle β and the constants a and b are fixed.

Proof By Lemma 2.1.4, using a unitary transformation and translation, we take z^0 to 0 and the tangent plane to ∂D at z^0 to the plane $T = \{w \in \mathbb{C}^n : \operatorname{Im} w_n = 0\}$. The surface ∂D will then be given in a neighborhood of 0 by a system of equations

$$\zeta_1 = w_1, \ldots, \zeta_{n-1} = w_{n-1}, \ \zeta_n = u_n + i\varphi(w),$$

where $w = (w_1, \ldots, w_{n-1}, u_n) \in T$, the function $\varphi(w)$ is of class $\mathscr{C}^1$ in a neighborhood W of 0 in the plane T, and $\varphi(w) = o(|w|)$ as $w \to 0$. We denote the projections $z^{\pm}$ onto the $\operatorname{Im} w_n$ axis by $\tilde{z}^{\pm}$. Then

$$|z^{\pm} - \tilde{z}^{\pm}| \leq |\tilde{z}^{\pm}| \tan \beta, \quad |z^{\pm}| \leq \frac{|\tilde{z}^{\pm}|}{\cos \beta},$$
$$a|\tilde{z}^+| \cos \beta \leq |\tilde{z}^-| \leq \frac{b|\tilde{z}^+|}{\cos \beta}. \tag{2.2.2}$$

We fix a ball B' in the plane T with the center at 0 and of radius ε such that

1. $B' \subset W$;
2. $|w - \tilde{z}^{\pm}| \leq C|\zeta(w) - z^{\pm}|$ for $w \in B'$, where C is the constant independent of the point $z^0 = 0$. Here $B' = B(z^0, \varepsilon) \cap T$ and $\Gamma = B(z^0, \varepsilon) \cap \partial D$.

Consider the difference

$$Mf(z^+) - Mf(z^-) = \int_{\partial D} (f(\zeta) - f(z^0)) U(\zeta, z^+)$$

$$- \int_{\partial D} (f(\zeta) - f(z^0)) U(\zeta, z^-) + f(z^0) \int_{\partial D} (U(\zeta, z^+) - U(\zeta, z^-)).$$

Since

$$\int_{\partial D} (U(\zeta, z^+) - U(\zeta, z^-)) = 1,$$

it is enough to show that

$$\lim_{z^{\pm} \to z^0} \int_{\partial D} (f(\zeta) - f(z^0))(U(\zeta, z^+) - U(\zeta, z^-)) = 0.$$

In the integral

$$\int_{\partial D\backslash\Gamma} (f(\zeta) - f(z^0))(U(\zeta, z^+) - U(\zeta, z^-)),$$

we can take the limit inside, since $z^0 \notin \partial D \setminus \Gamma$. It remains to consider this integral over the set Γ. From condition (2) on the choice of B' and the inequality

$$|\zeta(w)| \leq C_1|w| \leq C_1|w - \tilde{z}^\pm|$$

we obtain

$$\left| \frac{\bar\xi_k}{|\zeta - z^+|^{2n}} - \frac{\bar\xi_k}{|\zeta - z^-|^{2n}} \right|$$

$$= \left| \frac{1}{|\zeta - z^+|} - \frac{1}{|\zeta - z^-|} \right| \sum_{j=0}^{2n-1} \frac{|\xi_k| \, |\zeta - z^-|^{j+1-2n}}{|\zeta - z^+|^j}$$

$$= \left| |\zeta - z^+| - |\zeta - z^-| \right| \sum_{j=0}^{2n-1} \frac{|\xi_k| \, |\zeta - z^-|^{j-2n}}{|\zeta - z^+|^{j+1}}$$

$$\leq C_1 C^{2n} \sum_{j=0}^{2n-1} \frac{|w - \tilde{z}^-|^{j-2n}(|z^+| + |z^-|)}{|w - \tilde{z}^+|^j}. \qquad (2.2.3)$$

We may assume that $a_1 = a\cos\beta < 1$, then $|w - \tilde{z}^\pm| \geq |w - a_1\tilde{z}^+|$ in view of (2.2.2). Therefore from (2.2.3) we have that

$$\left| \frac{\bar\xi_k}{|\zeta - z^+|^{2n}} - \frac{\bar\xi_k}{|\zeta - z^-|^{2n}} \right| \leq \frac{d|\tilde{z}^+|}{|w - a_1\tilde{z}^+|^{2n}},$$

where d depends only on a, b, C, C_1, and β. In precisely the same way,

$$\left| \frac{\bar{z}_k^+}{|\zeta - z^+|^{2n}} - \frac{\bar{z}_k^-}{|\zeta - z^-|^{2n}} \right| \leq \frac{|z_k^+|}{|\zeta - z^+|^{2n}} + \frac{|z_k^-|}{|\zeta - z^-|^{2n}} \leq \frac{d_1|\tilde{z}|}{|w - a_1\tilde{z}^+|^{2n}}.$$

Finally, $d\sigma \leq d_2 dS$, where dS is the surface area element on the surface T, and d_2 is independent of z^0. Therefore

$$\left| \int_\Gamma (f(\zeta) - f(0))(U(\zeta, z^+) - U(\zeta, z^-)) \right| \leq d_3 \int_{B'} \frac{|f(\zeta(w)) - f(0)| \, |\tilde{z}^+|}{(|w|^2 + a_1^2|\tilde{z}^+|^2)^n} dS.$$

$$(2.2.4)$$

Since

$$\frac{|\bar{z}^+|}{(|w|^2 + a_1^2|\bar{z}^+|^2)^n}$$

is the Poisson kernel for the half-space and 0 is a Lebesgue point for $f(\zeta(w))$, the last expression tends to zero as $|\bar{z}^+| \to 0$ (see [77, Theorem 1.25]).

If f is continuous on ∂D, then for each $\delta > 0$, we choose a ball B' of radius ε such that $|f(\zeta(w)) - f(0)| < \delta$ for $w \in B'$ (where ε may be taken independent of the point $z^0 = 0$). Then from (2.2.4) we obtain

$$\left| \int_{\Gamma} (f(\zeta) - f(0))(U(\zeta, z^+) - U(\zeta, z^-)) \right|$$

$$\leq d_4 \delta \int_{B'} \frac{a_1|\bar{z}^+|dS}{(|w|^2 + a_1^2|\bar{z}^+|^2)^n} \leq d_4 \delta \int_{T} \frac{a_1|\bar{z}^+|dS}{(|w|^2 + a_1^2|\bar{z}^+|^2)^n},$$

and the last integral equals the constant independent of $\bar{z}^+$.   □

Theorem 2.2.1 for continuous functions can be found in [19].

Corollary 2.2.1 *Let* $f \in \mathscr{C}(\partial D)$. *If* M^+f *extends continuously to* $\overline{D}$, *then* M^-f *extends continuously to* $\mathbb{C}^n \setminus D$, *and vice versa.*

This corollary, given in [19], was also remarked by Harvey and Lawson in [30]. We now give an example to show that when f is continuous, the function Mf may fail to extend to certain points of the boundary ∂D. This example is contained in [19].

Example 2.2.1 Let D be a domain such that $\overline{D}$ is contained in a unit ball $B(0, 1)$, and ∂D contains a $(2n - 1)$-dimensional ball B' of radius $R < 1$ with the center at the point 0 in the plane $T = \{z \in \mathbb{C}^n : \operatorname{Im} z_n = 0\}$. We set $f(\zeta) = \dfrac{\zeta_n}{|\zeta| \ln |\zeta|}$ on ∂D, so that $f \in \mathscr{C}(\partial D)$. We will show that $Mf(z)$ is unbounded in any neighborhood of the origin.

Set $z = (0, \ldots, 0, iy_n)$, with $y_n > 0$. It suffices to show that the integral

$$I(z) = \int_{B'} f(\zeta)U(\zeta, z)$$

is unbounded in any neighborhood of the origin. Now $d\bar{\zeta}[k] \wedge d\zeta = 0$ on the set B' for $k \neq n$, so

$$I(z) = \frac{(n-1)!(-1)^{n-1}}{(2\pi i)^n} \int_{B'} \frac{\eta_n(\eta_n - iy_n)d\bar{\zeta}[n] \wedge d\zeta}{|\zeta| \ln |\zeta|(|\zeta|^2 + y_n^2)^n},$$

where $\eta_n = \operatorname{Re} \zeta_n$. As in Theorem 2.2.1,

$$\left| \int_{B'} f(\zeta) \frac{\eta_n y_n d\bar{\zeta}[n] \wedge d\zeta}{(|\zeta|^2 + y_n^2)^n} \right| \leq C \int_T \frac{y_n dS}{(|\zeta|^2 + y_n^2)^n} \leq C_1.$$

If we introduce polar coordinates in B', then $dS = |\zeta|^{2n-2} d|\zeta| \wedge d\omega$, where $d\omega$ is the surface area element on the unit sphere in $\mathbb{R}^{2n-1}$. Integrating over ω yields,

$$I_1 = \int_{B'} \frac{\eta_n^2 dS}{|\zeta| \, |\ln |\zeta|| \, (|\zeta|^2 + y_n^2)^n} = C_2 \int_0^R \frac{|\zeta|^{2n} d|\zeta|}{|\zeta| \, |\ln |\zeta|| \, (|\zeta|^2 + y_n^2)^n}$$

$$\geq C_2 \int_\varepsilon^R \frac{|\zeta|^{2n-1} d|\zeta|}{|\ln |\zeta|| \, (|\zeta|^2 + y_n^2)^n}.$$

However

$$\lim_{y_n \to 0} \int_\varepsilon^R \frac{|\zeta|^{2n-1} d|\zeta|}{|\ln |\zeta|| \, (|\zeta|^2 + y_n^2)^n} = \int_\varepsilon^R \frac{d|\zeta|}{|\zeta| \, |\ln |\zeta||} = -\ln |\ln R| + \ln |\ln \varepsilon|.$$

Fix $N > 0$. If we take ε sufficiently small, then $\ln |\ln \varepsilon| - \ln |\ln R| > 2N$, and so $I_1 > C_2 N$ for sufficiently small y_n.

2.2.2 Functions of Class $\mathscr{L}^p$

Again, let D be a bounded domain in $\mathbb{C}^n$ with a smooth boundary ∂D, and $f \in \mathscr{L}^p(\partial D)$ with $p \geq 1$. We denote the unit outer normal to ∂D at ζ by $\nu(\zeta)$.

Theorem 2.2.2 ([45]) *If $Mf(z)$ is an integral of the form (1.2.4), then*

$$\lim_{\varepsilon \to +0} \int_{\partial D} |Mf(z - \varepsilon \nu(z)) - Mf(z + \varepsilon \nu(z)) - f(z)|^p d\sigma = 0,$$

and in addition

$$\int_{\partial D} |Mf(z - \varepsilon \nu(z)) - Mf(z + \varepsilon \nu(z))|^p d\sigma \leq C \int_{\partial D} |f|^p d\sigma, \tag{2.2.5}$$

where the constant C is independent of f and ε (for sufficiently small ε, the point $z - \varepsilon \nu(z) \in D$, and $z + \varepsilon \nu(z) \in \mathbb{C}^n \setminus \overline{D}$). If $f \in \mathscr{L}^\infty(\partial D)$, then

$$\sup_{\partial D} |Mf(z - \varepsilon \nu(z)) - Mf(z + \varepsilon \nu(z))| \leq C \operatorname{ess\,sup}_{\partial D} |f|.$$

Proof We write $z^+ = z - \varepsilon v(z)$ and $z^- = z + \varepsilon v(z)$. For each point $\zeta \in \partial D$, we take a ball $B(\zeta, r)$ of radius r not depending on ζ such that, for $z \in \partial D \cap B(\zeta, r)$, we have

$$|\zeta - z^{\pm}|^2 \geq k(|w - \zeta|^2 + \varepsilon^2)$$

for $\varepsilon < \dfrac{r}{2}$ (here k is independent of ζ and ε), where w is the projection of z onto the tangent plane T_ζ to ∂D at ζ. This can always be done because

$$||\zeta - w| - |\zeta - z|| \leq |w - z| = o(|\zeta - w|)$$

as $w \to \zeta$ (see the proof of Theorem 2.2.1). We have

$$\int_{\partial D} |Mf(z^+) - Mf(z^-) - f(z)|^p d\sigma$$

$$= \int_{\partial D} d\sigma(z) \left| \int_{\partial D} (f(\zeta) - f(z))(U(\zeta, z^+) - U(\zeta, z^-)) \right|^p$$

$$\leq \int_{\partial D} d\sigma(z) \left(\int_{\partial D} |U(\zeta, z^+) - U(\zeta, z^-)| \right)^{p-1} \times$$

$$\times \int_{\partial D} |f(\zeta) - f(z)|^p \; |U(\zeta, z^+) - U(\zeta, z^-)|$$

by Jensen's inequality (see, for example [31, Sect. 2.2]) applied to the integral

$$\left(\int_{\partial D} |f(\zeta) - f(z)| \; |U(\zeta, z^+) - U(\zeta, z^-)| d\sigma \right)^p.$$

We estimated the integral $\int_{\partial D} |U(\zeta, z^+) - U(\zeta, z^-)|$ in Theorem 2.2.1 and showed it to be bounded by the constant not depending on ε, while the integral

$$\int_{\partial D} d\sigma(z) \int_{\partial D} |f(\zeta) - f(z)|^p |U(\zeta, z^+) - U(\zeta, z^-)|$$

$$\leq C_1 \sum_{m=1}^{n} \int_{\partial D} d\sigma(\zeta) \int_{\partial D} |f(\zeta) - f(z)|^p \left| \frac{\bar{\zeta}_m - \bar{z}_m^+}{|\zeta - z^+|^{2n}} - \frac{\bar{\zeta}_m - \bar{z}_m^-}{|\zeta - z^-|^{2n}} \right| d\sigma(z).$$

If $z \in B(\zeta, r) \cap \partial D$, then

$$\left| \frac{\bar{\xi}_m - \bar{z}_m}{|\zeta - z^+|^{2n}} - \frac{\bar{\xi}_m - \bar{z}_m}{|\zeta - z^-|^{2n}} \right|$$

$$= |\bar{\xi}_m - \bar{z}_m| \left| |\zeta - z^+| - |\zeta - z^-| \right| \sum_{j=0}^{2n-1} \frac{1}{|\zeta - z^+|^{j+1} |\zeta - z^-|^{2n-j}}$$

$$\leq \frac{6\varepsilon n}{k^n (|w - \varepsilon|^2 + \varepsilon^2)^n},$$

while

$$\left| \frac{\varepsilon \nu_m}{|\zeta - z^+|^{2n}} + \frac{\varepsilon \nu_m}{|\zeta - z^-|^{2n}} \right| \leq \frac{2\varepsilon}{k^n (|w - \zeta|^2 + \varepsilon^2)^n}.$$

Then

$$\int_{\partial D \cap B(\zeta, r)} |f(\zeta) - f(z)|^p \left| \frac{\bar{\xi}_m - \bar{z}_m^+}{|\zeta - z^+|^{2n}} - \frac{\bar{\xi}_m - \bar{z}_m^-}{|\zeta - z^-|^{2n}} \right| d\sigma(z)$$

$$\leq d \int_{T_\zeta \cap B(\zeta, r)} \frac{\varepsilon |f(\zeta) - f(z(w))|^p}{(|w - \zeta|^2 + \varepsilon^2)^n} dS(w) = d I_1.$$

Introducing the variable $t = \dfrac{w - \zeta}{\varepsilon}$ in $\mathbb{R}^{2n-1}$, we obtain that

$$I_1 = \int_{\{\varepsilon |t| < r\}} \frac{|f(\zeta) - f(z(\zeta + \varepsilon t))|^p}{(|t|^2 + 1)^n} dS(t),$$

and the integral

$$I_\varepsilon(t) = \int_{\partial D} |f(\zeta) - f(z(\zeta + \varepsilon t))|^p d\sigma(\zeta)$$

converges to zero as $\varepsilon \to +0$ for fixed t. Also $I_\varepsilon(t) \leq A \|f\|_{\mathscr{L}^p}^p$. Therefore

$$\int_{\partial D} d\sigma(\zeta) \int_{\{\varepsilon |t| < r\}} \frac{|f(\zeta) - f(z(\zeta + \varepsilon t))|^p}{(|t|^2 + 1)^n} dS(t)$$

$$= \int_{\{\varepsilon |t| < r\}} \frac{I_\varepsilon(t)}{(|t|^2 + 1)^n} dS(t) \leq \int_{\mathbb{R}^{2n-1}} \frac{I_\varepsilon^*(t)}{(|t|^2 + 1)^n} dS(t),$$

where $I_\varepsilon^*(t) = I_\varepsilon(t)$ inside the ball $\{t : \varepsilon|t| < r\}$, and $I_\varepsilon^*(t) = 0$ outside this ball. In the last integral, we may take the limit as $\varepsilon \to +0$ under the integral sign by Lebesgue's dominated convergence theorem.

It remains to consider the integral

$$\int_{\partial D} d\sigma(\zeta) \int_{\partial D \backslash B(\zeta, r)} |f(\zeta) - f(z)|^p \left| \frac{\bar{\zeta}_m - \bar{z}_m^+}{|\zeta - z^+|^{2n}} - \frac{\bar{\zeta}_m - \bar{z}_m^-}{|\zeta - z^-|^{2n}} \right| d\sigma(z).$$

Since $|\zeta - z| \geq r$, we have $|\zeta - z^\pm| \geq \left| |\zeta - z| - |z - z^\pm| \right| \geq r - \varepsilon > \dfrac{r}{2}$. Then

$$\left| \frac{\bar{\zeta}_m - \bar{z}_m}{|\zeta - z^+|^{2n}} - \frac{\bar{\zeta}_m - \bar{z}_m}{|\zeta - z^-|^{2n}} \right|$$

$$\leq |\bar{\zeta}_m - \bar{z}_m| \, |z^+ - z^-| \sum_{j=0}^{2n-1} \frac{1}{|\zeta - z^+|^{j+1} |\zeta - z^-|^{2n-j}} \leq d_1 \varepsilon,$$

while

$$\left| \frac{\varepsilon v_m}{|\zeta - z^+|^{2n}} + \frac{\varepsilon v_m}{|\zeta - z^-|^{2n}} \right| \leq D_2 \varepsilon,$$

that is,

$$\int_{\partial D} d\sigma(\zeta) \int_{\partial D \backslash B(\zeta, r)} |f(\zeta) - f(z)|^p \left| \frac{\bar{\zeta}_m - \bar{z}_m^+}{|\zeta - z^+|^{2n}} - \frac{\bar{\zeta}_m - \bar{z}_m^-}{|\zeta - z^-|^{2n}} \right| d\sigma(z)$$

$$\leq d_3 \varepsilon \left(\int_{\partial D} |f|^p d\sigma \right)^2.$$

Inequality (2.2.5) is proved analogously. □

2.3 Boundary Behavior of Derivatives of the Bochner–Martinelli Integral

2.3.1 Formulas for Finding Derivatives

Suppose D is a bounded domain with a piecewise-smooth boundary, $f \in \mathscr{C}^1(\partial D)$, and Mf is the Bochner–Martinelli integral (1.2.4).

Lemma 2.3.1 *Derivatives of Mf may be found by the formulas*

$$\frac{\partial (Mf)}{\partial z_m} = \int_{\partial D} \frac{\partial f}{\partial \zeta_m} U(\zeta, z) + (-1)^{n+m} \int_{\partial D} \sum_{s=1}^{n} \frac{\partial f}{\partial \bar{\zeta}_s} \frac{\partial g}{\partial \zeta_s} d\bar{\zeta} \wedge d\zeta[m], \quad (2.3.1)$$

$$\frac{\partial (Mf)}{\partial \bar{z}_m} = \int_{\partial D} \frac{\partial f}{\partial \bar{\zeta}_m} U(\zeta, z) + (-1)^{m} \int_{\partial D} \sum_{s=1}^{n} \frac{\partial f}{\partial \bar{\zeta}_s} \frac{\partial g}{\partial \zeta_s} d\bar{\zeta}[m] \wedge d\zeta, \quad (2.3.2)$$

where $g = g(\zeta, z)$ is the fundamental solution to Laplace's equation (see Sect. 1.1).

Proof We prove, for example, formula (2.3.1), formula (2.3.2) is proved analogously. Recall that

$$U(\zeta, z) = \sum_{k=1}^{n} (-1)^{k-1} \frac{\partial g}{\partial \zeta_k} (\zeta, z) \, d\bar{\zeta}[k] \wedge d\zeta.$$

Now

$$\frac{\partial (Mf)}{\partial z_m} = - \int_{\partial D} f(\zeta) \frac{\partial}{\partial \zeta_m} U(\zeta, z) = - \int_{\partial D} \frac{\partial}{\partial \zeta_m} (fU) + \int_{\partial D} \frac{\partial f}{\partial \zeta_m} U(\zeta, z),$$

however

$$(-1)^{k} \int_{\partial D} \frac{\partial}{\partial \zeta_m} \left(f \frac{\partial g}{\partial \zeta_k} \right) d\bar{\zeta}[k] \wedge d\zeta = (-1)^{n+m} \int_{\partial D} \frac{\partial}{\partial \bar{\zeta}_k} \left(f \frac{\partial g}{\partial \zeta_k} \right) d\bar{\zeta} \wedge d\zeta[m],$$

since

$$d \left(f \frac{\partial g}{\partial \zeta_k} \right) d\bar{\zeta}[k] \wedge d\zeta[m] = (-1)^{k-1} \frac{\partial}{\partial \bar{\zeta}_k} \left(f \frac{\partial g}{\partial \zeta_k} \right) d\bar{\zeta} \wedge d\zeta[m]$$

$$+ (-1)^{n+m} \frac{\partial}{\partial \zeta_m} \left(f \frac{\partial g}{\partial \zeta_k} \right) d\bar{\zeta}[k] \wedge d\zeta.$$

Consequently

$$\frac{\partial (Mf)}{\partial z_m} = \int_{\partial D} \frac{\partial f}{\partial \zeta_m} U(\zeta, z) + (-1)^{n+m} \sum_{k=1}^{n} \int_{\partial D} \frac{\partial}{\partial \bar{\zeta}_k} \left(f \frac{\partial g}{\partial \zeta_k} \right) d\bar{\zeta} \wedge d\zeta[m]$$

$$= \int_{\partial D} \frac{\partial f}{\partial \zeta_m} U(\zeta, z) + (-1)^{n+m} \sum_{k=1}^{n} \int_{\partial D} \frac{\partial f}{\partial \bar{\zeta}_k} \frac{\partial g}{\partial \zeta_k} d\bar{\zeta} \wedge d\zeta[m],$$

due to g being a harmonic function. ☐

Now consider a domain D with a boundary of class $\mathscr{C}^2$, and suppose that $f \in \mathscr{C}^1(\partial D)$. If $D = \{z \in \mathbb{C}^n : \rho(z) < 0\}$ and $\rho \in \mathscr{C}^2(\overline{D})$ with $d\rho \neq 0$ on ∂D, we denote

$$\rho_k = \frac{\partial \rho}{\partial z_k} \frac{1}{|\operatorname{grad} \rho|} \quad \text{and } \rho_{\bar{k}} = \bar{\rho}_k. \text{ The surface area element is then}$$

$$d\sigma = i^{-n} 2^{1-n} \sum_{k=1}^{n} (-1)^{n+k-1} \rho_{\bar{k}} d\zeta[k] \wedge d\bar{\zeta}\Big|_{\partial D}$$

$$= i^{-n} 2^{1-n} \sum_{k=1}^{n} (-1)^{k-1} \rho_k d\bar{\zeta}[k] \wedge d\zeta\Big|_{\partial D}$$

(see Lemma 2.1.1).

Lemma 2.3.2 *For $z \notin \partial D$, let*

$$\Phi(z) = i^n 2^{n-1} \int_{\partial D} f(\zeta) g(\zeta, z) d\sigma(\zeta)$$

be a single-layer potential. Then

$$\frac{\partial \Phi}{\partial z_m} = -\int_{\partial D} f(\zeta) \rho_m(\zeta) U(\zeta, z)$$

$$+ i^n 2^{n-1} \sum_{k=1}^{n} \int_{\partial D} \left[\rho_k \frac{\partial}{\partial \zeta_m}(f\rho_{\bar{k}}) - \rho_m \frac{\partial}{\partial \zeta_k}(f\rho_{\bar{k}}) \right] g(\zeta, z) \, d\sigma(\zeta), \qquad (2.3.3)$$

$$\frac{\partial \Phi}{\partial \bar{z}_m} = -\int_{\partial D} f(\zeta) \rho_{\bar{m}}(\zeta) U(\zeta, z)$$

$$+ i^n 2^{n-1} \sum_{k=1}^{n} \int_{\partial D} \left[\rho_k \frac{\partial}{\partial \bar{\zeta}_m}(f\rho_{\bar{k}}) - \rho_{\bar{m}} \frac{\partial}{\partial \zeta_k}(f\rho_{\bar{k}}) \right] g(\zeta, z) \, d\sigma(\zeta). \qquad (2.3.4)$$

Proof We have

$$\frac{\partial \Phi}{\partial z_m} = -\int_{\partial D} f(\zeta) \frac{\partial g}{\partial \zeta_m} \sum_{k=1}^{n} \rho_{\bar{k}} (-1)^{n+k-1} d\zeta[k] \wedge d\bar{\zeta}$$

$$= \sum_{k=1}^{n} (-1)^{n+k-1} \int_{\partial D} \frac{\partial}{\partial \zeta_m}(f\rho_{\bar{k}}) g(\zeta, z) \, d\zeta[k] \wedge d\bar{\zeta}$$

$$- \sum_{k=1}^{n} (-1)^{n+k-1} \int_{\partial D} \frac{\partial}{\partial \zeta_m}(f\rho_{\bar{k}} g) \, d\zeta[k] \wedge d\bar{\zeta}.$$

Just as in Lemma 2.3.1, we obtain

$$(-1)^{n+k} \int_{\partial D} \frac{\partial}{\partial \bar{\zeta}_m} (f\rho_{\bar{k}} g)\, d\zeta[k] \wedge d\bar{\zeta} = (-1)^{n+m} \int_{\partial D} \frac{\partial}{\partial \bar{\zeta}_k} (f\rho_{\bar{k}} g)\, d\zeta[m] \wedge d\bar{\zeta}.$$

Therefore

$$\frac{\partial \Phi}{\partial z_m} = \sum_{k=1}^{n} (-1)^{n+k-1} \int_{\partial D} \frac{\partial}{\partial \bar{\zeta}_m} (f\rho_{\bar{k}}) g(\zeta, z)\, d\zeta[k] \wedge d\bar{\zeta}$$

$$+ (-1)^{n+m} \sum_{k=1}^{n} \int_{\partial D} \frac{\partial}{\partial \bar{\zeta}_k} (f\rho_{\bar{k}}) g(\zeta, z)\, d\zeta[m] \wedge d\bar{\zeta} - i^n 2^{n-1} \sum_{k=1}^{n} \int_{\partial D} f\rho_{\bar{k}} \frac{\partial g}{\partial \bar{\zeta}_k}\, \rho_m d\sigma$$

$$= \sum_{k=1}^{n} \int_{\partial D} \left[(-1)^{n+k-1} \frac{\partial}{\partial \bar{\zeta}_m} (f\rho_{\bar{k}})\, d\zeta[k] \wedge d\bar{\zeta} \right.$$

$$\left. + (-1)^{n+m} \frac{\partial}{\partial \bar{\zeta}_k} (f\rho_{\bar{k}})\, d\zeta[m] \wedge d\bar{\zeta} \right] g(\zeta, z) - \int_{\partial D} f\rho_m U(\zeta, z).$$

Formula (2.3.4) is proved analogously.   □

Theorem 2.3.1 *If $\partial D \in \mathscr{C}^2$ and $f \in \mathscr{C}^2(\partial D)$, then the integral Mf extends to $\bar{D}$ and to $\mathbb{C}^n \setminus D$ as a function of class $\mathscr{C}^{1+\alpha}$ for $0 < \alpha < 1$. Moreover*

$$\frac{\partial (Mf)}{\partial z_m} = \int_{\partial D} \left(\frac{\partial f}{\partial \zeta_m} - \rho_m \sum_{k=1}^{n} \rho_k \frac{\partial f}{\partial \bar{\zeta}_k} \right) U(\zeta, z)$$

$$+ i^n 2^{n-1} \int_{\partial D} \psi_1(\zeta) g(\zeta, z)\, d\sigma(\zeta), \qquad (2.3.5)$$

where

$$\psi_1 = \sum_{s,k=1}^{n} \left[\rho_k \frac{\partial}{\partial \zeta_s} \left(\rho_m \rho_{\bar{k}} \frac{\partial f}{\partial \bar{\zeta}_s} \right) - \rho_m \frac{\partial}{\partial \zeta_k} \left(\rho_m \rho_{\bar{k}} \frac{\partial f}{\partial \bar{\zeta}_s} \right) \right],$$

and

$$\frac{\partial (Mf)}{\partial \bar{z}_m} = \int_{\partial D} \left(\frac{\partial f}{\partial \bar{\zeta}_m} - \rho_{\bar{m}} \sum_{k=1}^{n} \rho_k \frac{\partial f}{\partial \bar{\zeta}_k} \right) U(\zeta, z)$$

$$+ i^n 2^{n-1} \int_{\partial D} \psi_2(\zeta) g(\zeta, z)\, d\sigma(\zeta), \qquad (2.3.6)$$

where

$$\psi_2 = \sum_{s,k=1}^{n} \left[\rho_k \frac{\partial}{\partial \bar{\zeta}_s} \left(\rho_{\bar{m}} \rho_{\bar{k}} \frac{\partial f}{\partial \bar{\zeta}_s} \right) - \rho_{\bar{m}} \frac{\partial}{\partial \zeta_k} \left(\rho_{\bar{m}} \rho_{\bar{k}} \frac{\partial f}{\partial \bar{\zeta}_s} \right) \right].$$

Proof Formulas (2.3.5) and (2.3.6) follow from Lemmas 2.3.1 and 2.3.2 while the boundary behavior of the integral Mf follows from Theorem 2.1.1 and the properties of the single-layer potential.   □

Formulas (2.3.1)–(2.3.6) are essentially classical formulas of the potential theory.

2.3.2   Jump Theorem for Derivatives

Corollary 2.3.1 *If $\partial D \in \mathscr{C}^2$ and $f \in \mathscr{C}^2(\partial D)$, then the jump of the derivatives of Mf is given by*

$$\frac{\partial (M^+ f)}{\partial z_m} - \frac{\partial (M^- f)}{\partial z_m} = \frac{\partial f}{\partial z_m} - \rho_m \sum_{k=1}^{n} \frac{\partial f}{\partial \bar{z}_k} \rho_k, \quad z \in \partial D,$$

$$\frac{\partial (M^+ f)}{\partial \bar{z}_m} - \frac{\partial (M^- f)}{\partial \bar{z}_m} = \frac{\partial f}{\partial \bar{z}_m} - \rho_{\bar{m}} \sum_{k=1}^{n} \frac{\partial f}{\partial \bar{z}_k} \rho_k, \quad z \in \partial D.$$

(2.3.7)

If we are only concerned with the jump of derivatives (that is, with formula (2.3.7)), then we can weaken the conditions on ∂D and on f.

Let D be a bounded domain with a boundary of class $\mathscr{C}^1$ and $D = \{z \in \mathbb{C}^n : \rho(z) < 0\}$, where $\rho \in \mathscr{C}^1(\mathbb{C}^n)$, and $d\rho \neq 0$ on ∂D. If $z \in \partial D$, then $z^+ \in D$ and $z^- \notin \overline{D}$ and we denote points on the normal to ∂D at z such that $|z^+ - z| = |z^- - z|$.

Lemma 2.3.3 *Let*

$$\Phi_{m,\bar{k}}(z) = \int_{\partial D} \frac{\partial g(\zeta, z)}{\partial \zeta_m} d\bar{\zeta}[k] \wedge d\zeta, \quad z \notin \partial D,$$

$$\Phi_{\bar{m},k}(z) = \int_{\partial D} \frac{\partial g(\zeta, z)}{\partial \bar{\zeta}_m} d\zeta[k] \wedge d\bar{\zeta}, \quad z \notin \partial D.$$

Then

$$\lim_{z^\pm \to z} \left(\Phi_{m,\bar{k}}(z^+) - \Phi_{m,\bar{k}}(z^-) \right) = (-1)^{k-1} \rho_{\bar{k}} \rho_m,$$

$$\lim_{z^\pm \to z} \left(\Phi_{\bar{m},k}(z^+) - \Phi_{\bar{m},k}(z^-) \right) = (-1)^{k-1} \rho_k \rho_{\bar{m}},$$

and these limits are uniformly attained in z.

The proof is analogues to the proof of Theorem 2.2.1.

Theorem 2.3.2 (Aronov) *Suppose* $f \in \mathscr{C}^1(\partial D)$, *and* Mf *is the Bochner–Martinelli integral (1.2.4). Then*

$$\lim_{z^\pm \to z} \left(\frac{\partial(Mf(z^+))}{\partial \bar{z}_k^+} - \frac{\partial(Mf(z^-))}{\partial \bar{z}_k^-} \right) = \frac{\partial f}{\partial \bar{z}_k} - \rho_{\bar{k}} \sum_{m=1}^n \rho_m \frac{\partial f}{\partial \bar{z}_m}, \qquad (2.3.8)$$

$$\lim_{z^\pm \to z} \left(\frac{\partial(Mf(z^+))}{\partial z_k^+} - \frac{\partial(Mf(z^-))}{\partial z_k^-} \right) = \frac{\partial f}{\partial z_k} - \rho_k \sum_{m=1}^n \rho_m \frac{\partial f}{\partial \bar{z}_m}, \qquad (2.3.9)$$

and these limits are uniformly attained in $z \in \partial D$.

Proof By (2.3.2), we have

$$\frac{\partial(Mf)}{\partial \bar{z}_k} = \int_{\partial D} \frac{\partial f}{\partial \bar{\zeta}_k} U(\zeta, z) + (-1)^k \int_{\partial D} \sum_{m=1}^n \frac{\partial f}{\partial \bar{\zeta}_m} \frac{\partial g}{\partial \zeta_m} d\bar{\zeta}[k] \wedge d\zeta.$$

By Theorem 2.2.1, the jump of the first integral equals $\dfrac{\partial f}{\partial \bar{z}_k}$, and we represent the second integral in the form

$$\int_{\partial D} \frac{\partial f}{\partial \bar{\zeta}_m} \frac{\partial g}{\partial \zeta_m} d\bar{\zeta}[k] \wedge d\zeta$$

$$= \frac{\partial f}{\partial \bar{z}_m} \int_{\partial D} \frac{\partial g}{\partial \zeta_m} d\bar{\zeta}[k] \wedge d\zeta + \int_{\partial D} \left(\frac{\partial f}{\partial \bar{\zeta}_m} - \frac{\partial f}{\partial \bar{z}_m} \right) \frac{\partial g}{\partial \zeta_m} d\bar{\zeta}[k] \wedge d\zeta.$$

By Lemma 2.3.3, the jump of the first integral equals

$$(-1)^{k-1} \frac{\partial f}{\partial \bar{z}_m} \rho_{\bar{k}}(z)\rho_m(z),$$

and the jump of the second integral is zero (this is proved the same way as in Theorem 2.2.1). □

We obtain the following assertion from Theorem 2.3.2 by induction.

Corollary 2.3.2 *If* $\partial D \in \mathscr{C}^k$ *and* $f \in \mathscr{C}^m(\partial D)$, *where* $m \leq k$, *and* $M^+f \in \mathscr{C}^m(\overline{D})$, *then* $M^-f \in \mathscr{C}^m(\mathbb{C}^n \setminus D)$. *Conversely, if* $M^-f \in \mathscr{C}^m(\mathbb{C}^n \setminus D)$, *then* $M^+f \in \mathscr{C}^m(\overline{D})$.

Remark 2.3.1 Just as for Theorem 2.2.1, Theorem 2.3.2 can be obtained when f is differentiable on ∂D and all its derivatives are integrable on ∂D. Jump formulas (2.3.8) and (2.3.9) for derivatives will then hold at Lebesgue points of the derivatives of f.

Corollary 2.3.3 *If $\partial D \in \mathscr{C}^1$ and $f \in \mathscr{C}^1$, then the jump of the derivative*

$$\bar{\partial}_n(Mf) = \sum_{k=1}^{n} \frac{\partial(Mf)}{\partial \bar{z}_k} \rho_k \qquad (2.3.10)$$

is zero.

Expression (2.3.10) will be called a $\bar{\partial}$-normal derivative of Mf.

2.3.3   Jump Theorem for the $\bar{\partial}$-Normal Derivative

Corollary 2.3.3 shows that the jump of the $\bar{\partial}$-normal derivative $\bar{\partial}_n(Mf)$ of the Bochner–Martinelli integral is zero. It turns out that this assertion is valid even for continuous functions f if the boundary of the domain is assumed to be class $\mathscr{C}^2$ smooth.

In this case, we may take the defining function to be

$$\rho(z) = \begin{cases} - \inf\limits_{\zeta \in \partial D} |\zeta - z|, & z \in \overline{D}; \\[2mm] \inf\limits_{\zeta \in \partial D} |\zeta - z|, & z \in \mathbb{C}^n \setminus \overline{D}. \end{cases}$$

Then $D = \{z \in \mathbb{C}^n : \rho(z) < 0\}$. Moreover, when $\partial D \in \mathscr{C}^2$ we have the following (see, for example, [83, Sect. 2]):

1. There is a neighborhood V of ∂D such that $\rho \in \mathscr{C}^2(V)$;
2. $|\operatorname{grad} \rho| = \dfrac{1}{2}$ in V;
3. If $z^{\pm} \in V$ are the points on the normal to ∂D at z such that $|z^+ - z| = |z^- - z|$, then $\dfrac{\partial \rho}{\partial z_k}(z^{\pm}) = \dfrac{\partial \rho}{\partial z_k}(z)$ and $\dfrac{\partial \rho}{\partial \bar{z}_k}(z^{\pm}) = \dfrac{\partial \rho}{\partial \bar{z}_k}(z)$ for $k = 1, 2, \ldots, n$.

In this case $\rho_k = 2\dfrac{\partial \rho}{\partial z_k}$ and $\rho_{\bar{k}} = 2\dfrac{\partial \rho}{\partial \bar{z}_k}$. Hence

$$\bar{\partial}_n(Mf) = \sum_{k=1}^{n} \frac{\partial(Mf)}{\partial \bar{z}_k} \rho_k = 2 \sum_{k=1}^{n} \frac{\partial(Mf)}{\partial \bar{z}_k} \frac{\partial \rho}{\partial z_k}.$$

Theorem 2.3.3 ([45]) *If $f \in \mathscr{C}(\partial D)$, then the integral Mf of the form (1.2.4) satisfies*

$$\lim_{z^{\pm} \to z} \left(\bar{\partial}_n(Mf(z^+)) - \bar{\partial}_n(Mf(z^-)) \right) = 0.$$

This limit is attained uniformly with respect to $z \in \partial D$. If $\bar{\partial}_n(Mf(z^+))$ extends continuously to $\overline{D}$, then $\bar{\partial}_n(Mf(z^-))$ extends continuously to $\mathbb{C}^n \setminus D$, and vice versa.

Proof If f is constant, then $\bar{\partial}_n Mf \equiv 0$. Thus, we may assume that $f(z) = 0$ at the point $z \in \partial D$. By formula (2.1.3), the restriction of the kernel $U(\zeta, z)$ to ∂D has the form

$$\frac{(n-1)!}{\pi^n} \sum_{k=1}^n \frac{\partial \rho}{\partial \bar{\zeta}_k} \frac{(\bar{\zeta}_k - \bar{z}_k)}{|\zeta - z|^{2n}} \, d\sigma.$$

Consequently,

$$\bar{\partial}_n M(z^+) - \bar{\partial}_n M(z^-)$$

$$= -\frac{(n-1)!}{\pi^n} \int_{\partial D} f(\zeta) \sum_{k=1}^n \frac{\partial \rho(z)}{\partial z_k} \frac{\partial \rho(\zeta)}{\partial \bar{\zeta}_k} \left(\frac{1}{|\zeta - z^+|^{2n}} - \frac{1}{|\zeta - z^-|^{2n}} \right) d\sigma$$

$$+ \frac{n!}{\pi^n} \int_{\partial D} f(\zeta) \left[\left(\sum_{k=1}^n \frac{\partial \rho}{\partial z_k}(\bar{\zeta}_k - \bar{z}_k^+) \sum_{m=1}^n \frac{\partial \rho}{\partial \bar{\zeta}_m}(\bar{\zeta}_m - \bar{z}_m^+) \right) |\zeta - z^-|^{-2-2n} \right.$$

$$\left. - \left(\sum_{k=1}^n \frac{\partial \rho}{\partial z_k}(\bar{\zeta}_k - \bar{z}_k^-) \sum_{m=1}^n \frac{\partial \rho}{\partial \bar{\zeta}_m}(\bar{\zeta}_m - \bar{z}_m^-) \right) |\zeta - z^-|^{-2-2n} \right] d\sigma.$$

Denote the first integral by I_1 and the second one by I_2. Make a unitary transformation and translation so that z is taken to 0 and the tangent plane to ∂D at z is taken to the plane

$$T = \{ w \in \mathbb{C}^n : \operatorname{Im} w_n = 0 \}.$$

In a neighborhood of the origin, the boundary ∂D will be given by a system of equations

$$\zeta_1 = w_1, \ldots, \zeta_{n-1} = w_{n-1}, \quad \zeta_n = u_n + i\varphi(w),$$

where $w = (w_1, \ldots, w_{n-1}, u_n) \in T$. The function $\varphi(w)$ is class $\mathscr{C}^2$ in a neighborhood W of the origin, and $z^{\pm} = (0, \ldots, 0, \pm i y_n)$. The surface ∂D is the Lyapunov surface with the Hölder exponent equal to 1, so the following estimates hold [82, Sect. 22]:

$$|\varphi(w)| \leq C|w|^2, \qquad w \in W,$$

$$\left| \frac{\partial \varphi}{\partial u_j} \right| \leq C_1 |w|, \qquad j = 1, \ldots, n \qquad (2.3.11)$$

$$\left| \frac{\partial \varphi}{\partial v_j} \right| \leq C_1 |w|, \qquad j = 1, \ldots, n-1,$$

where $u_j = \operatorname{Re} w_j$ and $v_j = \operatorname{Im} w_j$. Since $\dfrac{\partial \varphi}{\partial w_j} = -\dfrac{\partial \rho}{\partial w_j} \Big/ \dfrac{\partial \rho}{\partial y_n}$, and $\left|\dfrac{\partial \rho}{\partial y_n}\right| \geq C_2 > 0$

for $w \in W$, it follows that

$$\left|\frac{\partial \rho}{\partial \zeta_k}(\zeta(w))\right| \leq C_3 |w|, \qquad \left|\frac{\partial \rho}{\partial \bar\zeta_k}(\zeta(w))\right| \leq C_3 |w| \tag{2.3.12}$$

for $w \in W$ and $k = 1, \dots, n-1$.

We note that the constants do not depend on the point z under consideration. Finally,

$$|\zeta(w)| \leq C_4 |w|. \tag{2.3.13}$$

We fix $\varepsilon > 0$, take a ball B' in the plane T with the center at the origin, and choose $a > 0$ such that

1. $B' \subset W$,
2. $|f(\zeta(w))| < \varepsilon$ for $w \in B'$,
3. $\{z \in \mathbb{C}^n : (z_1, \dots, z_{n-1}, \operatorname{Re} z_n) \in B', \ |\operatorname{Im} z_n| < a\} \subset W$,
4. $C(2|y_n| + C|w|^2) \leq d < 1$ for $|y_n| < a$ and $w \in B'$ (the constant C being borrowed from (2.3.11)).

Since $z^{\pm} = (0, \dots, 0, \pm i y_n)$, the identity

$$|\zeta(w) - z^{\pm}|^2 = |w|^2 + (\pm y_n - \varphi(w))^2$$

holds. Hence

$$|\zeta - z^{\pm}|^{-2} = |w - z^{\pm}|^{-2} \left(1 - (\pm 2\varphi y_n - \varphi^2)|w - z^{\pm}|^{-2}\right)^{-1}.$$

But

$$\frac{|\pm 2\varphi y_n - \varphi^2|}{|w - z^{\pm}|^2} \leq \frac{C|w|^2(2|y_n| - C|w|^2)}{|w|^2 + y_n^2} \leq C(2|y_n| + C|w|^2) \leq d < 1$$

for $|y_n| \leq a$ and $w \in B'$. Consequently

$$\left(1 - (\pm 2\varphi y_n - \varphi^2)|w - z^{\pm}|^{-2}\right)^{-1} = \sum_{k=0}^{\infty} \frac{(\pm 2\varphi y_n - \varphi^2)^k}{|w - z^{\pm}|^{2k}}$$

$$= 1 + (\pm 2\varphi y_n - \varphi^2)|w - z^{\pm}|^{-2} h(w, z)$$

and the function $h(w, z)$ is uniformly bounded for $w \in B'$ and $|y_n| \leq a$. Therefore

$$|\zeta - z^\pm|^{-2n} = |w - z^\pm|^{-2n}\big(1 + (\pm 2\varphi y_n - \varphi^2)|w - z^\pm|^{-2}h_1(w, z)\big), \qquad (2.3.14)$$

$$|\zeta - z^\pm|^{-2-2n} = |w - z^\pm|^{-2-2n}\big(1 + (\pm 2\varphi y_n - \varphi^2)|w - z^\pm|^{-2}h_2(w, z)\big) \qquad (2.3.15)$$

and the functions h_1 and h_2 are uniformly bounded for $w \in B'$ and $|y_n| \leq a$.

We set $\Gamma = \{\zeta \in \partial D : \zeta = \zeta(w),\ w \in B'\}$ and estimate the integral I_1 over the surface Γ. Using (2.3.14) and (2.3.15), we obtain

$$\big||\zeta - z^+|^{-2n} - |\zeta - z^-|^{-2n}\big| \leq 2(|2\varphi y_n| + \varphi^2)|w - z^+|^{-2-2n}|h_1|$$

$$\leq C_5(2|y_n| + C|w|^2)(|w|^2 + y_n^2)^{-n}.$$

Supposing that $d\sigma \leq C_6 dS$, where dS is the surface area element of the plane T, we have

$$|I_{1,\Gamma}| = \frac{(n-1)!}{\pi^n}\left| \int_\Gamma f(\zeta) \sum_{k=1}^n \frac{\partial \rho}{\partial z_k}\frac{\partial \rho}{\partial \bar{\zeta}_k}\big(|\zeta - z^+|^{-2n} - |\zeta - z^-|^{-2n}\big)\, d\sigma \right|$$

$$\leq \varepsilon C_7 \int_{B'} (2|y_n| + C|w|^2)(|w|^2 + y_n^2)^{-n} dS.$$

Now

$$\int_{B'} |y_n|(|w|^2 + y_n^2)^{-n} dS \leq \int_T |y_n|(|w|^2 + y_n^2)^{-n} dS = \text{const},$$

while

$$\int_{B'} |w|^2(|w|^2 + y_n^2)^{-n} dS \leq \int_{B'} (|w|^2 + y_n^2)^{1-n} dS.$$

Introducing polar coordinates in the ball B', we have $dS = |w|^{2n-2}d|w| \wedge d\omega$, where $d\omega$ is the surface area element in the unit sphere in $\mathbb{R}^{2n-1}$, so

$$\int_{B'} (|w|^2 + y_n^2)^{1-n} dS = \sigma_{2n-1}\int_0^R (|w|^2 + y_n^2)^{1-n} d|w| \leq R\sigma_{2n-1}.$$

Here R is the radius of the ball B', and σ_{2n-1} is the area of the unit sphere in $\mathbb{R}^{2n-1}$. Therefore $|I_{1,\Gamma}| \leq C_8\varepsilon$, where the constant C_8 is independent of z and y_n. Obviously the integral I_1 over the surface $\partial D \setminus \Gamma$ can be made as small as desired as $z^\pm \to 0$.

We now show that the form of the integral I_2 is not affected by the unitary transformation. Indeed, the distance does not change, so the functions ρ, $d\sigma$, and

$|\zeta - z|$ do not change either. Consider the expression

$$\sum_{k=1}^{n} \frac{\partial \rho}{\partial z_k} (\zeta_k - z_k).$$

Suppose the unitary transformation is given by the matrix $A = \|a_{jk}\|_{j,k=1}^{n}$, i.e., by

$$z_k' = \sum_{j=1}^{n} a_{jk} z_j, \quad k = 1, \ldots, n,$$

and the inverse transformation is given by the matrix $B = \|b_{jk}\|_{j,k=1}^{n}$. Then

$$\sum_{k=1}^{n} a_{kj} b_{sk} = \delta_{js},$$

where δ_{js} is the Kronecker symbol. Therefore

$$\sum_{k=1}^{n} \frac{\partial \rho}{\partial z_k} (\zeta_k - z_k) = \sum_{k,j,s=1}^{n} \frac{\partial \rho}{\partial z_j'} a_{kj} b_{sk} (\zeta_s' - z_s')$$

$$= \sum_{j,s=1}^{n} \frac{\partial \rho}{\partial z_j'} \delta_{js} (\zeta_s' - z_s') = \sum_{j=1}^{n} \frac{\partial \rho}{\partial z_j'} (\zeta_s' - z_s').$$

It can be shown in the same way that the sum

$$\sum_{k=1}^{n} \frac{\partial \rho}{\partial \bar{\zeta}_k} (\bar{\zeta}_k - \bar{z}_k)$$

does not change. Thus, the form of the integral I_2 is invariant under unitary transformation. Then

$$\sum_{k=1}^{n} \frac{\partial \rho}{\partial z_k} (0)(\zeta_k - z_k^{\pm}) \sum_{m=1}^{n} \frac{\partial \rho}{\partial \bar{\zeta}_m} (\bar{\zeta}_m - \bar{z}_m^{\pm})$$

$$= -\frac{i}{2} (\zeta_n - z_n^{\pm}) \sum_{m=1}^{n} \frac{\partial \rho}{\partial \bar{\zeta}_m} (\bar{\zeta}_m - \bar{z}_m^{\pm})$$

$$= -\frac{i}{2} \sum_{m=1}^{n-1} \frac{\partial \rho}{\partial \bar{\zeta}_m} \bar{\zeta}_m (\zeta_n - z_n^{\pm}) - \frac{i}{2} \frac{\partial \rho}{\partial \bar{\zeta}_n} (u_n^2 + \varphi^2 + y_n^2 \mp 2\varphi y_n).$$

We split the integral I_2 over the surface Γ into three integrals:

$$I_2' = \frac{in!}{\pi^n} \int_{B'} f(\zeta(w)) \left(2\varphi y_n \frac{\partial\rho}{\partial\bar{\zeta}_n} + \sum_{m=1}^{n-1} \frac{\partial\rho}{\partial\bar{\zeta}_m} \bar{\zeta}_m y_n \right) |w - z^+|^{-2-2n} d\sigma',$$

$$I_2^{\pm} = \pm \frac{in!}{2\pi^n} \int_{B'} f(\zeta(w)) \left(\frac{\partial\rho}{\partial\bar{\zeta}_n}(u_n^2 + \varphi^2 + y_n^2 \pm 2\varphi y_n) \right.$$

$$\left. + \sum_{m=1}^{n-1} \frac{\partial\rho}{\partial\bar{\zeta}_m} \bar{\zeta}_m(\zeta_n - z_n^{\pm}) \right) (\pm 2\varphi y_n - \varphi^2) h_2 |w - z^+|^{-4-2n} d\sigma',$$

where $d\sigma'$ is the image of $d\sigma$ under the mapping $w \to \zeta(w)$, and h_2 is defined in (2.3.15). Using (2.3.11)–(2.3.15), we find that

$$|I_2'| \leq M_1\varepsilon \int_{B'} (M_2|w|^3|y_n| + M_3|w|^2|y_n|)|w - z^+|^{-2-2n} dS$$

$$\leq M_4\varepsilon \int_T |y_n|(|w|^2 + y_n^2)^{-n} dS = M_5\varepsilon.$$

Now

$$|I_2^{\pm}|$$

$$\leq M_6\varepsilon \int_{B'} |w|^2(2|y_n| + C|w|^2)(M_7|w|^2 + M_8|w|^2|y_n| + M_9 y_n^2)(|w|^2 + y_n^2)^{-2-n} dS$$

$$\leq M_{10}\varepsilon \int_{B'} |y_n|(|w|^2 + y_n^2)^{-n} dS + M_{11}\varepsilon \int_{B'} (|w|^2 + y_n^2)^{1-n} dS \leq M_{12}\varepsilon.$$

The integral I_2 over $\partial D \setminus \Gamma$ also tends to zero. $\square$

Theorem 2.3.3 is an analogue of Lyapunov's theorem on the jump of the normal derivative of a double-layer potential. Just as in Theorem 2.2.1, it can be shown that for $f \in \mathscr{L}^1(\partial D)$, the difference

$$\bar{\partial}_n(Mf(z^+)) - \bar{\partial}_n(Mf(z^-)) \to 0$$

as $z^{\pm} \to z$ at Lebesgue points of f.

Remark 2.3.2   Theorem 2.3.3 does not hold for the derivative

$$\partial_n(Mf) = \sum_{k=1}^n \frac{\partial(Mf)}{\partial\bar{z}_k} \rho_{\bar{k}}.$$

2.4 The Hodge Operator

Let us define the Hodge star operator ($*$) for differential forms with respect to the Euclidean metric in $\mathbb{C}^n$ (see, for example, [84, Chap. 5, Sect. 1]). Consider a differential form of type (p, q)

$$\gamma = {\sum_{I,J}}' \gamma_{I,J}(z)\, dz_I \wedge d\bar{z}_J, \tag{2.4.1}$$

where $I = (i_1, \ldots, i_p)$ and $J = (j_1, \ldots, j_q)$ are the multi-indices of order p and q respectively, and $0 \le p, q \le n$. The prime on the summation sign indicates that the sum is taken over increasing multi-indices

$$1 \le i_1 < \ldots < i_p \le n, \quad 1 \le j_1 < \ldots < j_q \le n.$$

Let the differential forms dz_I and $d\bar{z}_J$ have the form $dz_I = dz_{i_1} \wedge \ldots \wedge dz_{i_p}$, $d\bar{z}_J = d\bar{z}_{j_1} \wedge \ldots \wedge d\bar{z}_{j_q}$. Then

$$*\gamma = \sum_{I,J} \gamma_{I,J}(z) * (dz_I \wedge d\bar{z}_J),$$

and

$$*(dz_I \wedge d\bar{z}_J) = 2^{p+q-n}(-1)^{np} i^n \sigma(I)\sigma(J) dz[J] \wedge d\bar{z}[I],$$

where the form $dz[J]$ is obtained from dz by eliminating the differentials $dz_{j_1}, \ldots,$ dz_{j_q}, and the symbols $\sigma(I)$ is defined by $dz_I \wedge dz[I] = \sigma(I)dz$. Thus, the form $*\gamma$ is a form of type $(n - q, n - p)$.

We now dwell on the basic properties of the Hodge operator.

Lemma 2.4.1 *If γ and φ are forms of type (p, q), then*

1. $* * \gamma = (-1)^{p+q}\gamma$,
2. $dz_I \wedge d\bar{z}_J \wedge *(\overline{dz_I \wedge d\bar{z}_J}) = 2^{p+q}dv$, *where dv is the volume form in $\mathbb{C}^n$,*
3. $* * (dz_I \wedge d\bar{z}_J) = (-1)^{p+q}dz_I \wedge d\bar{z}_J$,
4. $\overline{*\gamma} = *\bar{\gamma}$,
5. $*\gamma \wedge \bar{\varphi} = (-1)^{p+q}\gamma \wedge *\bar{\varphi}$.

This lemma is well known (see, for example, [84, Chap. 5]) and follows directly from the definition of the Hodge operator.

A scalar product (γ, φ) may be defined for (p, q)-forms γ and φ with coefficients of class $\mathscr{L}^2(D)$ by

$$(\gamma, \varphi) = \int_D \gamma \wedge *\bar{\varphi}.$$

This scalar product is called the Hodge product. Then $\|\gamma\| = \sqrt{(\gamma, \gamma)}$ is the Hodge norm.

By using the Hodge operator, it is easy to find the operators $\bar{\partial}^*$ and ∂^* formally dual to $\bar{\partial}$ and ∂, namely $\bar{\partial}^* = - * \partial *$ and $\partial^* = - * \bar{\partial}*$. Recall that for the form γ the operators $\bar{\partial}$ and ∂ are defined by the following equations:

$$\bar{\partial}\gamma = \sum_{k=1}^{n} {\sum_{I,J}}' \frac{\partial \gamma_{I,J}}{\partial \bar{z}_k} \, d\bar{z}_k \wedge dz_I \wedge d\bar{z}_J, \quad \partial\gamma = \sum_{k=1}^{n} {\sum_{I,J}}' \frac{\partial \gamma_{I,J}}{\partial z_k} \, dz_k \wedge dz_I \wedge d\bar{z}_J.$$

Let us find, for example, $\bar{\partial}^*$. If γ is a $(p, q - 1)$-form and φ is a (p, q)-form, γ and φ have smooth coefficients of class $\mathscr{L}^2(D)$ and φ has compact support in D, then $(\bar{\partial}\gamma, \varphi) = (\gamma, \bar{\partial}^*\varphi)$ and

$$(\bar{\partial}\gamma, \varphi) = \int_D \bar{\partial}\gamma \wedge *\bar{\varphi} = \int_D d\gamma \wedge *\bar{\varphi} = \int_D d(\gamma \wedge *\bar{\varphi}) + (-1)^{p+q} \int_D \gamma \wedge d * \bar{\varphi}$$

$$= (-1)^{p+q} \int_D \gamma \wedge \bar{\partial} * \bar{\varphi} = -\int_D f \wedge *(\overline{* \partial * \varphi}),$$

so $\bar{\partial}^* = - * \partial *$. In just the same way, we can see that $\partial^* = - * \bar{\partial}*$. The operator $\bar{\partial}^*$ carries forms of type (p, q) into forms of type $(p, q - 1)$. By definition, $\bar{\partial}^* = 0$ for forms of type $(p, 0)$.

Example 2.4.1 If f is a smooth function, then

$$*\bar{\partial}f = *\left(\sum_{k=1}^{n} \frac{\partial f}{\partial \bar{z}_k} \, d\bar{z}_k \right) = 2^{1-n} i^n \sum_{k=1}^{n} (-1)^{k-1} \frac{\partial f}{\partial \bar{z}_k} \, dz[k] \wedge d\bar{z} = 2^{1-n} i^n (-1)^n \mu_f.$$

Hence $\mu_f = i^n 2^{n-1} (*\bar{\partial}f)$.

Example 2.4.2 $U(\zeta, z) = 2^{n-1} i^n (*\partial g(\zeta, z))$.
We consider the operator

$$\Box = \bar{\partial}^* \bar{\partial} + \bar{\partial}\bar{\partial}^*,$$

which is known as the complex Laplacian. If φ is a function, then

$$\Box\varphi = \bar{\partial}^* \bar{\partial}\varphi = \bar{\partial}^* \sum_{k=1}^{n} \frac{\partial \varphi}{\partial \bar{z}_k} \, d\bar{z}_k = - * \partial \sum_{k=1}^{n} 2^{1-n} i^n (-1)^{k-1} \frac{\partial \varphi}{\partial \bar{z}_k} \, dz[k] \wedge d\bar{z}$$

$$= - * 2^{1-n} i^n \sum_{k=1}^{n} \frac{\partial^2 \varphi}{\partial \bar{z}_k \partial z_k} \, dz \wedge d\bar{z} = -2 \sum_{k=1}^{n} \frac{\partial^2 \varphi}{\partial \bar{z}_k \partial z_k} = -2\Delta\varphi,$$

i.e., $\square = -2\Delta$ for the functions and this identity continues to hold for the forms as well (see, for example, [28, p. 106]). Thus, in $\mathbb{C}^n$ the harmonic forms in the sense of $\square$ are forms with harmonic coefficients. It is also easy to show that $\square = \partial\partial^* + \partial^*\partial$.

2.5 Holomorphic Functions Represented by the Bochner–Martinelli Integral

2.5.1 Statement of the $\bar{\partial}$-Neumann Problem

Suppose $n > 1$, and $D = \{z \in \mathbb{C}^n : \rho(z) < 0\}$ is a bounded domain in $\mathbb{C}^n$ with a boundary of class $\mathscr{C}^1$, where ρ is the defining function. If $f \in \mathscr{C}^1(\overline{D})$, then denote

$$\bar{\partial}_n f = \sum_{k=1}^{n} \frac{\partial f}{\partial \bar{z}_k} \rho_k,$$

where $\rho_k = \dfrac{\partial \rho}{\partial z_k} \dfrac{1}{|\partial \rho|}$. The derivative $\bar{\partial}_n f$ is the $\bar{\partial}$-normal derivative of function f. We say that the tangential part $\bar{\partial}_\tau f$ of $\bar{\partial}f$ equals zero on ∂D if

$$\int_{\partial D} \bar{\partial}f \wedge \varphi = 0$$

for all forms $\varphi \in \mathscr{D}^{n,n-2}(\mathbb{C}^n)$. (Here $\mathscr{D}^{n,n-2}(\mathbb{C}^n)$ is the space of differential forms of type $(n, n-2)$ with coefficients of space $\mathscr{D}(\mathbb{C}^n)$.) But this means precisely that $f \wedge \partial \rho = 0$ on ∂D.

If we write the form as $\bar{\partial}f = \bar{\partial}_\tau f + \lambda \dfrac{\partial \rho}{|\partial \rho|}$, then $\lambda = \bar{\partial}_n f$. If we denote the outer unit normal to ∂D at z by $\nu(z)$, and $s(z) = i\nu(z)$, then

$$\bar{\partial}_n f = \frac{1}{2}\left(\frac{\partial f}{\partial \nu} + i\frac{\partial f}{\partial s} \right).$$

On the other hand from Example 2.4.1 and the equalities

$$d\bar{\zeta}[k] \wedge d\zeta\big|_{\partial D} = 2^{n-1} i^n (-1)^{k-1} \bar{\rho}_k \, d\sigma,$$

$$d\zeta[k] \wedge d\bar{\zeta}\big|_{\partial D} = 2^{n-1} i^n (-1)^{n+k-1} \rho_k \, d\sigma,$$

we have

$$\bar{\partial}_n f \, d\sigma = *\bar{\partial}f|_{\partial D} = 2^{1-n} i^n (-1)^n \mu_f\big|_{\partial D}.$$

So the normal part of the form $\bar\partial f$ has moved to the tangent part of the form $*\bar\partial f$.

If we consider the function

$$\tilde g(\zeta, z) = 2^{n-1} i^n g(\zeta, z) = -\frac{(n-2)!}{2\pi^n} \frac{1}{|\zeta - z|^{2n-2}},$$

then formula (1.2.1) can be rewritten as follows.

Corollary 2.5.1 *Let D be a bounded domain with a piecewise-smooth boundary, and let f be a harmonic function in D of class $\mathscr{C}^1(\overline{D})$. Then*

$$\int_{\partial D} f(\zeta) U(\zeta, z) - \int_{\partial D} \tilde g(\zeta, z) \bar\partial_n f(\zeta) d\sigma = \begin{cases} f(z), & z \in D, \\ 0, & z \notin \overline{D}. \end{cases} \tag{2.5.1}$$

We consider the $\bar\partial$-Neumann problem for functions.

2.5.1 For given function φ on ∂D, find a function f on $\overline{D}$ such that

$$\begin{cases} \bar\partial_n f = \varphi, & \text{on} \quad \partial D, \\ \Box f = 0, & \text{in} \quad D. \end{cases} \tag{2.5.2}$$

This problem is an exact analogue of an ordinary Neumann problem for harmonic functions.

Just as the ordinary Neumann problem, problem (2.5.2) is not always solvable. There is a necessary orthogonality condition to be satisfied. Indeed, if f is a harmonic function of class $\mathscr{C}^1(\overline{D})$, then $*\bar\partial f$ is a ∂-closed form in D, since

$$0 = \Box f = \bar\partial^* \bar\partial f = - * \partial(*\bar\partial f),$$

i.e., $\partial(*\bar\partial f) = 0$. Hence, if $\varphi = \bar\partial_n f$ on ∂D, and h is a holomorphic function on $\overline{D}$, then

$$\int_{\partial D} \varphi \bar h \, d\sigma = \int_{\partial D} \bar h(*\bar\partial f) = \int_D \partial(\bar h * \bar\partial f) = \int_D \bar h \partial(*\bar\partial f) = 0.$$

Thus, a necessary condition for solvability of (2.5.2) is the orthogonality condition

$$\int_D \varphi \bar h \, d\sigma = 0$$

for all $h \in \mathscr{O}(\overline{D})$.

Compare problem (2.5.2) with the following problem: for given function ψ, find a function f such that

$$\begin{cases} \bar{\partial}_n f = 0, & \text{on} \quad \partial D, \\ \Box f = \psi, & \text{in} \quad D. \end{cases} \tag{2.5.3}$$

If we ignore the smoothness of the functions, then (2.5.2) and (2.5.3) are equivalent. Indeed, the volume potential f_ψ is one of the solutions to the second equation in (2.5.3). Subtracting it from the solution of (2.5.3), we obtain

$$\Box(f - f_\psi) = 0 \quad \text{and} \quad \bar{\partial}_n(f - f_\psi) = \varphi$$

on ∂D, i.e., we have (2.5.2). Conversely, given (2.5.2), we take the single-layer potential $f_\varphi^\pm$ for φ and extend f_φ^- into D as a smooth function to obtain

$$\bar{\partial}_n(f - f_\varphi^+ + f_\varphi^-) = 0 \quad \text{on} \quad \partial D,$$

and

$$\Box(f - f_\varphi^+ + f_\varphi^-) = \psi,$$

i.e., we have (2.5.3).

Problem (2.5.2) is more adequate for studying the boundary properties of holomorphic functions. We will not dwell any further on the development of the inhomogeneous Neumann problem in this book. Several results on its solvability can be found in [45]. Here we will focus on the homogeneous problem.

2.5.2 The Homogeneous $\bar{\partial}$-Neumann Problem

We first consider the homogeneous $\bar{\partial}$-Neumann problem

$$\begin{cases} \bar{\partial}_n f = 0, & \text{on} \quad \partial D, \\ \Box f = 0, & \text{in} \quad D. \end{cases} \tag{2.5.4}$$

It is clear that holomorphic functions f satisfy (2.5.4). We will show that the converse is also true. First we reformulate the problem. Recall that Mf is the Bochner–Martinelli integral (1.2.4)

$$Mf(z) = \int_{\partial D} f(\zeta) U(\zeta, z), \quad z \notin \partial D.$$

Theorem 2.5.1 ([44]) *Let D be a domain such that a set $\mathbb{C}^n \setminus \overline{D}$ is connected, and let f be a harmonic function in D of class $\mathscr{C}^1(\overline{D})$. The following conditions are equivalent:*

1. $\bar{\partial}_n f = 0$ on ∂D;
2. $M^+ f = f$ in D;
3. $M^- f = 0$ in $\mathbb{C}^n \setminus \overline{D}$.

Proof Conditions (2) and (3) are equivalent by the jump theorem for the Bochner–Martinelli integral (see Corollary 1.2.4) and by the uniqueness theorem for harmonic functions.

If $\bar{\partial}_n f = 0$ on ∂D then formula (2.5.1) yields $M^+ f = f$ in D. If $M^+ f = f$ in D, then $M^- f = 0$ outside $\overline{D}$. Thus from (2.5.1) we obtain

$$\int_{\partial D} \frac{\bar{\partial}_n f(\zeta)}{|\zeta - z|^{2n-2}} \, d\sigma(\zeta) = 0$$

for all $z \notin \partial D$. Applying the Keldysh–Lavrent'ev theorem (see, for example, [56, p. 418]) on the density of fractions of the form

$$\frac{1}{|\zeta - z|^{2n-2}}$$

in the space $\mathscr{C}(\partial D)$, we obtain that $\bar{\partial}_n f = 0$ on ∂D. $\qquad\square$

Theorem 2.5.2 (Folland and Kohn [21]; Aronov and Kytmanov [10]) *Let f be a harmonic function in D of class $\mathscr{C}^1(\overline{D})$. The following conditions are equivalent:*

1. $\bar{\partial}_n f = 0$ on ∂D;
2. $M^+ f = f$ in D;
3. $M^- f = 0$ in $\mathbb{C}^n \setminus \overline{D}$;
4. f is holomorphic in D.

Proof It is sufficient to prove that condition (1) implies condition (4). Since the form $*\bar{\partial} f$ is ∂-closed, then

$$0 = \int_{\partial D} \bar{f}(*\bar{\partial} f) = \int_D \partial \bar{f} \wedge *\bar{\partial} f = 2^{1-n} i^n \int_D |\bar{\partial} f|^2 dz \wedge d\bar{z} = 2 \int_D |\bar{\partial} f|^2 dv.$$

Hence $\dfrac{\partial f}{\partial \bar{z}_k} = 0$ in D for all $k = 1, \ldots, n$, so $f \in \mathscr{O}(D)$. $\qquad\square$

Conditions (2), (3), (4) are equivalent without requirement that f is harmonic in D.

2.5.3 Holomorphic Functions Represented by the Bochner–Martinelli Integral

Let $n > 1$.

Theorem 2.5.3 ([45]) *If M^+f is holomorphic in D, $f \in \mathcal{C}^1(\partial D)$, and $\partial D \in \mathcal{C}^1$ is connected, then the boundary value of M^+f coincides with f.*

It is clear that Theorem 2.5.3 is not true when $n = 1$. This is not true either if ∂D is not connected: it suffices to set $f = 1$ on just one connected component of ∂D and $f = 0$ on the remaining components.

Consider a continuous function.

Theorem 2.5.4 ([46]) *Let D be a bounded domain with a connected boundary of class $\mathcal{C}^2$. A necessary and sufficient condition for the function $f \in \mathcal{C}(\overline{D})$ to be holomorphic in D is that $M^+f = f$ in D.*

Proof If $M^+f = f$ in D, then M^-f extends continuously to $\mathbb{C}^n \setminus D$, and $M^-f = 0$ on ∂D. By the uniqueness theorem for harmonic functions, $M^-f \equiv 0$. Then $\bar{\partial}_n(M^-f) \equiv 0$, so by Theorem 2.3.3, $\bar{\partial}_n(M^+f) = \bar{\partial}_n f$ extends continuously to $\overline{D}$, and $\bar{\partial}_n f = 0$ on ∂D. Let $\rho(z)$ be a defining function for D, and $D_\varepsilon = \{z \in D : \rho(z) < -\varepsilon\}, \varepsilon > 0$. Then

$$\int_{\partial D_\varepsilon} \bar{f}(*\bar{\partial}f) = 2^{1-n} i^n \int_{D_\varepsilon} \sum_{k=1}^n \left| \frac{\partial f}{\partial \bar{z}_k} \right|^2 dz \wedge d\bar{z} \to 0$$

when $\varepsilon \to +0$. Hence $\dfrac{\partial f}{\partial \bar{z}_k} = 0$ in D for all $k = 1, \dots, n$. $\square$

We get the corollaries from this theorem, Theorem 2.2.1 on the jump of the Bochner–Martinelli integral and Corollary 2.2.1.

Corollary 2.5.2 *If $f \in \mathcal{C}(\partial D)$ and $\partial D \in \mathcal{C}^2$, then a necessary and sufficient condition for f to extend into D as a function F of class $\mathcal{A}(D)$ is that $M^-f = 0$ outside $\overline{D}$.*

Corollary 2.5.3 *Suppose $n > 1$, $\partial D \in \mathcal{C}^2$ is connected, and $f \in \mathcal{C}(\partial D)$. If M^+f is holomorphic in D, then $M^+f \in \mathcal{C}(\overline{D})$, and $M^+f = f$ on ∂D.*

Proof Since M^+f is holomorphic in D, then $\bar{\partial}_n M^+f = 0$ in D. Using Theorem 2.3.3 we obtain $\bar{\partial}_n M^-f = 0$ on ∂D. Since $M^-f = O(|z|^{1-2n})$, and $\dfrac{\partial M^-f}{\partial \bar{z}_k} = O(|z|^{-2n})$ as $|z| \to \infty$, by applying Stokes' formula we find that

$$0 = \int_{\partial D} \overline{M^-f} \left(*\bar{\partial} M^-f \right) = -2^{1-n} i^n \int_{\mathbb{C}^n \setminus D} \sum_{k=1}^n \left| \frac{\partial M^-f}{\partial \bar{z}_k} \right|^2 dz \wedge d\bar{z}.$$

Consequently, $M^- f$ is holomorphic in $\mathbb{C}^n \setminus D$, and by Hartogs' theorem it extends holomorphically in $\mathbb{C}^n$. But then $M^- f \equiv 0$, since $M^- f \to 0$ as $|z| \to \infty$ (here we assume that $n > 1$ and ∂D is connected). Now applying Theorem 2.2.1 and Corollary 2.2.1, we obtain the required assertion $M^+ f = f$ on ∂D. □

2.5.4 Homogeneous Harmonic Polynomials Expansion of the Bochner–Martinelli Kernel

Let $B = B(0, 1)$ be a unit ball in $\mathbb{C}^n$ with the center at the origin, and let $S = S(0, 1)$ be its boundary. Consider a set of homogeneous harmonic polynomials that form a complete orthonormal system of functions (basis) in the space $\mathscr{L}^2(S)$ with respect to the Lebesgue measure $d\sigma$ on S. We denote these polynomials as $P_{k,s}$, where k is the degree of homogeneity, $k = 0, 1, 2, \ldots$, and $s = 1, 2, \ldots, \sigma(k)$, $\sigma(k) = \dfrac{2(n + k - 2)(k + 2n - 3)!}{k!(n - 2)!}$ is the number of linearly independent homogeneous polynomials of degree k (see, for example, [74, Chap. 10]). It is clear that $\sigma(k)$ is a polynomial (in k) of degree $(2n - 2)$ with the higher coefficient $-2(n - 2)!$.

Theorem 2.5.5 ([45, 46]) *If* $\{P_{k,s}\}$ *is a complete orthonormal system of homogeneous harmonic polynomials in the space* $\mathscr{L}^2(S)$, *then the Bochner–Martinelli kernel has the expansion*

$$U(\zeta, z) = -\sum_{k,s} \frac{P_{k,s}(z)}{n + k - 1}\left[*\partial \frac{\overline{P_{k,s}(\zeta)}}{|\zeta|^{2n+2k-2}} \right], \qquad (2.5.5)$$

where series (2.5.5) converges uniformly on compact sets in the domain

$$\{(\zeta, z) \in \mathbb{C}^{2n} : |\zeta| > |z|\}.$$

Likewise, we have the expansion

$$U(\zeta, z) = -\sum_{k,s} \frac{\overline{P_{k,s}(z)}}{(n + k - 1)|z|^{2n+2k-2}} * (\partial P_{k,s}(\zeta)), \qquad (2.5.6)$$

where series (2.5.6) converges uniformly on compact sets in the domain

$$\{(\zeta, z) \in \mathbb{C}^{2n} : |\zeta| < |z|\}.$$

Proof We denote the restriction of the polynomial $P_{k,s}$ on S by $Y_{k,s}$. Then $\{Y_{k,s}\}$ is an orthonormal basis in $\mathscr{L}^2(S)$, consisting of spherical functions. Let $\zeta \in S, z \in B$

and

$$\frac{1}{|\zeta - z|^{2n-2}} = \sum_{k,s} c_{k,s}\overline{Y}_{k,s}$$

(z being fixed here), where

$$c_{k,s}(z) = \int_S \frac{Y_{k,s}(\zeta)}{|\zeta - z|^{2n-2}}\, d\sigma(\zeta).$$

If we express $|\zeta - z|^{2-2n}$ in terms of the Poisson kernel $P(\zeta, z)$ for the ball, where

$$P(\zeta, z) = \frac{(n-1)!}{2\pi^n}\frac{1 - |z|^2}{|\zeta - z|^{2n}},$$

then

$$c_{k,s}(z) = \frac{2\pi^n}{(n-1)!}\int_S P(\zeta, z)\frac{1 - \langle \zeta, \bar{z}\rangle - \langle \bar{\zeta}, z\rangle + |z|^2}{1 - |z|^2}\, Y_{k,s}(\zeta)\, d\sigma(\zeta).$$

It is easy to verify that the functions

$$\zeta_j P_{k,s}(\zeta) - \frac{1}{n+k-1}\frac{\partial P_{k,s}(\zeta)}{\partial \bar{\zeta}_j}\,(|\zeta|^2 - 1)$$

and

$$\bar{\zeta}_j P_{k,s}(\zeta) - \frac{1}{n+k-1}\frac{\partial P_{k,s}(\zeta)}{\partial \zeta_j}\,(|\zeta|^2 - 1)$$

give harmonic extension of the functions $\zeta_j Y_{k,s}$ and $\bar{\zeta}_j Y_{k,s}$ in B from the sphere S. Therefore

$$c_{k,s}(z) = \frac{2\pi^n}{(n-1)!}\frac{1}{1-|z|^2}\left[(1 + |z|^2)P_{k,s}(z) - 2|z|^2 P_{k,s}(z) + \right.$$

$$\left. + \frac{|z|^2 - 1}{n+k-1}\sum_{j=1}^n \left(\bar{z}_j\frac{\partial P_{k,s}}{\partial \bar{z}_j} + z_j\frac{\partial P_{k,s}}{\partial z_j}\right)\right] = \frac{2\pi^n}{(n-2)!(n+k-2)}\, P_{k,s}(z).$$

Then

$$\frac{1}{|\zeta - z|^{2n-2}} = \frac{2\pi^n}{(n-2)!}\sum_{k,s}\frac{1}{n+k-1}P_{k,s}(z)\,\overline{Y}_{k,s}(\zeta). \qquad (2.5.7)$$

Series (2.5.7) converges in ζ in the sense of $\mathscr{L}^2(S)$ and z uniformly on compact subsets of B. The harmonic extension on ζ in series (2.5.7) on B is given by

$$\left(|\zeta| \left|\frac{\zeta}{|\zeta|^2} - z\right|\right)^{2-2n} = \frac{2\pi^n}{(n-2)!} \sum_{k,s} \frac{1}{n+k-1} P_{k,s}(z) \overline{P}_{k,s}(\zeta). \tag{2.5.8}$$

Applying the Kelvin transformation in ζ to both sides in (2.5.8), we find

$$\frac{1}{|\zeta - z|^{2n-2}} = \frac{2\pi^n}{(n-2)!} \sum_{k,s} \frac{P_{k,s}(z)}{n+k-1} \frac{\overline{P}_{k,s}(\zeta)}{|\zeta|^{2n+2k-2}},$$

where the series converges uniformly on compact sets in $\{(z, \zeta) \in \mathbb{C}^{2n} : |\zeta| > |z|\}$. Since

$$U(\zeta, z) = -\frac{(n-2)!}{2\pi^n} * \partial_\zeta |\zeta - z|^{2-2n}$$

(Example 2.4.2), we obtain the required equality (2.5.5).

Similarly, by exchanging ζ and z in formula (2.5.7), we obtain

$$\frac{1}{|\zeta - z|^{2n-2}} = \frac{2\pi^n}{(n-2)!} \sum_{k,s} \frac{1}{n+k-1} P_{k,s}(\zeta) \overline{Y}_{k,s}(z), \tag{2.5.9}$$

where $\zeta \in B, z \in S$. Harmonic extension of the left- and right-hand sides of (2.5.9) on z in B results in

$$\left(|z| \left|\frac{z}{|z|^2} - \zeta\right|\right)^{2-2n} = \frac{2\pi^n}{(n-2)!} \sum_{k,s} \frac{1}{n+k-1} P_{k,s}(\zeta) \overline{P}_{k,s}(z). \tag{2.5.10}$$

Applying the Kelvin transformation in z to (2.5.10), we obtain

$$\frac{1}{|\zeta - z|^{2n-2}} = \frac{2\pi^n}{(n-2)!} \sum_{k,s} \frac{P_{k,s}(\zeta)}{n+k-1} \frac{\overline{P}_{k,s}(z)}{|z|^{2n+2k-2}},$$

where the series converges uniformly on compact sets in $\{(z, \zeta) \in \mathbb{C}^{2n} : |\zeta| < |z|\}$. From this and Example 2.4.2 we obtain expansion (2.5.6). $\qquad\square$

Corollary 2.5.4 *Suppose $\partial D \in \mathscr{C}^2$ is connected, and $f \in \mathscr{C}(\partial D)$. A necessary and sufficient condition for f to extend into D as a function $F \in \mathscr{A}(D)$ is that*

$$\int_{\partial D} f(*\partial P_{k,s}) = 0$$

for all k, s.

Proof The function $M^- f$ is harmonic outside $\overline{D}$, so to prove $M^- f = 0$ outside $\overline{D}$ it suffices to prove that $M^- f = 0$ outside some ball $B(0, R) \supset \overline{D}$. When $z \notin B(0, R)$, the function $|\zeta - z|^{2-2n}$ is harmonic in ζ in $\overline{B}(0, R)$, and therefore the kernel $U(\zeta, z)$ can be represented by uniformly convergent series (2.5.6) on $\overline{B}(0, R)$. We thus obtain the required equality

$$\int_{\partial D} f(\zeta) U(\zeta, z) = 0$$

for $z \notin \overline{B}(0, R)$.   □

When $n = 1$, Corollary 2.5.4 reduces to the classical criterion for the existence of a holomorphic extension, which consists in orthogonality of f to the monomials z^k, $k = 0, 1, \ldots$, since in this case $P_{k,s,t} = az^k + b\overline{z}^s$, and so $*\partial P_{k,s,t} = cz^{k-1} dz$, $k \geq 1$.

2.6 Boundary Behavior of the Integral (of Type) of the Logarithmic Residue

Suppose D is a bounded domain in $\mathbb{C}^n$ with a piecewise-smooth boundary ∂D and $w = \psi(z)$ is a holomorphic mapping on $\overline{D}$ in $\mathbb{C}^n$ having a finite number of zeros E_ψ on $\overline{D}$. Similarly to Sect. 1.4, consider the multiplicity μ_a of zero a of this map.

For a point $z \in E_\psi \cap \partial D$ we consider a ball $B(z, R)$, that does not contain any other zeros of ψ, and use $\tau_\psi(z)$ to denote

$$\tau_\psi(z) = \lim_{r \to +0} \frac{\mathscr{L}^{2n-1}[S(0, r) \cap \psi(B(z, R) \cap D)]}{\mathscr{L}^{2n-1}[S(0, r)]},$$

where $\mathscr{L}^{2n-1}$ is $(2n - 1)$-dimensional Lebesgue measure. In other words, we consider the solid angle of the tangent cone of the image $\psi(B(z, R) \cap D)$ at the point 0 rather than the solid angle of the tangent cone to the domain D at the point z. (The definition of the tangent cone can be found in [20, Sect. 3.1.21].)

By Sard's theorem, for $z \in E_\psi$ and a sufficiently small neighborhood of V_z of the point z the set $B_\psi(z, r) = \{\zeta \in V_z : |\psi(\zeta)| < r\}$ is relatively compact in V_z, and the set $S_\psi(z, r) = \{\zeta \in V_z : |\psi(\zeta)| = r\}$ is a smooth $(2n-1)$-dimensional cycle (for almost all sufficiently small $r > 0$).

We define the principal value p.v.$^\psi$ of the integral of a measurable function φ at the point $z \in E_\psi$ on neighborhood S of the point z of the surface ∂D as follows:

$$\text{p.v.}^\psi \int_S \varphi(\zeta) \, d\mathscr{L}^{2n-1}(\zeta) = \lim_{r \to +0} \int_{S \setminus B_\psi(z, r)} \varphi(\zeta) \, d\mathscr{L}^{2n-1}(\zeta).$$

This definition differs from the conventional definition of the Cauchy principal value p.v. in that we remove the curved ball $B_\psi(z, r)$ rather than the ball neighborhood of z.

Consider the kernel of $U(\psi(\zeta))$ that is used in the multidimensional logarithmic residue in Sect. 1.4. It is obtained by substituting $w = \psi(z)$ from the Bochner–Martinelli kernel $U(w)$, defined by formula (1.4.1). The kernel $U(\psi(\zeta))$ is a closed differential form of type $(n, n-1)$ on $\overline{D}$ with singularities at $a \in E_\psi$.

We now formulate the main result of this section.

Theorem 2.6.1 ([65]) *If a holomorphic function f in D satisfies the Holder condition with the exponent $\gamma > 0$ on $\overline{D}$ (i.e., $f \in \mathscr{C}^\gamma(\overline{D})$), then*

$$\text{p.v.} \int_{\partial D} f(\zeta) U(\psi(\zeta)) = \sum_{a \in E_\psi \cap D} \mu_a f(a) + \sum_{a \in E_\psi \cap \partial D} \tau_\psi(a) \mu_a f(a).$$

This formula is the formula of a multidimensional logarithmic residue with singularities on the boundary. If the zeros of mapping ψ do not lie on the boundary, this formula turns into the ordinary logarithmic residue formula [9, Sect. 3]. In the case of simple zeros $a \in \partial D$, it gives the theorem from [68]. Furthermore, this theorem is a generalization of Theorem 20.7 in [45], where additional conditions are imposed on the boundary ∂D and the map ψ.

For the proof we need the following Theorem 3.2.5 from [20].

Suppose that the map $\psi : \mathbb{R}^m \to \mathbb{R}^n$ is the Lipschitz one and $m \leq n$. Then

$$\int_A g(\psi(x)) J_m \psi(x) \, d\mathscr{L}^m(x) = \int_{\mathbb{R}^n} g(y) N(\psi|A, y) \, d\mathscr{H}^m(y), \qquad (2.6.1)$$

if the set A is $\mathscr{L}^m$-measurable, $g : \mathbb{R}^n \to \overline{\mathbb{R}}$ and $N(\psi|A, y) < \infty$ for $\mathscr{H}^m$ for almost all y.

Here $J_m \psi(x)$ is the m-dimensional Jacobian of the mapping ψ, $\mathscr{L}^m$ is the m-dimensional Lebesgue measure, $\mathscr{H}^m$ is the m-dimensional Hausdorff measure, $N(\psi|A, y)$ is the function of multiplicity of the mapping ψ, i.e., the number of inverse images $\psi^{-1}(y)$, lying in A.

First we prove the theorem for the principal value p.v.$^\psi$.

Lemma 2.6.1 *Under the hypotheses of Theorem 2.6.1 the equality*

$$\text{p.v.}^\psi \int_{\partial D} f(\zeta) U(\psi(\zeta)) = \sum_{a \in E_\psi \cap D} \mu_a f(a) + \sum_{a \in E_\psi \cap \partial D} \tau_\psi(a) \mu_a f(a)$$

holds.

Proof In the domain

$$D_r = D \setminus \bigcup_{a \in E_\psi \cap \partial D} B_\psi(a, r)$$

by the multidimensional logarithmic residue formula (1.4.2) we have

$$\int_{\partial D_r} f(\zeta) U(\psi(\zeta)) = \sum_{a \in E_\psi \cap D} \mu_a f(a),$$

and

$$\text{p.v.}^\psi \int_{\partial D} f(\zeta) U(\psi(\zeta)) = \lim_{r \to +0} \int_{\partial D \setminus \bigcup_{a \in E_\psi \cap \partial D} B_\psi(a,r)} f(\zeta) U(\psi(\zeta)).$$

So that

$$\int_{\partial D_r} f(\zeta) U(\psi(\zeta)) = \int_{\partial D \setminus \bigcup_a B_\psi(a,r)} f(\zeta) U(\psi(\zeta)) - \sum_a \int_{S_\psi(a,r) \cap D} f(\zeta) U(\psi(\zeta)),$$

and

$$\int_{S_\psi(a,r) \cap D} f(\zeta) U(\psi(\zeta))$$

$$= \int_{S_\psi(a,r) \cap D} (f(\zeta) - f(a)) U(\psi(\zeta)) + f(a) \int_{S_\psi(a,r) \cap D} U(\psi(\zeta)). \qquad (2.6.2)$$

Next, we use the Lojasiewicz inequality [61, p. 73]

$$|\zeta - a| \le C |\psi(\zeta)|^\alpha \qquad (2.6.3)$$

for some positive numbers α and C and points ζ from a sufficiently small neighborhood a. We show that the first integral in (2.6.2) tends to zero as $r \to +0$. Using the Holder condition for the function f, equality (2.6.1) and inequality (2.6.3), we obtain

$$\int_{S_\psi(a,r) \cap D} |f(\zeta) - f(a)| \frac{|\psi_k|}{|\psi(\zeta)|^{2n}} |d\overline{\psi}[k] \wedge d\psi|$$

$$\le C_1 \int_{S_\psi(a,r) \cap D} |\psi(\zeta)|^{\gamma\alpha + 1 - 2n} |d\overline{\psi}[k] \wedge d\psi|$$

$$\le C_1 \mu_a \int_{S(0,r) \cap \psi(D)} |w|^{\gamma\alpha + 1 - 2n} \, d\mathcal{H}^{2n-1}(w)$$

$$\le C_2 \int_{S(0,r)} |w|^{\gamma\alpha + 1 - 2n} \, d\mathcal{L}^{2n-1}(w),$$

since the mapping ψ is smooth, therefore $\mathscr{H}^{2n-1}(\psi(S)) \leq C_3 \mathscr{L}^{2n-1}(S)$, and the last integral obviously tends to zero as $r \to +0$. For the second integral in (2.6.2) we apply equality (2.6.1) and obtain

$$\lim_{r \to +0} \int_{S_\psi(a,r) \cap D} U(\psi(\zeta)) = \lim_{r \to +0} \mu_a \int_{S(0,r) \cap \psi(D)} U(w) = \mu_a \tau_\psi(a),$$

since

$$\int_{S(0,r) \cap \psi(D)} U(w) = \frac{\mathscr{L}^{2n-1}[S(0,r) \cap \psi(D)]}{\mathscr{L}^{2n-1}[S(0,r)]}$$

by Lemma 2.1 from [45]. $\square$

Now let $\psi = (\psi_1, \ldots, \psi_n)$ be a holomorphic mapping from $\mathbb{C}^n$ to $\mathbb{C}^n$, that consists of entire functions and has the only zero at the origin. Multiplicity of zero of the mapping ψ will be denoted by μ.

As in (1.4.5), we denote the integrals by the formula

$$\int_{\partial D_\zeta} f(\zeta) U(\psi(\zeta - z)) = \begin{cases} M_\psi^+ f(z), & z \in D, \\ M_\psi^- f(z), & z \notin \overline{D}. \end{cases} \tag{2.6.4}$$

Lemma 2.6.2 *If $f \in \mathscr{C}^\gamma(\partial D)$, $\gamma > 0$, then the integrals $M_\psi^\pm f$ extend continuously to ∂D and $M_\psi^+ f(z) - M_\psi^- f(z) = \mu f(z)$ on ∂D.*

Proof We extend f in a neighborhood V of the boundary of the domain D to a function, satisfying the Hölder condition with the exponent γ in this neighborhood. We now prove that the functions

$$\int_{\partial D_\zeta} (f(\zeta) - f(z)) U(\psi(\zeta - z))$$

are continuous in V. To do this, we need to show that integrals of the form

$$\int_{S_\zeta} (f(\zeta) - f(z)) \frac{\overline{\psi_k(\zeta - z)}}{|\psi(\zeta - z)|^{2n}} d\overline{\psi}[k] \wedge d\psi$$

converge absolutely (here S is the neighborhood of z on the surface ∂D). Inequality (2.6.3), when applied to the $\psi(\zeta - z)$, and the Hölder condition of f yields

$$|f(\zeta) - f(z)| \leq c|\zeta - z|^\gamma \leq c_1 |\psi(\zeta - z)|^{\gamma\alpha}$$

for ζ from a sufficiently small neighborhood of z. Using (2.6.1), as we did when proving Lemma 2.6.1, we obtain

$$\int_{S_\zeta} |f(\zeta) - f(z)| \frac{|\psi_k(\zeta - z)|}{|\psi(\zeta - z)|^{2n}} \left| d\overline{\psi}[k] \wedge d\psi \right|$$

$$\leq c_1 \int_{S_\zeta} |\psi(\zeta - z)|^{\gamma\alpha + 1 - 2n} \left| d\overline{\psi}[k] \wedge d\psi \right| \leq c_1 \mu \int_{\psi(S)} |w|^{\gamma\alpha + 1 - 2n} \, d\mathcal{H}^{2n-1}(w)$$

$$\leq c_2 \int_S |w|^{\gamma\alpha + 1 - 2n} \, d\mathcal{L}^{2n-1}(w),$$

and the last integral obviously converges.

The formula

$$\int_{\partial D} U(\psi(\zeta - z)) = \begin{cases} \mu, & z \in D, \\ 0, & z \notin \overline{D}, \end{cases}$$

completes the proof.   □

Let us return to the original mapping ψ.

Lemma 2.6.3 *For the function* $f \in \mathscr{C}^\gamma(\partial D)$, $\gamma > 0$ *the equality*

$$\text{p.v.}^\psi \int_S f(\zeta) U(\psi(\zeta)) = \text{p.v.} \int_S f(\zeta) U(\psi(\zeta))$$

holds.

This lemma generalizes the statement from [68] about equality of the principal values for the case of simple zeros of the mapping ψ.

Proof As shown in Lemma 2.6.2, the integral

$$\int_S (f(\zeta) - f(z)) U(\psi(\zeta))$$

converges absolutely, so the principal values are equal to the given integral. Hence, we only need to prove that

$$\text{p.v.}^\psi \int_S U(\psi(\zeta)) = \text{p.v.} \int_S U(\psi(\zeta)).$$

We transform the integral (r is small enough), taking $S = \partial D \cap B(z, R)$,

$$\int_{S \setminus B_\psi(z,r)} U(\psi(\zeta)) = \int_{\partial(D \cap B(z,R) \setminus B_\psi(z,r))} U(\psi(\zeta)) - \int_{D \cap S(z,R)} U(\psi(\zeta))$$

$$+ \int_{D \cap S_\psi(z,r)} U(\psi(\zeta)) = - \int_{D \cap S(z,R)} U(\psi(\zeta)) + \int_{D \cap S_\psi(z,r)} U(\psi(\zeta))$$

according to the multidimensional logarithmic residue formula. Therefore, it remains to prove that

$$\lim_{r \to +0} \int_{D \cap S_\psi(z,r)} U(\psi(\zeta)) = \lim_{r \to +0} \int_{D \cap S(z,r)} U(\psi(\zeta)).$$

By Theorem 3.2.5 in [20] (equality (2.6.1)) we have

$$\int_{D \cap S_\psi(z,r)} U(\psi(\zeta)) = \mu_z \int_{\psi(D) \cap S(0,r)} U(w),$$

$$\int_{D \cap S(z,r)} U(\psi(\zeta)) = \mu_z \int_{\psi(D \cap S(z,r))} U(w).$$

Therefore, we need to show that

$$\lim_{r \to +0} \int_{\psi(D) \cap S(0,r)} U(w) = \lim_{r \to +0} \int_{\psi(D \cap S(z,r))} U(w).$$

In this equality the tangent cone of Π to $\psi(D)$ at 0 can be chosen instead of $\psi(D)$. We show that

$$\int_{\Pi \cap S(0,r_1)} U(w) = \int_{\Pi \cap \psi(S(z,r_2))} U(w).$$

Consider the domain G bounded by the surfaces $\Pi \cap S(0, r_1)$, $\Pi \cap \psi(S(z, r_2))$ and the part of the conical surface $M \cap \partial\Pi$ (r_1 and r_2 are chosen so that the ball $B(0, r_1)$ contains the surface $\psi(S(z, r_2))$). By the Bochner–Martinelli formula we have

$$\int_{\partial G} U(w) = 0,$$

therefore

$$\int_{\Pi \cap S(0,r_1)} U(w) - \int_{\Pi \cap \psi(S(z,r_2))} U(w) = \int_M U(w).$$

We show that

$$\int_M U(w) = 0.$$

We pass from the complex coordinates w to real ones $w_j = \xi_j + i\xi_{n+j}, j = 1, \ldots, n$. Then (see [68] or [45])

$$\operatorname{Re} U(w) = \frac{(n-1)!}{2\pi^n} \sum_{k=1}^{2n} (-1)^{k-1} \frac{\xi_k}{|\xi|^{2n}} d\xi[k],$$

$$\operatorname{Im} U(w) = -\frac{(n-2)!}{4\pi^n} d \left(\sum_{k=1}^{n} \frac{1}{|\xi|^{2n-2}} d\xi[k, n+k] \right), \quad n > 1,$$

and when $n = 1$

$$\operatorname{Im} U(w) = -\frac{d \ln |\xi|^2}{4\pi}.$$

The restriction of the differential form $\operatorname{Re} U(w)$ onto the conical surface M (at smooth points of M) equals 0. In fact, let M be given by a zero set of a homogeneous real-valued function φ: $M = \{\xi : \varphi(\xi) = 0\}$. Then at the points of smoothness M the restriction of $d\xi[k]$ to M equals $(-1)^{k-1}\gamma_k d\sigma$, where $\gamma_k = \dfrac{\partial \varphi}{\partial \xi_k} \dfrac{1}{|\operatorname{grad} \varphi|}$ are the direction cosines of the normal and $d\sigma$ is an element of the surface M. Then

$$\sum_{k=1}^{2n} (-1)^{k-1} \frac{\xi_k}{|\xi|^{2n}} d\xi[k] \bigg|_M$$

$$= \sum_{k=1}^{2n} \xi_k \frac{\partial \varphi}{\partial \xi_k} \frac{1}{|\operatorname{grad} \varphi||\xi|^{2n}} d\sigma = l\varphi \frac{1}{|\operatorname{grad} \varphi||\xi|^{2n}} d\sigma = 0$$

by virtue of Euler's formula for homogeneous functions (l is the degree of homogeneity of φ); the $(2n-1)$-dimensional measure of the set of points of non-smoothness is zero.

Integration over M will be done with real lines on M of the form

$$l_b = \{\xi : \xi_j = b_j t, j = 1, \ldots, 2n, t \in \mathbb{R}\},$$

where $|b| = 1$. For a fixed $b \in S(0, 1)$ the variable t changes from some $r_2(b)$ to r_1. The function $r_2(b)$ is measurable. Thus, M is a bundle over the cycle $\partial \Pi \cap S(0, 1)$.

In these variables

$$\operatorname{Im} U(w) = c_n d\left(\frac{dt}{t} \wedge \sum_{k,j} \pm b_k db[j, k, n+k]\right) = c_n \frac{dt}{t} \wedge \sum_{k=1}^{n} db[k, n+k],$$

as the form containing a product of more than $2n - 2$ differentials db_j is equal to zero on $S \cap \partial\Pi$. Then

$$\int_M \operatorname{Im} U(w) = c_n \int_{S(0,1)\cap\partial\Pi} \ln \frac{r_1}{r_2(b)} \sum_{k=1}^{n} db[k, n+k].$$

Variables b_k, b_{n+k} at almost all points of $S \cap \partial\Pi$ are the functions of the remaining variables $b_j, j \neq k, n+k$. Therefore, the last integral takes the form

$$\int_{S(0,1)\cap\partial\Pi} \sum_{k=1}^{n} \ln \Phi_k(b_1, \ldots, [k], \ldots, [n+k], \ldots, b_{2n}) \, db[k, n+k]$$

$$= \int_{S(0,1)\cap\Pi} d\left(\sum_{k=1}^{n} \ln \Phi_k(b_1, \ldots, [k], \ldots, [n+k], \ldots, b_{2n}) \, db[k, n+k]\right) = 0$$

by Stokes's formula. □

Proof of Theorem 2.6.1 follows from Lemmas 2.6.1 and 2.6.3. □

Lemma 2.6.2 adds strength to Theorem 1 from [49], which was proved for smooth functions.

2.7   On the Holomorphicity of Functions Represented by the Logarithmic Residue Formula

Suppose D is a bounded domain in $\mathbb{C}^n$ with a piecewise-smooth boundary ∂D. We want to prove the converse assertion to Corollary 1.4.2: if equality (1.4.4) is valid for a function f, then this function is holomorphic in D.

Theorem 2.7.1 ([48]) *If $f \in \mathscr{C}^1(\partial D)$ satisfies the condition*

$$M_{\psi}^{-} f(z) = 0, \quad z \notin \overline{D}, \tag{2.7.1}$$

then f extends holomorphically into D as a function $F \in \mathscr{C}^1(\overline{D})$.

For the proof we will need some of the properties of determinants made up of differential forms. Consider n-dimensional vectors $\theta^1, \ldots, \theta^m$ consisting of exterior differential forms. We introduce determinants of order n:

$$\mathbf{D}_{\nu_1,\dots,\nu_m}\left(\theta^1,\dots,\theta^m\right), \tag{2.7.2}$$

where the first ν_1 columns are the vectors θ^1, the second ν_2 columns are the vectors θ^2, and so on, and the last ν_m columns are the vectors θ^m, $\nu_1 + \cdots + \nu_m = n$. The properties of these determinants can be found in [8, Sect. 1]. Consider the vector

$$\eta = \frac{\bar{\psi}}{|\psi|^2} = \left(\frac{\bar{\psi}_1}{|\psi|^2},\dots,\frac{\bar{\psi}_n}{|\psi|^2}\right),$$

then $\langle \eta, \psi \rangle = \eta_1\psi_1 + \cdots + \eta_n\psi_n = 1$ outside the zeros of the map ψ. The kernel $\omega(\psi)$ takes the form

$$\omega(\psi) = \frac{1}{(2\pi i)^n}\mathbf{D}_{1,n-1}(\eta,\bar{\partial}_\zeta\eta) \wedge d\psi$$

$$= \frac{1}{(2\pi i)^n}\frac{1}{|\psi|^{2n}}\mathbf{D}_{1,n-1}(\bar{\psi},\bar{\partial}_\zeta\bar{\psi}) \wedge d\psi. \tag{2.7.3}$$

By virtue of the properties of determinants in the expression $\mathbf{D}_{1,n-1}$, the first column can be replaced by any column such that $\langle \tau, \psi \rangle = 1$. Indeed, the determinant

$$\mathbf{D}_{1,n-1}(\eta - \tau, \bar{\partial}_\zeta\eta) = 0$$

due to the lines $\langle \eta - \tau, \psi \rangle = 0$ and $\langle \bar{\partial}_\zeta\eta, \psi \rangle = \bar{\partial}_\zeta\langle \eta, \psi \rangle = 0$ being linearly dependent.

Let $\Phi = \psi_1\varphi_1 + \cdots + \psi_n\varphi_n$, where the functions $\varphi_j = \varphi_j(\zeta, z), j = 1,\dots,n$ are holomorphic in $\mathbb{C}^n \times \mathbb{C}^n$, and $\tau = \frac{1}{\Phi}(\varphi_1,\dots,\varphi_n)$, then $\langle \tau, \psi \rangle = 1$ outside the zeros of the function Φ.

We introduce the determinant

$$U_\Phi = \frac{1}{(2\pi i)^n}\mathbf{D}_{1,1,n-2}(\tau,\eta,\bar{\partial}_\zeta\eta) \wedge d\psi.$$

Lemma 2.7.1 *The equality $\bar{\partial}_\zeta U_\Phi = \omega$ holds outside the zeros of Φ.*

The assertion of Lemma 2.7.1 can be checked by direct computation using the equation $\bar{\partial}_\zeta\tau = 0$.

Lemma 2.7.2 *For all $j = 1, 2,\dots,n$ we have the equations*

$$\frac{\partial}{\partial\bar{z}_j} U_\Phi = \frac{n-1}{(2\pi i)^n}\mathbf{D}_{1,1,n-2}\left(\tau,\frac{\partial\eta}{\partial\bar{z}_j},\bar{\partial}_\zeta\eta\right) \wedge d\psi + \bar{\partial}_\zeta\xi_j,$$

where

$$\xi_j = \frac{n-2}{(2\pi i)^n} D_{1,1,1,n-3}\left(\tau, \eta, \frac{\partial \eta}{\partial \bar{z}_j}, \bar{\partial}_\zeta \eta\right) \wedge d\psi$$

($\xi_j = 0$ *for* $n = 2$).

Proof Indeed

$$\frac{\partial}{\partial \bar{z}_j} U_\Phi = \frac{1}{(2\pi i)^n} D_{1,1,n-2}\left(\tau, \frac{\partial \eta}{\partial \bar{z}_j}, \bar{\partial}_\zeta \eta\right) \wedge d\psi$$

$$+ \frac{n-2}{(2\pi i)^n} D_{1,1,1,n-3}\left(\tau, \eta, \frac{\partial}{\partial \bar{z}_j}\bar{\partial}_\zeta \eta, \bar{\partial}_\zeta \eta\right) \wedge d\psi.$$

On the other hand

$$\bar{\partial}_\zeta D_{1,1,1,n-3}\left(\tau, \eta, \frac{\partial \eta}{\partial \bar{z}_j}, \bar{\partial}_\zeta \eta\right)$$

$$= -D_{1,1,n-2}\left(\tau, \frac{\partial \eta}{\partial \bar{z}_j}, \bar{\partial}_\zeta \eta\right) + D_{1,1,1,n-3}\left(\tau, \eta, \bar{\partial}_\zeta \frac{\partial \eta}{\partial \bar{z}_j}, \bar{\partial}_\zeta \eta\right).$$

Hence we have the required result. □

Lemma 2.7.3 *The following formulas*

$$\frac{\partial}{\partial \bar{z}_j}\omega(\psi) = \frac{n-1}{(2\pi i)^n} \bar{\partial}_\zeta D_{1,1,n-2}\left(\eta, \frac{\partial \eta}{\partial \bar{z}_j}, \bar{\partial}_\zeta \eta\right) \wedge d\psi \qquad (2.7.4)$$

hold for all $\zeta \neq z$.

Proof From Lemmas 2.7.1 and 2.7.2 we have

$$\frac{\partial}{\partial \bar{z}_j}\omega(\psi) = \frac{n-1}{(2\pi i)^n} \bar{\partial}_\zeta D_{1,1,n-2}\left(\tau, \frac{\partial \eta}{\partial \bar{z}_j}, \bar{\partial}_\zeta \eta\right) \wedge d\psi.$$

The determinant

$$D_{1,1,n-2}\left(\tau - \eta, \frac{\partial \eta}{\partial \bar{z}_j}, \bar{\partial}_\zeta \eta\right) = 0$$

by virtue of the linear dependence of the lines

$$\langle \tau - \eta, \psi \rangle = 0, \quad \left\langle \frac{\partial \eta}{\partial \bar{z}_j}, \psi \right\rangle = \frac{\partial}{\partial \bar{z}_j}\langle \eta, \psi \rangle = 0, \quad \langle \bar{\partial}_\zeta \eta, \psi \rangle = 0,$$

from which (2.7.4) follows. □

Lemma 2.7.3 shows that the derivatives with respect to $\bar{z}_j$ of form $\omega(\psi)$ are $\bar{\partial}$-exact differential forms with the point singularities $\zeta = z$.

Remark 2.7.1 Using the homogeneity property of the determinant, formula (2.7.4) can be rewritten as

$$\frac{\partial}{\partial \bar{z}_j} \omega(\psi) = \frac{n-1}{(2\pi i)^n} \bar{\partial}_\zeta \left[\frac{1}{|\psi|^{2n}} \mathbf{D}_{1,1,n-2} \left(\bar{\psi}, \overline{\frac{\partial \psi}{\partial z_j}}, \overline{\partial_\zeta \psi} \right) \wedge d\psi \right].$$

Lemma 2.7.4 *The following equalities*

$$\frac{\partial}{\partial \bar{z}_j} \omega(\psi) = -\frac{1}{(2\pi i)^n} \bar{\partial}_\zeta \left[\sum_{s=1}^{n} \frac{\partial}{\partial z_s} \left(\frac{1}{|\psi|^{2n-2}} \mathbf{D}_{1,1,n-2} \left(A^s, \overline{\frac{\partial \psi}{\partial z_j}}, \overline{\partial_\zeta \psi} \right) \wedge d\zeta \right) \right],$$

hold, where A^s is the i-th column of cofactors $A^s k$ to the elements of the Jacobi matrix $\left\| \dfrac{\partial \psi_s}{\partial z_k} \right\|_{s,k=1}^{n}$.

Proof It is not hard to check that

$$\sum_{s=1}^{n} \frac{\partial}{\partial z_s} A_k^s = 0$$

for all $k = 1, 2, \ldots, n$. Then

$$\sum_{s=1}^{n} \frac{\partial}{\partial z_s} \left[\frac{1}{|\psi|^{2n-2}} \mathbf{D}_{1,1,n-2} \left(A^s, \overline{\frac{\partial \psi}{\partial z_j}}, \overline{\partial_\zeta \psi} \right) \wedge d\zeta \right]$$

$$= \sum_{s=1}^{n} \frac{\partial}{\partial z_s} \left(\frac{1}{|\psi|^{2n-2}} \right) \mathbf{D}_{1,1,n-2} \left(A^s, \overline{\frac{\partial \psi}{\partial z_j}}, \overline{\partial_\zeta \psi} \right) \wedge d\zeta$$

$$= -(n-1) \sum_{s=1}^{n} \frac{\sum_{k=1}^{n} \bar{\psi}_k \frac{\partial \psi_k}{\partial z_s}}{|\psi|^{2n}} \sum_{m=1}^{n} (-1)^{m-1} A_m^s \mathbf{D}_{1,n-2}^m \left(\overline{\frac{\partial \psi}{\partial z_j}}, \overline{\partial_\zeta \psi} \right) \wedge d\zeta$$

$$= -(n-1) \sum_{k=1}^{n} (-1)^{k-1} \frac{\bar{\psi}_k}{|\psi|^{2n}} \mathbf{D}_{1,n-2}^k \left(\overline{\frac{\partial \psi}{\partial z_j}}, \overline{\partial_\zeta \psi} \right) \wedge d\psi$$

$$= -(n-1) \frac{1}{|\psi|^{2n}} \mathbf{D}_{1,1,n-2} \left(\bar{\psi}, \overline{\frac{\partial \psi}{\partial z_j}}, \overline{\partial_\zeta \psi} \right) \wedge d\psi,$$

were $\mathbf{D}_{1,n-2}^m \left(\dfrac{\partial \psi}{\partial z_j}, \overline{\partial_\zeta \psi} \right)$ is the determinant obtained from the determinant

$\mathbf{D}_{1,1,n-2} \left(\bar\psi, \dfrac{\partial \psi}{\partial z_j}, \overline{\partial_\zeta \psi} \right)$ by deleting the first column and the n-th line.   □

Proof of Theorem 2.7.1 Consider the differential form

$$\beta = \sum_{k=1}^{n} \beta_k \, d\bar z[k] \wedge dz,$$

where

$$\bar\beta_k = \frac{(-1)^k}{(2\pi i)^n} \int_{\partial D} f \bar\partial_\zeta \left[\frac{1}{|\psi|^{2n-2}} \mathbf{D}_{1,1,n-2} \left(A^k, \overline{\frac{\partial \psi}{\partial z_j}}, \overline{\partial_\zeta \psi} \right) \wedge d\zeta \right].$$

Then $\beta_k \in \mathscr{C}(\mathbb{C}^n)$ and they are real-analytic outside the boundary of the domain D. Indeed, the form

$$\bar\partial_\zeta \left[\frac{1}{|\psi|^{2n-2}} \mathbf{D}_{1,1,n-2} \left(A^k, \overline{\frac{\partial \psi}{\partial z_j}}, \overline{\partial_\zeta \psi} \right) \wedge d\zeta \right]$$

has absolutely integrable coefficients in D by Remark 1.4.1. Then by Stokes' formula

$$\int_{\partial D} f \bar\partial_\zeta \left[\frac{1}{|\psi|^{2n-2}} \mathbf{D}_{1,1,n-2} \left(A^k, \overline{\frac{\partial \psi}{\partial z_j}}, \overline{\partial_\zeta \psi} \right) \wedge d\zeta \right]$$

$$= \int_D \bar\partial_\zeta f \wedge \bar\partial_\zeta \left[\frac{1}{|\psi|^{2n-2}} \mathbf{D}_{1,1,n-2} \left(A^k, \overline{\frac{\partial \psi}{\partial z_j}}, \overline{\partial_\zeta \psi} \right) \wedge d\zeta \right],$$

and the last integral is a continuous function in $\mathbb{C}^n$, $k, j = 1, 2, \ldots, n$.

From the condition of Theorem 2.7.1, the function $M_\psi^- f(z) = 0$ outside $\overline{D}$, then $\dfrac{\partial}{\partial \bar z_j} M_\psi^- f(z) = 0$. By Lemmas 2.7.3 and 2.7.4, we obtain that the form β is $\bar\partial$-closed outside $\overline{D}$. Now we need the solution to the $\bar\partial$-problem in a ball $B_R(z^0) = \{z \in \mathbb{C}^n : |z - z^0| < R\}$, discussed in [9, Sect. 25] for the case of strongly convex domains.

Let $\rho = |z - z^0|^2 - R^2$, then $B_R(z^0) = \{z \in \mathbb{C}^n : \rho(z) < 0\}$. Consider the vector-valued function

$$P(z) = (P_1, \ldots, P_n) = \left(\frac{\partial \rho}{\partial z_1}, \ldots, \frac{\partial \rho}{\partial z_n} \right) = (\overline{z_1 - z_1^0}, \ldots, \overline{z_n - z_n^0})$$

and the functions

$$\Phi(z, w) = \sum_{j=1}^{n} P_j(z)(z_j - w_j) = |z|^2 - \langle \bar{z}, w \rangle - \langle \bar{z}^0, z \rangle + \langle \bar{z}^0, w \rangle,$$

$$\check{\Phi}(z, w) = \Phi(z, w) - \rho(z) = R^2 - \langle \bar{z}, w \rangle + \langle \bar{z}^0, w \rangle - |z^0|^2 + \langle \bar{z}, z^0 \rangle.$$

We introduce the vector-valued functions

$$u(z, w) = \frac{P(z)}{\Phi(z, w)},$$

$$u_\lambda(z, w) = (1 - \lambda)\frac{\bar{z} - \bar{w}}{|z - w|^2} + \lambda u(z, w), \qquad \lambda \in [0, 1],$$

and

$$\varphi_k(z, w) = \left[1 - \left(\frac{\Phi(z, w)}{\check{\Phi}(z, w)} \right)^{2n} \right]^k = \left(-\frac{\rho(z)}{\check{\Phi}(z, w)} \right)^k \left(\sum_{j=0}^{2n-1} \left(\frac{\Phi(z, w)}{\check{\Phi}(z, w)} \right)^j \right)^k.$$

If $z, w \in \bar{B}_R(z^0)$ and $z \neq w$, then $\Phi(z, w) \neq 0$.

Define the operators G^k to be

$$(G^k \beta)(w) = -\int_{B_R(z^0)} \beta(z)\, \varphi_k(z, w) U_{n,n-2}(z, w)$$

$$+ \int_{B_R(z^0) \times [0,1]} \beta(z) \wedge \bar{\partial}_z \varphi_k(z, w) \wedge W_{n,n-2}(u_\lambda, z, w), \qquad (2.7.5)$$

where

$$U_{n,n-2}(z, w) = \frac{n-1}{(2\pi i)^n} \mathbf{D}_{1,n-2,1}(t(z, w), \bar{\partial}_w t, \bar{\partial}_z t) \wedge dw,$$

$$W_{n,n-2}(z, w) = -\frac{1}{(2\pi i)^n} \mathbf{D}_{1,n-2,1}(u_\lambda, \bar{\partial}_w u_\lambda, \bar{\partial}_z u_\lambda) \wedge dw,$$

and $t(z, w) = \dfrac{\bar{z} - \bar{w}}{|z - w|^2}$ (see [9, Sect. 25]). Then by Theorems 25.6 from [9] at $k \geq 2$, we get

$$\bar{\partial}(G^k \beta) = \beta$$

in a ball $B_R(z^0)$ not intersecting $\bar{D}$, and the coefficients of $G^k \beta$ are continuous on the closure of $B_R(z^0)$ and real-analytic in this ball.

Consider the form $\gamma(w, z0, R) = G^k \beta$ in an arbitrary ball $B_R(z^0)$. We show that for $w \notin \partial D$, this form is real-analytic for w, z^0, R. Let $w^0 \notin \partial D$. Take a ball $B_\varepsilon(w^0)$ that does not intersect the boundary D, and break each of the integrals in formula (2.7.5) into two sets: $B_\varepsilon(w^0)$ and $B_R(z^0) \setminus B_\varepsilon(w^0)$. The integrals over $B_R(z^0) \setminus B_\varepsilon(w^0)$ are obviously real-analytic in $B_\varepsilon(w^0)$ due to the real-analytic kernels and absence of singularities. In the integrals over $B_\varepsilon(w^0)$ we replace the variables $z - w \to z'$.

Since the denominators responsible for singularity in the kernels depend on $z - w$, we get the integrals

$$\int_{B_\varepsilon(w-z)} A(z, z^0, R, w, z') \frac{\bar{z}'_j}{|z'|^{2n}} d\bar{z}' \wedge dz'$$

with coefficients depending real-analytically on the variables. Expanding these coefficients in a series in $z - w$ and integrating, we obtain the desired result.

Thus, in the ball $B_R(z^0) \subset \mathbb{C}^n \setminus \overline{D}$ the equality

$$\bar{\partial}_z \gamma(z, z^0, R) = \beta(z) \tag{2.7.6}$$

holds. Because of being real-analytic, equality (2.7.6) will also hold in the case when the ball $B_R(z^0)$ intersects the boundary D. Taking R and z^0 such that $B_R(z^0) \supset \overline{D}$, we obtain that on the set $B_R(z^0) \setminus \overline{D}$ the form γ yields a solution to the $\bar{\partial}$-problem for the form β.

Furthermore, the form γ has real-analytic coefficients in D, and equality (2.7.6) on the boundary ∂D is satisfied by virtue of continuity. The coefficients of β are generalized potentials, for which the maximum modulus theorem holds (see [56, Chaps. 6, 8, 9]), and the coefficients of the form γ are integrals of the form β, i.e., the uniform limits of such potentials. Then the maximum modulus theorem is also true for the coefficients of the form γ. Therefore, if Eq. (2.7.6) holds on the boundary D, then it is also valid in the domain D. Consequently, the form β is $\bar{\partial}$-closed in D. This means that the function $F = \frac{1}{\mu} M_\psi^+ f$ is holomorphic in D and gives continuation of f in D. This continuation F belongs to $\mathscr{C}(\overline{D})$, but since $f \in \mathscr{C}^1(\partial D)$, then $F \in \mathscr{C}^1(\overline{D})$. $\qquad\square$

Corollary 2.7.1 *Let D be a bounded domain in $\mathbb{C}^n$ with a connected smooth boundary. If the integral $M_\psi^- f(z) = 0$ outside $\overline{D}$ for the function $f \in \mathscr{C}^\gamma(\partial D)$ then the function f extends holomorphically to domain D.*

Proof repeats the proof of Theorem 1 from [49] using Lemma 2.6.2 instead of Corollary 1 from [49]. $\qquad\square$

Theorem 2.7.2 ([48]) *Let the function $M_\psi^+ f(z)$ be a holomorphic function in D for a function $f \in \mathscr{C}^1(\partial D)$. Then the function $F = \frac{1}{\mu} M_\psi^+ f$ is a holomorphic continuation of f from the boundary of the domain D into D.*

Proof repeats the proof of Theorems 2.7.1, we only have to use the fact that $M_{\psi}^{-}f(z) \to 0$ as $|z| \to +\infty$. Therefore, having the holomorphicity of functions $M_{\psi}^{-}f$ outside $\overline{D}$, we get $M_{\psi}^{-}f \equiv 0$. $\qquad\qquad\square$

Corollary 2.7.2 *If for the function* $f \in \mathscr{C}^1(\partial D)$ *the integral* $\dfrac{1}{\mu}M_{\psi}^{+}f$ *gives a continuous extension of* f *from the boundary of* D *into* D, *then the extension* $\dfrac{1}{\mu}M_{\psi}^{+}f$ *is holomorphic.*

Proof By Corollary 1.4.3 the function $M_{\psi}^{-}f(z) = 0$ on ∂D and $M_{\psi}^{-}f(z) \to 0$ as $|z| \to +\infty$. Then $M_{\psi}^{-}f \equiv 0$, since the integral $M_{\psi}^{-}f$ is a generalized potential [56, Chaps. 6, 8, 9]. $\qquad\qquad\square$

Remark 2.7.2 Theorem 2.7.2, Corollaries 2.7.1 and 2.7.2 are also valid for functions $f \in \mathscr{C}^{\gamma}(\partial D)$.

Chapter 3
On the Multidimensional Boundary Analogues of the Morera Theorem

Abstract This chapter contains some results related to the analytic continuation of functions given on the boundary of a bounded domain $D \subset \mathbb{C}^n$, $n > 1$, to this domain. The subject is not new. Results about the continuation of the Hartogs–Bochner theorem are well known and have already became classical. They are the subject of many monographs and surveys (see, for example, Aizenberg and Yuzhakov, Khenkin, Rudin, and many others). Here we will discuss boundary multidimensional variants of the Morera theorem. We desire to show how integral representations can be applied to the study of analytic continuation of functions. The same question about continuation connected with the direction about gluing discs can also be applied to the above Morera theorems based on the theory of extremal discs, developed by Lempert. However, since it is based on other ideas and methods, it does not fit into our book devoted to integral representations and their applications.

3.1 Functions with the Morera Property Along Complex and Real Planes

Let D be a bounded domain in $\mathbb{C}^n$ ($n > 1$) with a connected smooth boundary ∂D of class $\mathscr{C}^2$.

Definition 3.1.1 We say that a continuous function f on ∂D ($f \in \mathscr{C}(\partial D)$) satisfies the *Morera property (condition)* along a complex plane l of dimension k, $1 \le k \le n - 1$, if

$$\int_{\partial D \cap l} f(\zeta)\beta(\zeta) = 0 \qquad (3.1.1)$$

for any differential form β of type $(k, k - 1)$ with constant coefficients.

© Springer International Publishing Switzerland 2015

A.M. Kytmanov, S.G. Myslivets, *Multidimensional Integral Representations*,

DOI 10.1007/978-3-319-21659-1_3

It is assumed that the plane l transversely intersects the boundary of the domain D. If l is a complex line intersecting ∂D transversally, then the *Morera property* along l consists of the equality

$$\int_{\partial D \cap l} f(z+bt)dt = \int_{\partial D \cap l} f(z_1 + b_1 t, \ldots, z_n + b_n t)dt = 0$$

for the given parametrization $\zeta = z + bt$ $(z, b \in \mathbb{C}^n, t \in \mathbb{C})$ of the complex line l.

Clearly, the boundary values of functions $F \in \mathscr{A}(D)$ satisfy this property. Moreover, the same is true for continuous CR-functions f on ∂D. Recall that

Definition 3.1.2 A function $f \in \mathscr{C}(\partial D)$ is called a *CR-function* on ∂D if

$$\int_{\partial D} f(\zeta)\, \bar{\partial}\alpha(\zeta) = 0 \tag{3.1.2}$$

for all exterior differential forms α of type $(n, n-2)$ with coefficients of class $\mathscr{C}^\infty$ in $\overline{D}$.

Conditions (3.1.2) are called the *tangential Cauchy–Riemann equations*.

The Hartogs–Bochner theorem, which is now classical, tells us that *any continuous function f on ∂D is a CR-function if and only if it is holomorphically extended to D up to a certain function $F \in \mathscr{A}(D)$ (the boundary of D is connected)*.

In [26], the following inverse problem was considered: let a function $f \in \mathscr{C}(\partial D)$ satisfy the Morera property (3.1.1) along any complex k-plane l intersecting ∂D transversely. Is it true that f is a CR-function on ∂D?

Obviously, the greater the dimension k of the complex plane, the weaker the Morera property along complex k-planes. Therefore, if the Morera property holds along all complex lines, so it does along all complex hyperplanes. The following theorem is the first sufficiently general assertion on the solution of this problem.

Theorem 3.1.1 (Globevnik and Stout [26]) *Let $1 \leq k \leq n-1$, and let a function $f \in \mathscr{C}(\partial D)$ satisfy the Morera property (3.1.1) along any complex k-plane l intersecting ∂D transversely, then f is a CR-function on ∂D (and, therefore, it is holomorphically continued to D by the Hartogs–Bochner theorem).*

Proof Let f satisfy the condition of the theorem. We show that f satisfies conditions (3.1.2), i.e., it is a CR-function on ∂D. Without loss of generality, we can assume that

$$\alpha = A(\zeta)d\zeta \wedge d\bar{\zeta}[n-1, n],$$

where $A(\zeta)$ is a smooth function in $\mathbb{C}^n$ with compact support, $d\zeta = d\zeta_1 \wedge \ldots \wedge d\zeta_n$, and $d\bar{\zeta}[n-1, n]$ is obtained from $d\bar{\zeta}$ by removing the differentials $d\bar{\zeta}_{n-1}$, and $d\bar{\zeta}_n$.

Let us represent A as an inverse Fourier transform. If we denote the Fourier transform of A by $\hat{A}$ and set $(\zeta, \bar{z}) = \zeta_1 \bar{z}_1 + \ldots + \zeta_n \bar{z}_n$, then

$$A(\zeta) = c \int_{\mathbb{C}^n} \hat{A}(z) e^{i \operatorname{Re}(\zeta, \bar{z})} dz \wedge d\bar{z}$$

for some constant c. Then

$$\int_{\partial D} f \bar{\partial} \alpha = \int_{\partial D} f \bar{\partial} A(\zeta) \wedge d\zeta \wedge d\bar{\zeta}[n-1, n]$$

$$= c \int_{\partial D} f(\zeta) \left\{ \int_{\mathbb{C}^n} \hat{A}(z) \bar{\partial}_\zeta e^{i \operatorname{Re}(\zeta, \bar{z})} dz \wedge d\bar{z} \right\} d\zeta \wedge d\bar{\zeta}[n-1, n].$$

Changing the order of integration, we obtain

$$\int_{\partial D} f \bar{\partial} \alpha = c \int_{\mathbb{C}^n} \hat{A}(z) \left\{ \int_{\partial D} f(\zeta) \bar{\partial}_\zeta e^{i \operatorname{Re}(\zeta, \bar{z})} d\zeta \wedge d\bar{\zeta}[n-1, n] \right\} dz \wedge d\bar{z}.$$

For a fixed z, the inner integral equals 0. To see this, we make a non-singular linear change of variables with respect to ζ so that in the new coordinates $w = (w_1, \ldots, w_n)$, we have $w_1 = (\zeta, \bar{z})$. Then if $f^*(w) = f(\zeta(w))$ and D^* is the domain in the new variables w, we obtain

$$\int_{\partial D} f(\zeta) \bar{\partial}_\zeta e^{i \operatorname{Re}(\zeta, \bar{z})} d\zeta \wedge d\bar{\zeta}[n-1, n]$$

$$= c' \int_{\partial D^*} f^*(w) e^{(i/2)(w_1 + \bar{w}_1)} d\bar{w}_1 \wedge dw \wedge \Omega(\bar{w}), \qquad (3.1.3)$$

where c' is some constant and $\Omega(\bar{w})$ stands for the form $d\bar{\zeta}[n-1, n]$ in the variables w. Therefore,

$$\Omega(\bar{w}) = \sum_{1 \le j < k \le n} \beta_{jk} d\bar{w}[j, k]$$

for some constants β_{jk}. Since

$$d\bar{w}_1 \wedge \Omega = \sum_{2 \le k \le n} \beta_{1k} d\bar{w}[k],$$

the integral on the right-hand side of formula (3.1.3) is equal to the sum of integrals of the form

$$I_k = \int_{\partial D^*} f^*(w)e^{(i/2)(w_1+\bar{w}_1)}dw \wedge d\bar{w}[k], \qquad k \geq 2.$$

Each of the integrals I_k equals 0. Indeed, consider the integral I_n. Let Π : $\mathbb{C}^n \to \mathbb{C}^{n-k}$ be the projection $\Pi(w_1, \ldots, w_n) = w' = (w_1, \ldots, w_{n-k})$. By Fubini's theorem, we have

$$I_n = \int_{\Pi(D^*)} e^{(i/2)(w_1+\bar{w}_1)} \left\{ \int_{\Pi^{-1}(w') \cap \partial D^*} f^*(w)dw_{n-k+1} \wedge \ldots \wedge dw_n \wedge \right.$$

$$\left. \wedge d\bar{w}_{n-k+1} \ldots \wedge d\bar{w}_{n-1} \right\} dw_1 \wedge \ldots \wedge dw_{n-k} \wedge d\bar{w}_1 \wedge \ldots \wedge d\bar{w}_{n-k}.$$

For almost all $w' \in \Pi(D^*)$, the inner integral equals zero:

$$\int_{\partial D \cap l} f(\zeta)\eta(\zeta) = 0$$

by the condition of the theorem, where η is the form

$$dw_{n-k+1} \wedge \ldots \wedge dw_n \wedge d\bar{w}_{n-k+1} \wedge \ldots \wedge d\bar{w}_{n-1}$$

in the old variables ζ. That is, η is a $(k, k-1)$-form with constant coefficients.   □

A more exact analysis shows that Theorem 3.1.1 holds for real planes. By definition, the *CR*-dimension of a real plane l in $\mathbb{C}^n$ is the dimension of the maximal complex plane belonging to l. Denote by $\dim_R l$ and $\dim_{CR} l$ the real dimension of the plane l and the *CR*-dimension of l, respectively. Then, obviously,

$$\max(0, \dim_R l - n) \leq \dim_{CR} l \leq \left[\frac{\dim_R l}{2} \right].$$

A continuous function f on ∂D satisfies the Morera condition along a real k-dimensional plane l of *CR*-dimension p that transversally intersects the boundary ∂D if

$$\int_{\partial D \cap l} f\beta = 0$$

for all $(k-p, p-1)$-differential forms β with constant coefficients.

Theorem 3.1.2 (Govekar [27]) *Let* $2 \leq k \leq 2n-1$ *and* $\max(1, k-n) \leq p \leq [k/2]$. *A continuous function f on ∂D is a CR-function if and only if f satisfies the Morera* *property along any k-dimensional plane l of CR-dimension p that intersects ∂D transversely.*

In particular, for real hypersurfaces, the previous theorem yields the following assertion.

Theorem 3.1.3 (Govekar [27]) *A function $f \in \mathscr{C}(\partial D)$ is a CR-function on ∂D if and only if*

$$\int_{\partial D \cap l} f\beta = 0$$

for all real hyperplanes l intersecting ∂D transversely and for all differential forms β of type $(n, n - 2)$ with constant coefficients.

For complex k-planes l, we have $\dim_R l = 2k$ and $p = k$, therefore, Theorem 3.1.2 transforms into Theorem 3.1.1.

By definition, a complex k-wave function is a function f in $\mathbb{C}^n$ depending on k variables under a certain linear change of variables. As was shown in [26], linear combinations of complex waves form dense sets in spaces of smooth functions. This allows us to substantially reduce the sets of planes along which the Morera property ensures holomorphic extension of the function.

Let $G(n, k)$ be the Grassmann manifold of complex k-planes in $\mathbb{C}^n$, passing through 0.

Theorem 3.1.4 (Globevnik and Stout [26]) *Let $1 \le k \le n - 1$, and let W be an open set in $G(n, k)$. If a function $f \in \mathscr{C}(\partial D)$ satisfies the Morera property along each complex k-plane l of the form $z + \Sigma$, $\Sigma \in W$, intersecting ∂D transversely, then f is a CR-function on ∂D.*

Theorem 3.1.5 (Globevnik and Stout [26]) *Let $1 \le k \le n - 1$, and K be a compact convex set in D. Assume that the function $f \in \mathscr{C}(\partial D)$ satisfies the Morera property along all complex k-planes not intersecting K. Then f is a CR-function on the boundary of D.*

Theorem 3.1.6 (Globevnik and Stout [26]) *Let B be a ball lying in $\mathbb{C}^n \setminus \overline{D}$. Assume that a function $f \in \mathscr{C}(\partial D)$ satisfies the Morera property along all complex k-planes intersecting B. Then f is a CR-function on the boundary of D.*

3.2 Functions with the Morera Property Along Complex Lines

Let D be a bounded domain in $\mathbb{C}^n$ $(n > 1)$ with a connected smooth boundary ∂D of class $\mathscr{C}^2$. The classical Hartogs' theorem asserts that any function f is holomorphic in the domain D if its restriction to any complex line parallel to one of the coordinate complex lines is holomorphic.

The following natural question arises: for which sets of complex one-dimensional cross-sections of the domain does the existence of holomorphic continuations along the cross-sections imply the existence of a holomorphic continuation to the whole domain?

A set of all complex lines intersecting a given domain with a twice smooth boundary is sufficient. For the case of a complex ball, this was proved by Agranovskii and Val'skii in [4], Nagel and Rudin in [66], and Grinberg [29]. For an arbitrary domain, an analogues result was proved by Stout [78]. Let us formulate it here.

Consider one-dimensional complex lines l of the form

$$l = \{\zeta \in \mathbb{C}^n : \zeta_j = z_j + b_j t, \; j = 1, \ldots, n, \; t \in \mathbb{C}\} \tag{3.2.1}$$

passing through a point $z \in \mathbb{C}^n$ in the direction of a vector $b \in \mathbb{CP}^{n-1}$ (the direction of b is determined with an accuracy of up to multiplication by a complex number $\lambda \neq 0$). By Sard's theorem, for almost all $z \in \mathbb{C}^n$ and almost all $b \in \mathbb{CP}^{n-1}$, the intersection $l \cap \partial D$ is a finite set of piecewise-smooth curves (except for the degenerate case where $\partial D \cap l = \varnothing$). Let us give the following definition.

Definition 3.2.1 The function $f \in \mathscr{C}(\partial D)$ has the *one-dimensional holomorphic extension property along complex line l* of the form (3.2.1) if for any line l such that $\partial D \cap l \neq \varnothing$, there exists a function F having the following properties:

1. $F \in \mathscr{C}(\overline{D} \cap l)$,
2. $F = f$ on the set $\partial D \cap l$,
3. The function F is holomorphic at interior (with respect to the topology of l) points of the set $\overline{D} \cap l$.

An analogues definition can be made for complex k-planes. Clearly, if a function f satisfies the holomorphic extension property along all complex k-planes, then it satisfies this property along complex lines. Therefore, in what follows, we restrict ourselves to consideration of this case.

Theorem 3.2.1 (Stout [78]) *If a function $f \in \mathscr{C}(\partial D)$ has the one-dimensional holomorphic extension property along complex lines of the form (3.2.1), then f is holomorphically extended into D.*

A more narrow set of complex lines sufficient for the holomorphic continuation was considered by Agranovskii and Semenov [3]. Consider an open set $V \subset D$ and a set $\mathfrak{L}_V$ of complex lines intersecting this set.

Theorem 3.2.2 (Agranovskii and Semenov [3]) *If a function $f \in \mathscr{C}(\partial D)$ has the one-dimensional holomorphic extension property along lines from the set $\mathfrak{L}_V$ for a certain open set $V \subset D$, then the function f is holomorphically extended into D.*

The strengthening of the previous results consists of assertions dealing with the boundary analogues of the Morera theorem, since they are completely implied by them. We now formulate the assertion belonging to Globevnik and Stout [26] (a particular case of Theorem 3.1.1).

Theorem 3.2.3 (Globevnik and Stout [26]) *Let a function $f \in \mathscr{C}(\partial D)$, and for almost all $z \in \mathbb{C}^n$ and almost all $b \in \mathbb{CP}^{n-1}$, let*

$$\int_{\partial D \cap l} f(z + bt) \, dt = \int_{\partial D \cap l} f(z_1 + b_1 t, \dots, z_n + b_n t) \, dt = 0. \qquad (3.2.2)$$

Then the function f is holomorphically extended into D up to a function $F \in \mathscr{C}(\overline{D})$. (If $\partial D \cap l = \varnothing$, then the integral in (3.2.2) is assumed to be equal to zero.)

We note that without the connectivity condition of the boundary of the domain, Theorem 3.2.3 is obviously false.

In [26], the problem of finding *sufficient* sets of complex lines $\mathfrak{L} = \{l\}$ for which condition (3.2.2) for $l \in \mathfrak{L}$ implies a holomorphic extension of the function f to D was posed. For example, is a set $\mathfrak{L}_V$ of lines l intersecting a certain open set $V \subset D$ such a sufficient set? In paper [3] Agranovskii and Semenov give an affirmative answer to this question; Theorem 3.2.3 is generalized there. In paper [47] Kytmanov and Myslivets obtained a statement from which Theorems 3.2.1, 3.2.2, 3.2.3 follow.

Theorem 3.2.4 ([47, 49]) *Let k be a fixed non-negative integer and let a function $f \in \mathscr{C}(\partial D)$. If, for almost all $z \in \mathbb{C}^n$ and almost all $b \in \mathbb{CP}^{n-1}$, the condition*

$$\int_{\partial D \cap l} f(z_1 + b_1 t, \dots, z_n + b_n t) t^k \, dt = 0 \qquad (3.2.3)$$

holds, then f is holomorphically extended to D.

For $k = 0$, we obtain Theorem 3.2.3.

Theorem 3.2.5 ([47, 49]) *For a fixed k and a function $f \in \mathscr{C}(\partial D)$, let condition (3.2.3) hold for almost all lines l (of the form (3.2.1)) intersecting an open set $V \subset D$ (or an open set $V \subset \mathbb{C}^n \setminus \overline{D}$), then the function f is holomorphically extended into D.*

Proof of the theorem is performed for the case of the set $V \subset D$.

Let $U(\zeta, z)$ be a Bochner–Martinelli kernel of the form (1.1.1). Consider complex lines l of the form (3.2.1) passing through z in the direction of the vector $b \in \mathbb{CP}^{n-1}$.

Lemma 3.2.1 *The Bochner–Martinelli kernel in the coordinates t and b has the form*

$$U(\zeta, z) = \lambda(b) \wedge \frac{dt}{t},$$

where $\lambda(b)$ is some differential form of the type $(n-1, n-1)$ in $\mathbb{CP}^{n-1}$ independent of t.

Proof Assume that $z = 0$. Express the Bochner–Martinelli kernel in the variables t and b. We obtain

$$U(\zeta, 0) = \frac{(n-1)!}{(2\pi i)^n} \frac{\sum\limits_{k=1}^{n}(-1)^{k-1}\bar\zeta_k d\bar\zeta[k] \wedge d\zeta}{|\zeta|^{2n}}$$

$$= \frac{(-1)^{n-1}(n-1)!}{(2\pi i)^n} \frac{\sum\limits_{k=1}^{n}(-1)^{k-1}\bar b_k d\bar b[k] \wedge \sum\limits_{k=1}^{n}(-1)^{k-1}b_k db[k]}{|b|^{2n}} \wedge \frac{dt}{t}$$

$$= \lambda(b) \wedge \frac{dt}{t}.$$

$\square$

Lemma 3.2.2 *If condition (3.2.3) holds for a point $z \in \mathbb{C}^n \setminus \partial D$ and for almost all $b \in \mathbb{CP}^{n-1}$, then*

$$\int_{\partial D_\zeta} (\zeta - z)^\alpha f(\zeta)\, U(\zeta, z) = 0, \tag{3.2.4}$$

where $\alpha = (\alpha_1, \ldots, \alpha_n)$ is an arbitrary multi-index such that

$$\|\alpha\| = \alpha_1 + \cdots + \alpha_n = k + 1, \quad \text{and} \quad (\zeta - z)^\alpha = (\zeta_1 - z_1)^{\alpha_1} \cdots (\zeta_n - z_n)^{\alpha_n}.$$

Proof We use the representation of the Bochner–Martinelli kernel from Lemma 3.2.1. Then by Fubini's theorem, we have

$$\int_{\partial D} (\zeta - z)^\alpha f(\zeta)\, U(\zeta, z) = \int_{\mathbb{CP}^{n-1}} \lambda(b) \int_{\partial D \cap l} b^\alpha t^k f(z + bt)\, dt = 0$$

by condition (3.2.3). $\square$

Lemma 3.2.3 *Let condition (3.2.4) hold for points $z \in V$, then the function f is holomorphically extended into D.*

Proof If (3.2.4) holds for $z \in V$, then it also holds for all points $z \in D$ by the real-analyticity of the integral in (3.2.4). Let us rewrite (3.2.4) in a different form. Introduce the following differential forms $U_s(\zeta, z)$, considered for the first time by Martinelli [63] (see also [45, Chap. 2]):

$$U_s(\zeta, z) = \frac{(-1)^s(n-2)!}{(2\pi i)^n}\left(\sum_{j=1}^{s-1}(-1)^j\frac{\bar\zeta_j - \bar z_j}{|\zeta - z|^{2n-2}} d\bar\zeta[j, s]\right.$$

$$\left. + \sum_{j=s+1}^{n}(-1)^{j-1}\frac{\bar\zeta_j - \bar z_j}{|\zeta - z|^{2n-2}} d\bar\zeta[s, j]\right) \wedge d\zeta. \tag{3.2.5}$$

It is easy to verify that

$$\bar{\partial}\left(\frac{1}{\zeta_s - z_s} U_s(\zeta, z)\right) = U(\zeta, z)$$

for $\zeta_s \neq z_s$, $s = 1, \ldots, n$. Then condition (3.2.4) can be written in the form

$$\int_{\partial D} f(\zeta)\bar{\partial}\left((\zeta - z)^\beta U_s(\zeta, z)\right) = 0, \quad z \in D, \tag{3.2.6}$$

for all monomials $(\zeta - z)^\beta$ with $\|\beta\| = k$.

Let us show that condition (3.2.6) also holds for the monomials $(\zeta - z)^\gamma$ with $\|\gamma\| < k$. Indeed, consider such a monomial $(\zeta - z)^\gamma$ and $\|\gamma\| = k - 1$. Then (3.2.6) holds for monomials of the form

$$(\zeta - z)^\gamma (\zeta_m - z_m), \quad m = 1, \ldots, n,$$

since the degree of these monomials is equal to k.

We have

$$\frac{\partial}{\partial \zeta_m}\left((\zeta - z)^\gamma (\zeta_m - z_m) U_s(\zeta, z)\right) = (\gamma_m + 1)(\zeta - z)^\gamma U_s(\zeta, z)$$

$$- (n-1)(\zeta - z)^\gamma \frac{(\zeta_m - z_m)(\bar{\zeta}_m - \bar{z}_m)}{|\zeta - z|^2} U_s(\zeta, z). \tag{3.2.7}$$

Summing relations (3.2.7) with respect to m, we obtain

$$\sum_{m=1}^{n} \frac{\partial}{\partial \zeta_m}\left((\zeta - z)^\gamma (\zeta_m - z_m) U_s(\zeta, z)\right) = (\|\gamma\| + 1)(\zeta - z)^\gamma U_s(\zeta, z). \tag{3.2.8}$$

Since condition (3.2.6) can be differentiated in z and the derivatives in z and ζ of the integrand are equal, (3.2.8) implies that the degree of the monomial in (3.2.6) can be reduced by 1. Sequentially reducing this degree, we arrive at the conditions

$$\int_{\partial D} f(\zeta)\bar{\partial} U_s(\zeta, z) = 0, \quad z \in D, s = 1, \ldots, n,$$

i.e.,

$$\int_{\partial D} (\zeta_s - z_s) f(\zeta) U(\zeta, z) = 0, \quad z \in D, s = 1, \ldots, n. \tag{3.2.9}$$

Applying the Laplace operator

$$\Delta = \frac{\partial^2}{\partial z_1 \partial \bar{z}_1} + \cdots + \frac{\partial^2}{\partial z_n \partial \bar{z}_n},$$

to the left-hand side of relation (3.2.9), we obtain

$$\frac{\partial}{\partial \bar{z}_s} \int_{\partial D} f(\zeta)\, U(\zeta - z) = 0, \ z \in D, \ s = 1, \ldots, n.$$

Here, we have used the harmonicity of the kernel $U(\zeta - z)$ and the identity

$$\Delta(fh) = h\Delta f + f\Delta h + \sum_{j=1}^{n} \frac{\partial f}{\partial \bar{z}_j} \frac{\partial h}{\partial z_j} + \sum_{j=1}^{n} \frac{\partial f}{\partial z_j} \frac{\partial h}{\partial \bar{z}_j}.$$

Therefore, the Bochner–Martinelli-type integral of f of the form (1.2.4)

$$Mf(z) = \int_{\partial D} f(\zeta)\, U(\zeta, z)$$

is a function holomorphic in the domain D. Therefore, taking $F(z) = Mf(z)$ and applying Corollary 2.5.3, according to which, in this case, $F \in \mathscr{C}(\overline{D})$ and the boundary value of the function F coincides with f on ∂D, we obtain the desired extension of the function $f(z)$. $\qquad\square$

The proposition is also true in the case where the open set $V \subset \mathbb{C}^n \setminus \overline{D}$. Instead of Corollary 2.5.3 we need to apply Corollary 2.5.2.

Theorems 3.2.4 and 3.2.5 are consequences of Lemmas 3.2.2 and 3.2.3.

Corollary 3.2.1 *Let A be an algebraic hypersurface in $\mathbb{C}^n$. If condition (3.2.3) for a function f holds for almost all complex lines l intersecting A, then the function f is holomorphically extended into D.*

Proof Since almost every complex line l intersects A, condition (3.2.3) holds for almost all $z \in \mathbb{C}^n$. $\qquad\square$

3.3 Holomorphic Extension Along Complex Curves and Analogues of the Morera Theorem

3.3.1 Holomorphic Extension Along Complex Curves

Consider classes of complex curves $l_{z,b}$ of the following types:

Type 1: algebraic curves

$$l_{z,b} = \left\{ \zeta \in \mathbb{C}^n : \zeta_1 = z_1 + t^{k_1}, \ \zeta_j = z_j + b_j t^{k_j}, j = 2, \ldots, n, \ t \in \mathbb{C} \right\},$$

where the constants $k_j \in \mathbb{N}$ are fixed, $j = 1, \ldots, n$, and the vector $b = (1, b_2, \ldots, b_n) \in \mathbb{C}^n$;

Type 2: complex curves of the form

$$l_{z,b} = \left\{ \zeta \in \mathbb{C}^n : \zeta_1 = z_1 + t, \ \zeta_j = z_j + b_j t^{k_j} \chi_j(t), j = 2, \ldots, n, \ t \in \mathbb{C} \right\},$$

where $\chi_j(t)$ are the entire holomorphic functions in the variable t, and, moreover these functions do not vanish at any point, $j = 2, \ldots, n$;

Type 3: complex curves of the form

$$l_{z,b} = \left\{ \zeta \in \mathbb{C}^n : \zeta_1 = z_1 + t^{k_1}, \ \zeta_j = z_j + b_j t^{k_j} \chi_j(t^{k_1}), j = 2, \ldots, n, \ t \in \mathbb{C} \right\}, \tag{3.3.1}$$

where $\chi_j(\tau)$ are the entire complex functions of the variable τ that do not vanish at any point, $j = 2, \ldots, n$.

The case of algebraic curves was examined in [48], Type 2 in [48], and Type 3 in [64]. We note that for $k_1 = 1$, the curves of the second type are obtained from curves of the form (3.3.1) and for $\chi_j \equiv 1$ $(j = 2, \ldots, n)$, we obtain algebraic curves; therefore, we will consider curves of the form (3.3.1).

The third class of curves also contains curves of the form

$$l_{z,b} = \left\{ \zeta \in \mathbb{C}^n : \zeta_1 = z_1 + \varphi_1(t), \zeta_j = z_j + b_j \varphi_j(t), j = 2, \ldots, n, \ t \in \mathbb{C} \right\},$$

where $\varphi_j(t)$ are the entire functions in the variable t having one zero of the first order at the point $t = 0$. Indeed, in this case, we can introduce a different parametrization taking the first function φ_1 as the parameter t.

If we fix a point $z \in \mathbb{C}^n$, then for any point ζ such that $z_1 \neq \zeta_1$, there exists a curve $l_{z,b}$ passing through ζ (subject to appropriate choice of the vector b). For a fixed z, all curves $l_{z,b}$ intersect at the point 0. If they also intersect at another points, it is easy to show that the j-coordinates of the vectors b for these points are obtained from each other by rotation by an angle that is a multiple of $\dfrac{2\pi k_j}{k_1}$. Therefore, to uniquely find the vector b, we assume that the argument b_j satisfies the condition

$$0 \leq \arg b_j < 2\pi r_j, \quad j = 2, \ldots, n, \tag{3.3.2}$$

where r_j is the fractional part of the number $\dfrac{k_j}{k_1}$ (if k_j is dividable by k_1, then no conditions are imposed on $\arg b_j$).

In fact, $l_{z,b}$ is a parameterization of the following complex curves given in an explicit form:

$$\{\zeta : \zeta_j = z_j + b_j(\zeta_1 - z_1)^{\frac{k_j}{k_1}} \chi_j(\zeta_1 - z_1), \quad j = 2, \ldots, n\}.$$

Therefore, for a fixed z, we obtain fibering of $\mathbb{C}^n \setminus \{\zeta : \zeta_1 = z_1\}$ into the curves $l_{z,b}$ for vectors b satisfying condition (3.3.2). Then Sard's theorem shows that for almost all b satisfying this condition, the intersection of $l_{z,b}$ with the boundary ∂D is either empty or is a union of a finite set of closed piecewise-smooth curves. Perhaps, class (3.3.1) is the most general class of curves having such properties.

Introduce the following holomorphic functions:

$$\psi_1(\zeta) = \zeta_1^{p_1}, \quad \psi_j = \frac{\zeta_j^{p_j}}{\chi_j^{p_j}(\zeta_1)}, \quad j = 2, \ldots, n, \tag{3.3.3}$$

where the natural numbers p_j are chosen so that $p_1 k_1 = \ldots = p_n k_n = p$. These functions are holomorphic in $\mathbb{C}^n$ and have only one common zero, the origin of multiplicity $\mu = p_1 \cdots p_n$. Consider the kernel $\omega(\zeta) = U(\zeta, 0)$ and $\omega(\psi(\zeta - z)) = U(\psi(\zeta - z), 0)$ in the new coordinates t, b. The symbol $*$ will mean transition from variables ζ to the new variables (t, b).

Lemma 3.3.1 *In the coordinates t, b the kernel $\omega(\psi(\zeta - z))$ has the form*

$$\omega(\psi^*(t, b)) = \frac{dt}{t} \wedge \lambda(b), \tag{3.3.4}$$

where

$$\lambda(b) = \frac{p(n-1)!}{(2\pi i)^n} \frac{(-1)^{n-1} d\bar{b}_2^{p_2} \wedge \cdots \wedge d\bar{b}_n^{p_n} \wedge db_2^{p_2} \wedge \cdots \wedge db_n^{p_n}}{\left(1 + \sum\limits_{j=2}^{n} |b_j|^{2p_j}\right)^n}.$$

Proof Indeed, $\psi_1^*(t, b) = t^p$, $\psi_j^*(t, b) = b_j^{p_j} t^p$, $j = 2, \ldots, n$. Therefore the module

$$|\psi(\zeta - z)|^2 = |\psi^*(t, b)|^2 = |t|^{2p}\left(1 + \sum\limits_{j=2}^{n} |b_j|^{2p_j}\right).$$

We have

$$d\psi^* = d\psi_1^* \wedge \cdots \wedge d\psi_n^* = pt^{pn-1} dt \wedge db_2^{p_2} \wedge \cdots \wedge db_n^{p_n},$$

and the form

$$\sum_{k=1}^{n} (-1)^{k-1} \psi_k^* d\psi^*[k] = t^p d\psi_2^* \wedge \cdots \wedge d\psi_n^* + \sum_{k=2}^{n} (-1)^{k-1} b_k^{p_k} t^p d\psi^*[k]$$

$$= t^{pn} db_2^{p_2} \wedge \cdots \wedge db_n^{p_n} + pt^{pn-1} \left(\sum_{j=2}^{n} (-1)^{j-2} b_j^{p_j} dt \wedge db_2^{p_2} \wedge \cdots [j] \cdots \wedge db_n^{p_n} \right.$$

$$\left. + \sum_{k=2}^{n} (-1)^{k-1} b_k^{p_k} t^{pn-1} dt \wedge db_2^{p_2} \wedge \cdots [k] \cdots \wedge db_n^{p_n} \right) = t^{pn} db_2^{p_2} \wedge \cdots \wedge db_n^{p_n}.$$

This completes the proof.   □

This statement generalizes Lemma 3.2.1.

Definition 3.3.1 A function $f \in \mathscr{C}(\partial D)$ has the *one-dimensional holomorphic extension property along complex curves* of the form $l_{z,b}$ if for any curve $l_{z,b}$ such that $\partial D \cap l_{z,b} \neq \varnothing$, there exists a function $F_{z,b}(t)$ having the following properties:

1. $F_{z,b} \in \mathscr{C}(\overline{D} \cap l_{z,b})$,
2. $F_{z,b} = f$ on the set $\partial D \cap l_{z,b}$,
3. The function $F_{z,b}$ is holomorphic with respect to t in interior (in the topology of $l_{z,b}$) points of the set $\overline{D} \cap l_{z,b}$.

Therefore, this definition is completely analogues to that of functions with the one-dimensional holomorphic extension property along lines.

Proposition 3.3.1 *If a function $f \in \mathscr{C}(\partial D)$ has the one-dimensional holomorphic extension property along complex curves of the form (3.3.1), then*

$$\int_{\partial D} f(\zeta) \, \omega(\psi(\zeta - z)) = 0,$$

for all $z \notin \overline{D}$, for the functions $(\psi_1, \ldots, \psi_n)$ of the form (3.3.3).

Proof By Sard's theorem, almost all complex curves $l_{z,b}$ intersect the boundary of the domain D along piecewise-smooth curves. Therefore Fubini's theorem and Lemma 3.3.1 give the equality

$$\int_{\partial D} f(\zeta) \, \omega(\psi(\zeta - z)) = \int_{\mathbb{C}^{n-1}} \lambda(b) \int_{\partial D \cap l_{z,b}} f^*(t, b) \frac{dt}{t}.$$

However the inner integral is zero if $z \notin \overline{D}$, since f has the one-dimensional holomorphic extension property.   □

Theorem 3.3.1 ([48, 64]) *Let $\partial D \in \mathscr{C}^2$, and a function $f \in \mathscr{C}(\partial D)$ has the one-dimensional holomorphic extension property along complex curves $l_{z,b}$, then f is holomorphically extended into D.*

Theorem 3.3.1 is a direct consequence of Theorem 2.7.1 and Proposition 3.3.1. This assertion generalizes Stout's theorem 3.2.1 on functions with the one-dimensional holomorphic extension property along complex lines.

Let us consider sufficient families of curves $l_{z,b}$ the holomorphic extension along which can ensure the holomorphic extension to the domain D. The first such family comprises the curves $l_{z,b}$ with the point z belonging to a certain open set $V \subset \mathbb{C}^n \setminus \overline{D}$, and b being any vector. In this case, the integral in Proposition 3.3.1 is equal to zero in V, and, therefore, it is equal to zero everywhere (by its real-analyticity) outside $\overline{D}$ and Theorem 3.3.1 is applicable.

Consider an open set $V \subset D$, and denote the set of curves $l_{z,b}$, intersecting this set by $\mathfrak{L}_V$.

Theorem 3.3.2 ([48]) *Let a bounded domain D with a smooth connected boundary be such that ∂D is the Shilov boundary for the function algebra $\mathscr{O}(D) \cap \mathscr{C}^1(\overline{D})$ (for example, D is a strictly pseudo-convex domain). If a function $f \in \mathscr{C}^1(\partial D)$ has the one-dimensional holomorphic extension property along complex curves from the set $\mathfrak{L}_V$ for a certain open set $V \subset D$, then the function f is holomorphically extended into D.*

Proof Let a function $\varphi \in \mathscr{O}(D) \cap \mathscr{C}^1(\overline{D})$, then φf has also the one-dimensional property of holomorphic extension along complex curves of the set $\mathfrak{L}_V$. We have the equality

$$\int_{\partial D} \varphi(\zeta) f(\zeta) \omega(\psi(\zeta - z))$$

$$= \int_{\mathbb{C}^{n-1}} \lambda(b) \int_{\partial D \cap l_{z,b}} \varphi^*(t, b) f^*(t, b) \frac{dt}{t} = \varphi(z) \int_{\partial D} f(\zeta) \omega(\psi(\zeta - z)),$$

i.e.,

$$\int_{\partial D} \varphi(\zeta) f(\zeta) \omega(\psi(\zeta - z)) = \varphi(z) \int_{\partial D} f(\zeta) \omega(\psi(\zeta - z)), \quad z \in V.$$

By virtue of the real-analyticity of this integral this equation is satisfied everywhere in D. Denote the integrals by $M_{\psi}^{\pm} f$ (as in formula (1.4.5)). Then

$$M_{\psi}^{+}[\varphi f] = \varphi M_{\psi}^{+} f \tag{3.3.5}$$

in domain D. For points $z \in D$, the relations: $M_\psi^+ 1 = \mu$ and $M_\psi^+ \varphi = \mu\varphi$ hold, because φ is holomorphic in D. Then from (3.3.5) we obtain

$$M_\psi^+ [\varphi(\zeta)(f(\zeta) - f(z))] = \varphi(z)M_\psi^+ [f(\zeta) - f(z)], \quad z \in D. \tag{3.3.6}$$

Here we took a continuation of f to a function of class $\mathscr{C}^1(\overline{D})$. Since $\varphi f \in \mathscr{C}^1(\overline{D})$, then $|f(\zeta) - f(z)| \leq c|\zeta - z|$ and the integral $M_\psi^+ [f(\zeta) - f(z)]$ converges absolutely for $z \in \partial D$ (see the proof of Theorem 1.4.1). Therefore, in Eq. (3.3.6) we can go to the limit for z tending to ∂D. And, therefore, equality (3.3.6) holds for points $z \in \partial D$.

By the hypothesis of Theorem 3.3.2, the boundary ∂D is a closure of the points for algebra $\mathscr{O}(D) \cap \mathscr{C}^1(\overline{D})$. Let $z \in \partial D$ be a peak point and φ be the peak function, i.e., $\varphi(z) = 1$ and $|\varphi(\zeta)| < 1$ for points $\zeta \neq z$.

Consider the functions φ^k and apply equality (3.3.6) to them. We obtain

$$M_\psi^+ [\varphi^k(\zeta)(f(\zeta) - f(z))] = \varphi^k(z)M_\psi^+ [f(\zeta) - f(z)]. \tag{3.3.7}$$

By Lebesgue's theorem and the inequality $|\varphi^k| \leq 1$ there exists a limit in (3.3.7) when $k \to \infty$. We have

$$M_\psi^+ [\varphi^k(\zeta)(f(\zeta) - f(z))] \to 0,$$

then formula (3.3.6) yields

$$M_\psi^+ [f(\zeta) - f(z)] = 0$$

for the peak points $z \in \partial D$. And since they are dense in ∂D and this integral is a continuous function, we obtain that the boundary values of the function $M_\psi^+ f$ coincide with μf. This means that the function $F = \dfrac{1}{\mu} M_\psi^+ f$ is holomorphic in D by Corollary 2.7.2 and is an extension of the function f. $\square$

3.3.2 Some Integral Criteria of Holomorphic Extension of Functions

As in Sect. 2.7, we consider the map $\psi = (\psi_1, \ldots, \psi_n)$ and the differential form $\omega(w)$.

Proposition 3.3.2 *If $\partial D \in \mathscr{C}^d$ $(d \geq 1)$, then every function $f \in \mathscr{C}^l(\partial D)$, $0 \leq l \leq d$ is the limit in the metric $\mathscr{C}^l(\partial D)$ of linear combinations of fractions of the form*

$$\sum_{k=1}^{n} \frac{A_k^s(\zeta - z)\overline{A}_k^m(\zeta - z)}{|\psi(\zeta - z)|^{2n-2}}, \quad s, m = 1, 2, \ldots, n, \tag{3.3.8}$$

where $z \in \partial D$, and ζ is a fixed point not lying on ∂D. Here A_k^s are the cofactors to

the elements $\dfrac{\partial \psi_k}{\partial \zeta_s}$ in the Jacobi matrix of the map ψ. Instead of fractions of the form

(3.3.8) we can as well take fractions of the form

$$
\frac{1}{|\psi(\zeta - z)|^{2n}} \left(\sum_{r=1}^{n} \psi_r(\zeta - z) A_r^s(\zeta - z) \right) \left(\sum_{p=1}^{n} \overline{\psi_p(\zeta - z) A_p^m(\zeta - z)} \right). \qquad (3.3.9)
$$

Proof Consider the determinants

$$
\mathbf{D}_{v_1, \ldots, v_m} \left(\theta^1, \ldots, \theta^m \right)
$$

of the form (2.7.2).

Lemma 3.3.2 *The kernel $\omega(\psi(\zeta - z))$ can be represented as*

$$
\omega(\psi) = \frac{1}{(n-1)(2\pi i)^n} \sum_{s=1}^{n} \frac{\partial}{\partial z_s} \left(\frac{1}{|\psi|^{2n-2}} \mathbf{D}_{1, n-1} \left(A^s, \overline{\partial_\zeta \psi} \right) \right) \wedge d\zeta,
$$

where A^s is a column of cofactors A_k^s, $k = 1, 2, \ldots, n$.

Proof From formula (2.7.3) we have

$$
\omega(\psi) = \frac{1}{(2\pi i)^n} \frac{1}{|\psi|^{2n}} \mathbf{D}_{1, n-1} \left(\bar{\psi}, \overline{\partial_\zeta \psi} \right) \wedge d\psi.
$$

Since

$$
\sum_{s=1}^{n} \frac{\partial A_k^s}{\partial z_s} = 0, \quad k = 1, \ldots, n,
$$

we obtain

$$
\sum_{s=1}^{n} \frac{\partial}{\partial z_s} \left(\frac{1}{|\psi|^{2n}} \mathbf{D}_{1, n-1} \left(A^s, \overline{\partial_\zeta \psi} \right) \right) \wedge d\zeta
$$

$$
= \sum_{s=1}^{n} \frac{\partial}{\partial z_s} \left(\frac{1}{|\psi|^{2n-2}} \right) \mathbf{D}_{1, n-1} \left(A^s, \overline{\partial_\zeta \psi} \right) \wedge d\zeta
$$

$$
= -(n-1) \sum_{s=1}^{n} \frac{\sum_{k=1}^{n} \bar{\psi}_k \frac{\partial \psi_k}{\partial z_s}}{|\psi|^{2n}} \sum_{j=1}^{n} (-1)^{j-1} A_j^s \, \mathbf{D}_{n-1}^j (\overline{\partial_\zeta \psi}) \wedge d\zeta
$$

$$= (n-1) \sum_{s=1}^{n} \frac{\bar{\psi}_s}{|\psi|^{2n}} (-1)^{s-1} \mathbf{D}_{n-1}^s (\overline{\partial_\zeta \psi}) \wedge d\psi$$

$$= \frac{(n-1)}{|\psi|^{2n}} \mathbf{D}_{1,n-1} \left(\bar{\psi}, \overline{\partial_\zeta \psi} \right) \wedge d\psi = (n-1)(2\pi i)^n \omega(\psi).$$

Here $\mathbf{D}_{n-1}^j (\overline{\partial_\zeta \psi})$ is the determinant of the $(n-1)$-th order, that is obtained from $\mathbf{D}_{1,n-1} \left(A^s, \overline{\partial_\zeta \psi} \right)$ by deleting the first column and the j-th row.   □

Lemma 3.3.3 *The kernel $U(\psi, 0)$ can be represented as*

$$\omega(\psi) = \frac{(n-2)!}{(2\pi i)^n} \sum_{s=1}^{n} \frac{\partial}{\partial z_s} \sum_{m=1}^{n} \frac{(-1)^{m-1}}{|\psi|^{2n-2}} \left(\sum_{k=1}^{n} A_k^s \overline{A}_k^m \right) d\bar{\zeta}[m] \wedge d\zeta.$$

Proof follows from Lemma 3.3.2 and the identity

$$\mathbf{D}_{1,n-1} \left(A^s, \overline{\partial_\zeta \psi} \right) = (n-1)! \sum_{m=1}^{n} (-1)^{m-1} \left(\sum_{k=1}^{n} A_k^s \overline{A}_k^m \right) d\bar{\zeta}[m].$$

□

From formula (1.4.2), Lemmas 3.3.2 and 3.3.3 we can easily obtain the proof of Proposition 3.3.2. Indeed, consider a sufficiently small neighborhood V of the boundary of the domain ∂D (where all functions of $\psi_j(\zeta - z)$ are defined). The function f continues in V to a function of class $\mathscr{C}^l$ with compact support in V. Approximating f in V in the metric $\mathscr{C}^l$ by functions of class $\mathscr{C}^\infty$, we can assume that the function f is infinitely differentiable. In the neighborhood V we apply formula (1.4.2). Then we obtain

$$- \int_{V_\zeta} \bar{\partial} f(\zeta) \wedge \omega \left(\psi(\zeta - z) \right) = \mu f(z), \quad z \in \partial D.$$

Making the replacement $\zeta = z + w$, we have

$$- \int_{\mathbb{C}^n} \bar{\partial} f(z + w) \wedge \omega \left(\psi(w) \right) = \mu f(z), \quad z \in \partial D.$$

In this equation we can find the derivatives up to order l with respect to the variable z and $\bar{z}$ by differentiating under the integral sign (by virtue of absolute convergence of the integral).

Therefore, choosing a sufficiently small neighborhood V' of the boundary ∂D, we find that the integral over V' can be made arbitrarily small in the metric $\mathscr{C}^l$. And, in the integral over $V \setminus V'$ let us replace the integrand by the integral sums, and the derivatives by the difference ratios (using Lemmas 3.3.2 and 3.3.3). The resulting fractions are arbitrarily close to the function f in the metric $\mathscr{C}^l$. The density of (3.3.9)-type fractions follows directly from representation (2.7.3) using fractions of the form (3.3.8). $\qquad\square$

Corollary 3.3.1 *If $f \in \mathscr{C}(\partial D)$ satisfies the moment conditions*

$$\int_{\partial D} f(\zeta)\bar{\partial}_\zeta \left(\sum_{k=1}^n \frac{A_k^s(\zeta - z)\overline{A}_k^m(\zeta - z)}{|\psi(\zeta - z)|^{2n-2}} \right) \wedge d\bar{\zeta}[j, p] \wedge d\zeta = 0 \qquad (3.3.10)$$

for all $z \notin \partial D$ and all $j, s, m, p = 1, \ldots, n$, then f is holomorphically extended into D up to a function $F \in \mathscr{C}(\overline{D})$.

Proof From Proposition 3.3.2 we obtain that

$$\int_{\partial D} f(\zeta)\bar{\partial}_\zeta(\alpha(\zeta) \wedge d\bar{\zeta}[j, p] \wedge d\zeta) = 0$$

for all smooth functions $\alpha(\zeta)$ defined in the neighborhood of the boundary ∂D. Hence f is a *CR*-function on ∂D. Since ∂D is connected, then the function f extends holomorphically to D. $\qquad\square$

Corollary 3.3.1 is one of the variants of the Hartogs–Bochner theorem. We note that in this statement equality (3.3.10) can be demanded to be satisfied only for points z in some open set $V \subset D$ or $V \subset \mathbb{C}^n \setminus \overline{D}$.

As we will see later, the condition of Morera's theorem turns into the following orthogonality condition:

$$\int_{\partial D_\zeta} f(\zeta)\Phi(\zeta - z)U(\psi(\zeta - z), 0) = 0, \quad z \notin \partial D, \qquad (3.3.11)$$

for the function $f \in \mathscr{C}(\partial D)$ and a function Φ of the form

$$\Phi(w) = \varphi_1(w)\psi_1(w) + \ldots + \varphi_n(w)\psi_n(w), \qquad (3.3.12)$$

where the functions $\varphi_j(w)$ (as well as functions $\psi_j(w)$) are holomorphic in some neighborhood of compact K_D, or φ_j are meromorphic and such that the form $\Phi\omega\psi$ has no singularities for $\zeta \neq z$. First we study condition (3.3.11) for a special choice of a function Φ of the form (3.3.12).

Lemma 3.3.4 *Equality (3.3.11) can be rewritten as*

$$\int_{\partial D_\zeta} f(\zeta)\bar{\partial}_\zeta \left[\frac{1}{|\psi(\zeta - z)|^{2n-2}} \, \mathbf{D}_{1,1,n-2}\left(\varphi(\zeta - z), \overline{\psi(\zeta - z)}, \overline{\partial_\zeta \psi} \right) \wedge d\psi \right] = 0,$$

where $z \notin \partial D$, and φ is a column of functions $\varphi_j, j = 1, \ldots, n$.

Proof Consider columns $\tau = \dfrac{\varphi}{\Phi}$ and

$$\eta = \frac{\bar{\psi}}{|\psi|^2} = \left(\frac{\bar{\psi}_1}{|\psi|^2}, \ldots, \frac{\bar{\psi}_n}{|\psi|^2}\right).$$

By Lemma 2.7.1

$$\omega(\psi(\zeta - z)) = \bar{\partial}_\zeta U_\Phi = \frac{1}{(2\pi i)^n} \bar{\partial}_\zeta \mathbf{D}_{1,1,n-2}\left(\tau(\zeta - z), \eta(\zeta - z), \bar{\partial}_\zeta \eta(\zeta - z)\right) \wedge d\psi$$

outside zeros of the function $\Phi(\zeta - z)$. Using the homogeneity property of the determinant $\mathbf{D}$ of the differential forms, we obtain

$$\Phi(\zeta - z)\omega(\psi(\zeta - z)) = \frac{1}{(2\pi i)^n} \times$$

$$\times \bar{\partial}_\zeta \left[\frac{1}{|\psi(\zeta - z)|^{2n-2}} \mathbf{D}_{1,1,n-2}\left(\varphi(\zeta - z), \overline{\psi(\zeta - z)}, \overline{\partial_\zeta \psi(\zeta - z)}\right)\right] \wedge d\psi.$$

$\square$

Lemma 3.3.5 *Equality (3.3.11) can be rewritten as*

$$\sum_{s=1}^{n} \int_{\partial D} f(\zeta)\bar{\partial}_\zeta \left[\frac{\partial}{\partial z_s}\left(\frac{1}{|\psi|^{2n-4}}\right) \mathbf{D}_{1,1,n-2}\left(\varphi, A^s, \overline{\partial_\zeta \psi}\right) \wedge d\zeta\right] = 0, \qquad (3.3.13)$$

if $z \notin \partial D$, and $n > 2$.

Proof is the same as in Lemma 3.3.2.

$\square$

We show that for a special choice of the functions φ_j in condition (3.3.13) the derivatives with respect to z_s can be taken out from under the integral sign. Let the vector-column φ have the form

$$\varphi(w) = \frac{1}{J} A^k,$$

where J is the determinant of the Jacobian matrix of the map ψ ($J \not\equiv 0$), and A^k is the column of the cofactors A^k_m, $m = 1, \ldots, n$, $k = 1, \ldots, n$.

Lemma 3.3.6 *Condition (3.3.11) can be rewritten as*

$$\sum_{s \neq k} \frac{\partial}{\partial z_s} \int_{\partial D} f(\zeta)\bar{\partial}_\zeta \left[\frac{1}{|\psi|^{2n-4}} \mathbf{D}_{1,1,n-2}\left(\frac{A^k}{J}, A^s, \overline{\partial_\zeta \psi}\right) \wedge d\zeta\right] = 0, \qquad (3.3.14)$$

if $z \notin \partial D$.

Although the determinant of J may be 0 on some surface as we shall see from the proof of Lemma 3.3.6, the determinants **D**, standing under the integral sign in (3.3.14), have no singularities.

Proof Let $\varphi = \dfrac{1}{J}A^1$. To prove the lemma it is sufficient to show (by Eq. (3.3.13)), that

$$\sum_{s=2}^{n} \frac{\partial}{\partial z_s} \mathbf{D}_{1,1,n-2}\left(\frac{A^1}{J}, A^s, \overline{\partial\psi}\right) = 0.$$

According to Laplace's theorem, we have

$$\sum_{s=2}^{n} \frac{\partial}{\partial z_s} \mathbf{D}_{1,1,n-2}\left(\frac{A^1}{J}, A^s, \overline{\partial\psi}\right) = \sum_{s=2}^{n} \frac{\partial}{\partial z_s} \sum_{p<r} \frac{(-1)^{p+r}}{J} \begin{vmatrix} A_p^1 & A_p^s \\ A_r^1 & A_r^s \end{vmatrix} \mathbf{D}_{n-2}^{p,r}(\overline{\partial\psi}),$$

where $\mathbf{D}_{n-2}^{p,r}$ is the determinant obtained from $\mathbf{D}_{1,1,n-2}$ by deleting the first two columns and rows with numbers p, r. By the property of the determinants of the cofactors we find that

$$\begin{vmatrix} A_p^1 & A_p^s \\ A_r^1 & A_r^s \end{vmatrix} = J A_{p,r}^{1,s}, \tag{3.3.15}$$

where $A_{p,r}^{1,s}$ are the cofactors in the Jacobi matrix of the map ψ, which stand at the intersection of the first and s-th columns, and the p-th and r-th rows.

On the other hand (as is easy to show)

$$\sum_{s=2}^{n} \frac{\partial}{\partial z_s} A_{p,r}^{1,s} = 0. \tag{3.3.16}$$

From this and from (3.3.15) we obtain the desired result. Moreover, (3.3.15) implies that the determinants $\mathbf{D}_{1,1,n-2}$ in equality (3.3.14) do not have singularities.   □

The proof of this lemma shows that equality (3.3.14) is equivalent to

$$\sum_{\substack{s \neq k}}^{n} \frac{\partial}{\partial z_s} \int_{\partial D} f(\zeta)\bar{\partial}_\zeta \beta_{k,s} \wedge d\zeta = 0 \quad \text{for} \quad z \notin \partial D, \tag{3.3.17}$$

where

$$\beta_{k,s} = \frac{1}{|\psi(\zeta - z)|^{2n-4}} \sum_{p<r}\sum_{l<m} (-1)^{p+r} A_{p,r}^{k,s}(\zeta - z)\overline{A_{p,r}^{l,m}(\zeta - z)} d\bar{\zeta}[l, m].$$

Therefore, we need to verify the density of linear combinations of fractions of a more general form than in Proposition 3.3.2 in the class $\mathscr{C}^k(\partial D)$.

Proposition 3.3.3 *Let $n > 2$ and $\partial D \in \mathscr{C}^d$. Linear combinations of fractions of the form*

$$Q_{l,s,m,k}(\zeta - z) = \sum_{1 \leq p < r \leq n} \frac{A_{p,r}^{l,s}(\zeta - z)\overline{A_{p,r}^{m,k}(\zeta - z)}}{|\psi(\zeta - z)|^{2n-4}},$$

$z \notin \partial D$, $\zeta \in \partial D$, $k, s, m, l, = 1, \ldots, n$, *are dense in the space $\mathscr{C}^u(\partial D)$, $0 \leq u \leq d$.*

Proof From identity (3.3.16) we obtain the following (for all $s, p, r = 1, \ldots, n$):

$$\sum_{l<s} \frac{\partial}{\partial z_l} A_{p,r}^{l,s} - \sum_{l>s} \frac{\partial}{\partial z_l} A_{p,r}^{s,l} = 0. \tag{3.3.18}$$

Indeed, we replace in identity (3.3.16) the variable z_1 with z_s, and z_s with z_1. Then we put the first columns in the resulting expression back in their place, and obtain (3.3.18).

Let $p < r$, the identity

$$\sum_{l<s} \frac{\partial \psi_q}{\partial z_l} A_{p,r}^{l,s} - \sum_{l>s} \frac{\partial \psi_q}{\partial z_l} A_{p,r}^{s,l} = \begin{cases} A_r^s, & \text{if } q = p, \\ -A_p^s, & \text{if } q = r, \\ 0, & \text{if } q \neq p, r \end{cases} \tag{3.3.19}$$

holds. This identity is obtained using the rules for decomposition of the determinant by one of the lines and the signs of cofactors $A_{p,r}^{l,s}$ and A_p^s. Using Eqs. (3.3.18) and (3.3.19), we have (for fixed s, m, k)

$$\sum_{l<s} \frac{\partial}{\partial z_l} Q_{l,s,m,k} - \sum_{l>s} \frac{\partial}{\partial z_l} Q_{s,l,m,k}$$

$$= -(n-2) \frac{\displaystyle\sum_{p<r} \sum_q \left(\sum_{l<s} \bar{\psi}_q \frac{\partial \psi_q}{\partial z_l} A_{p,r}^{l,s} - \sum_{l>s} \bar{\psi}_q \frac{\partial \psi_q}{\partial z_l} A_{p,r}^{s,l} \right) \overline{A_{p,r}^{m,k}}}{|\psi|^{2n-2}}$$

$$= -(n-2) \frac{\displaystyle\sum_{p<r} \left(\bar{\psi}_p A_r^s - \bar{\psi}_r A_p^s \right) \overline{A_{p,r}^{m,k}}}{|\psi|^{2n-2}} = R_{s,m,k}.$$

Then again using Eqs. (3.3.18) and (3.3.19), we obtain (for fixed s, k)

$$
\sum_{m<k} \frac{\partial}{\partial \bar{z}_m} R_{s,m,k} - \sum_{m>k} \frac{\partial}{\partial \bar{z}_m} R_{s,k,m}
$$

$$
= -\frac{(n-2)}{|\psi|^{2n-2}} \sum_{p<r} \left[A_r^s \left(\sum_{m<k} \frac{\overline{\partial \psi_p}}{\partial z_m} \overline{A}_{p,r}^{m,k} - \sum_{m>k} \frac{\overline{\partial \psi_p}}{\partial z_m} \overline{A}_{p,r}^{k,m} \right) \right.
$$

$$
\left. - A_p^s \left(\sum_{m<k} \frac{\overline{\partial \psi_r}}{\partial z_m} \overline{A}_{p,r}^{m,k} - \sum_{m>k} \frac{\overline{\partial \psi_r}}{\partial z_m} \overline{A}_{p,r}^{k,m} \right) \right]
$$

$$
+ \frac{(n-2)(n-1)}{|\psi|^{2n}} \sum_{p<r} (\bar{\psi}_p A_r^s - \bar{\psi}_r A_p^s) \left[\sum_{m<k} \overline{A}_{p,r}^{m,k} \sum_q \psi_q \frac{\overline{\partial \psi_q}}{\partial z_m} \right.
$$

$$
\left. - \sum_{m>k} \overline{A}_{p,r}^{k,m} \sum_q \psi_q \frac{\overline{\partial \psi_q}}{\partial z_m} \right] = -\frac{(n-2)}{|\psi|^{2n-2}} \sum_{p<r} \left(A_r^s \overline{A}_r^k + A_p^s \overline{A}_p^k \right)
$$

$$
+ \frac{(n-2)(n-1)}{|\psi|^{2n}} \sum_{p<r} (\bar{\psi}_p A_r^s - \bar{\psi}_r A_p^s) \left(\psi_p \overline{A}_r^k - \psi_r \overline{A}_p^k \right)
$$

$$
= -\frac{(n-2)(n-1)}{|\psi|^{2n}} \left(\sum_r \psi_r A_r^s \right) \left(\sum_p \bar{\psi}_p \overline{A}_p^k \right).
$$

Replacing the derivatives by the difference ratio and applying Proposition 3.3.2, we obtain the desired result.   □

Theorem 3.3.3 ([64]) *Let $\partial D \in \mathscr{C}^2$, and $f \in \mathscr{C}(\partial D)$. If*

$$
\int_{\partial D_\zeta} f(\zeta) \Phi_k(\zeta - z) \omega(\psi(\zeta - z)) = 0 \tag{3.3.20}
$$

for all $z \notin \partial D$, $k = 1, \dots, n$, where

$$
\Phi_k(\zeta - z) = \frac{1}{J(\zeta - z)} \sum_{s=1}^n A_s^k(\zeta - z) \psi_s(\zeta - z),
$$

then the function f is holomorphically extended into D up to a function $F \in \mathscr{C}(\overline{D})$.

Condition (3.3.20) (as shown by formula (2.7.3)) is equivalent to the following:

$$
\int_{\partial D_\zeta} f(\zeta) \frac{\sum_{s=1}^n A_s^k(\zeta - z) \psi_s(\zeta - z)}{|\psi(\zeta - z)|^{2n}} \mathbf{D}_{1,n-1}(\bar{\psi}, \overline{\partial_\zeta \psi}) \wedge d\zeta = 0,
$$

however, in spite of the Jacobian J in the denominator, the integrand in formula (3.3.20) has no singularities at $\zeta \neq z$.

Proof Let $n > 2$. We write the condition of the theorem in the form of equality (3.3.17). Using Lemma 3.3.6 and Proposition 3.3.3, we approximate the function $|\zeta - z|^{4-2n}$ by linear combinations of fractions in this proposition from the class $\mathscr{C}^2(\partial D)$, $\zeta \in \partial D$, and z is fixed and does not lie on ∂D. Then for $z \notin \partial D$ from (3.3.20) we obtain

$$\sum_{s \neq k} \frac{\partial}{\partial z_s} \int_{\partial D} f(\zeta) \bar{\partial}_\zeta \left[\frac{1}{|\zeta - z|^{2n-4}} \mathbf{D}_{1,1,n-2} \left(\tilde{A}^k, \tilde{A}^s, \overline{\partial_\zeta(\zeta - z)} \right) \wedge d\zeta \right] = 0,$$

where $\tilde{A}^k$ are the respective cofactors in the identity mapping $\tilde{\psi}(\zeta - z) = \zeta - z$.

Lemmas 3.3.4 and 3.3.5 show that this condition can be written as

$$\int_{\partial D} f(\zeta)(\zeta_k - z_k) U(\zeta, z) = 0, \quad z \notin \partial D, \quad k = 1, \ldots, n.$$

Applying to the left side of the last equation the Laplace operator

$$\Delta = \frac{\partial^2}{\partial z_1 \partial \bar{z}_1} + \ldots + \frac{\partial^2}{\partial z_n \partial \bar{z}_n}$$

and using harmonicity of the coefficients of the Bochner–Martinelli kernel, we have

$$\frac{\partial}{\partial \bar{z}_k} \int_{\partial D} f(\zeta) U(\zeta, z) = 0, \quad z \notin \partial D, \quad k = 1, \ldots, n.$$

Thus, the Bochner–Martinelli integral of the function f is a holomorphic function outside the boundary of D. Since ∂D is connected and this integral tends to 0 as $|z| \to \infty$, then it equals zero outside D. By Corollary 1.2.4 about the jump of the Bochner–Martinelli integral we obtain that the desired holomorphic extension is given by the Bochner–Martinelli integral. □

3.3.3 Analogues of the Morera Theorem

Let D be a bounded domain in $\mathbb{C}^n$ ($n > 1$) with a connected smooth boundary ∂D of class $\mathscr{C}^1$. Assume that for a function $f \in \mathscr{C}(\partial D)$ integrals over $\partial D \cap l$ are equal to zero for all complex curves l in a certain class. Our goal is to answer the question whether f extends holomorphically to D as a function of n complex variables. This question was investigated for the case of complex lines by Globevnik and Stout in [26], Semenov and Agranovskii in [3], and by us in [47].

As in Sect. 3.3.1, we consider a class of complex curves $l_{z,b}$ of the form (3.3.1)

$$l_{z,b} = \{\zeta \in \mathbb{C}^n : \zeta_1 = z_1 + t^{k_1},\ \zeta_j = z_j + b_j t^{k_j} \chi_j(t^{k_1}), j = 2,\ldots,n,\ t \in \mathbb{C}\}$$

with the same properties. Similarly, we introduce holomorphic functions of the form (3.3.3):

$$\psi_1(\zeta) = \zeta_1^{p_1},\quad \psi_j = \frac{\zeta_j^{p_j}}{\chi_j^{p_j}(\zeta_1)},\quad j = 2,\ldots,n,$$

having the properties as in Sect. 3.3.1. For these functions, Theorem 3.3.3 holds. By Lemma 3.3.1 in the coordinates $t,\ b$ the kernel $\omega(\psi(\zeta - z))$ will have the form (3.3.4):

$$\omega(\psi^*(t,b)) = \frac{dt}{t} \wedge \lambda(b),$$

where

$$\lambda(b) = \frac{p(n-1)!}{(2\pi i)^n} \frac{(-1)^{n-1} d\bar{b}_2^{p_2} \wedge \cdots \wedge d\bar{b}_n^{p_n} \wedge db_2^{p_2} \wedge \cdots \wedge db_n^{p_n}}{\left(1 + \sum\limits_{j=2}^n |b_j|^{2p_j}\right)^n}.$$

Here $\psi^*(t,b)$ is the composition of the map $\psi(\zeta - z)$ and the map $\zeta - z$, defining the curves $l_{z,b}$.

To see how condition (3.3.20) in Theorem 3.3.3 will change for the mapping ψ, in particular, we find the functions Φ_j in the new coordinates $t,\ b$. Denote the functions $\dfrac{1}{\chi_j^{p_j}}$ by $\gamma_j, j = 2,\ldots,n$. We have

$$J(\zeta - z) = p_1 \cdots p_n (\zeta_1 - z_1)^{p_1-1} \cdots (\zeta_n - z_n)^{p_n-1} \gamma_2(\zeta_1 - z_1) \cdots \gamma_n(\zeta_1 - z_1).$$

The column vectors A^s take the form

$$A^1 = \left(p_2 \cdots p_n (\zeta_2 - z_2)^{p_2-1} \cdots (\zeta_n - z_n)^{p_n-1} \gamma_2 \cdots \gamma_n, 0, \ldots, 0\right),$$

$$A^2 = \left(-p_3 \cdots p_n (\zeta_2 - z_2)^{p_2}(\zeta_3 - z_3)^{p_3-1} \cdots (\zeta_n - z_n)^{p_n-1} \gamma_2' \gamma_3 \cdots \gamma_n,\right.$$
$$\left. p_1 p_3 \cdots p_n (\zeta_1 - z_1)^{p_1-1}(\zeta_3 - z_3)^{p_3-1} \cdots (\zeta_n - z_n)^{p_n-1} \gamma_1 \gamma_3 \cdots \gamma_n, 0, \ldots, 0\right),$$

$$\cdots,$$

$$A^n = \left(-p_2 \cdots p_{n-1}(\zeta_2 - z_2)^{p_2-1} \cdots (\zeta_{n-1} - z_{n-1})^{p_{n-1}-1}(\zeta_n - z_n)^{p_n} \gamma_2 \cdots \gamma_{n-1}\gamma_n',\right.$$
$$\left. 0, \ldots, p_1 \cdots p_{n-1}(\zeta_1 - z_1)^{p_1-1} \cdots (\zeta_{n-1} - z_{n-1})^{p_{n-1}-1} \gamma_1 \cdots \gamma_{n-1}\right).$$

Calculating the function Φ_j, we obtain

$$\Phi_1(\zeta - z) = \frac{\zeta_1 - z_1}{p_1},$$

$$\Phi_2(\zeta - z) = \frac{\zeta_2 - z_2}{p_2} - \frac{(\zeta_1 - z_1)(\zeta_2 - z_2)\gamma_2'}{\gamma_2 p_1 p_2},$$

$$\ldots,$$

$$\Phi_n(\zeta - z) = \frac{\zeta_n - z_n}{p_n} - \frac{(\zeta_1 - z_1)(\zeta_n - z_n)\gamma_n'}{\gamma_n p_1 p_n}.$$

In the coordinates t, b, these functions take the form

$$\Phi_1^*(t, b) = \frac{t}{p}(t^{k_1})',$$

$$\Phi_2^*(t, b) = \frac{b_2 t}{p}\left(t^{k_2}\chi_2(t^{k_1})\right)',$$

$$\ldots,$$

$$\Phi_n^*(t, b) = \frac{b_n t}{p}\left(t^{k_n}\chi_n(t^{k_1})\right)'.$$

(3.3.21)

Theorem 3.3.4 ([64]) *Let $\partial D \in \mathscr{C}^2$, and a function $f \in \mathscr{C}(\partial D)$ satisfy the conditions*

$$\int_{\partial D \cap l_{z,b}} f^*(t, b)\, d(t^{k_j}\chi_j) = 0$$

for all $j = 1, \ldots, n$, of almost all points z, lying in a neighborhood of $\overline{D}$, and almost all vectors b, satisfying condition (3.3.2), then the function f is holomorphically extended into D up to a function F of the class $\mathscr{C}(\overline{D})$ (we believe that for $j = 1$ the function $\chi_1 = 1$).

Proof From Theorem 3.3.3, the kind (3.3.4) of the differential form $\omega(\psi)$ and the kind (3.3.21) of the functions Φ_j in the coordinates t, b, as well as from Fubini's theorem, it follows that

$$\int_{\partial D_\zeta} f(\zeta)\Phi_j(\zeta - z)\omega(\psi(\zeta - z)) = \int_{\mathbb{C}^{n-1}} \lambda(b) \int_{\partial D \cap l_{z,b}} f^*(t, b)\, d(t^{k_j}\chi_j) = 0.$$

Then the function f extends holomorphically into D. $\qquad\square$

Theorem 3.3.4 is a generalization of the boundary Morera theorem given in [26] (see Theorem 3.2.3), where the case of complex lines $l_{z,b}$ is considered. If all functions $\chi_j \equiv 1$, $j = 1, \ldots, n$, then Theorem 3.3.4 becomes one of the boundary version of the Morera theorem for algebraic curves.

3.4   Morera Theorem in Classical Domains

In this section, we consider the boundary variant of the Morera theorem for classical domains. The starting point of this theorem is the result of Nagel and Rudin [66], which says that if a function f is continuous on the boundary of a ball in $\mathbb{C}^N$ and the integral

$$\int_0^{2\pi} f(\psi(e^{i\varphi}, 0 \ldots, 0)) \, e^{i\varphi} \, d\varphi = 0$$

for all (holomorphic) automorphisms ψ of the ball, then the function f is holomorphically extended to the ball.

An alternative proof of the theorem of Nagel and Rudin was given by Kosbergenov in [37]. It allows this assertion to be generalized for the case of classical domains.

In [2] Agranovskii gives a description of Möbius-invariant spaces of continuous functions in classical domains of tubular type, i.e., those classical domains for which the real dimension of the Shilov boundary coincides with the complex dimension of the domain. In [2] by using this description, the assertion which essentially coincides with Theorem 1.3.2 for classical domains of tubular type was proved.

3.4.1   Classical Domains

We recall certain definitions and introduce notations needed for further discussion. By a classical domain $D \subset \mathbb{C}^N$ (see [32, p. 9]), we mean an irreducible bounded symmetric domain of several complex variables of one of the following four types:

1. The domain D_I is formed by matrices Z consisting of m rows and n columns (entries of matrices are complex numbers) and satisfying the condition

$$I^{(m)} - ZZ^* > 0.$$

Here, $I^{(m)}$ is the identity matrix of order m, $Z^* = \overline{Z}'$ is the matrix complex-conjugate to the transposed matrix Z', and, as usual, the inequality $H > 0$ for a Hermitian matrix H means that this matrix is positive definite.

2. The domain D_{II} is formed by symmetric (square) matrices Z of order n satisfying the condition

$$I^{(n)} - Z\bar{Z} > 0.$$

3. The domain D_{III} is formed by skew-symmetric matrices Z of order n satisfying the condition

$$I^{(n)} + Z\bar{Z} > 0.$$

4. The domain D_{IV} is formed by n-dimensional vectors $z = (z_1, \ldots, z_n)$ satisfying the condition

$$|zz'|^2 + 1 - 2\bar{z}\bar{z}' > 0, \qquad |zz'| < 1.$$

The complex dimension of these four types of domains is equal to mn, $\dfrac{n(n+1)}{2}$, $\dfrac{n(n-1)}{2}$, and n, respectively. These domains are complete circular convex domains. In our case, the domain D means a domain of one of the types presented above. Let S be the Shilov boundary for the domain D (see [32, p. 10]).

1. S_I is formed by matrices U consisting of m rows and n columns with the condition that

$$UU^* = I^{(m)}.$$

In particular, for $m = n$, the manifold S_I coincides with the set of all unitary matrices $U(n)$.
2. S_{II} is formed by all symmetric unitary matrices of order n.
3. S_{III} is defined in different ways depending on the evenness or oddness of n. If n is even, then S_{III} is formed by all skew-symmetric unitary matrices of order n. If n is odd, then S_{III} is formed by all matrices of the form UFU', where U is an arbitrary unitary matrix and

$$F = \begin{pmatrix} 0 & 1 \\ -1 & 0 \end{pmatrix} \dotplus \begin{pmatrix} 0 & 1 \\ -1 & 0 \end{pmatrix} \dotplus \ldots \dotplus 0.$$

4. S_{IV} is formed by vectors of the form $e^{i\varphi} x$, where x is the real n-dimensional vector satisfying the condition $xx' = 1, 0 \leq \varphi \leq 2\pi$.

The manifolds S_I, S_{II}, S_{III}, S_{IV} have the real dimension $m(2n - m)$, $\dfrac{n(n+1)}{2}$, $\dfrac{n(n-1)}{2} + (1 + (-1)^n)\dfrac{(n-1)}{2}$, and, n, respectively. All these manifolds are generic CR-manifolds, and, moreover, in the case where their dimension is equal to the (complex) dimension of the space $\mathbb{C}^N$, these manifolds are totally real (i.e., have no complex tangent vectors).

3.4.2   Morera Theorem in Classical Domains

We define the class $\mathcal{H}^1(D)$ as a class of all functions f, holomorphic in D such that

$$\sup_{0<r<1} \int_S |f(r\zeta)| d\mu < +\infty,$$

were $r\zeta = (r\zeta_1, \ldots, r\zeta_N)$, and $d\mu$ is the normalized Lebesgue measure on the manifold S, which is a Haar measure, and, therefore, it is invariant with respect to rotations.

For any function f in D and any $\zeta \in S$, consider a cut-function f_ζ in $\Delta = \{t \in \mathbb{C} : |t| < 1\}$ of the following form: $f_\zeta(t) = f(t\zeta)$. This cut-function allows us to relate certain N-dimensional properties of the function f to one-dimensional properties of f_ζ.

Fix a point $\lambda_0 \in S$ ($\lambda_0 = (\lambda_1^0, \ldots, \lambda_N^0)$) and consider the following embedding of a disk Δ in the domain D:

$$\{\zeta \in \mathbb{C}^N : \zeta_j = t\lambda_j^0, j = 1, \ldots, N, |t| < 1\}. \tag{3.4.1}$$

Under this embedding, the boundary T of the disk Δ moves to a circle lying on S. If ψ is an arbitrary (holomorphic) automorphism of the domain D (i.e., a biholomorphic self-map of the domain D), then the set of the form (3.4.1) passes to a certain analytic disk with the boundary on S under the action of this automorphism.

Theorem 3.4.1 ([38]) *If a function $f \in \mathscr{C}(S)$ satisfies the condition*

$$\int_T f(\psi(t\lambda_0)) \, dt = 0 \tag{3.4.2}$$

for all automorphisms ψ of the domain D, then the function f is holomorphically extended into D up to a function F of class $\mathscr{C}(\overline{D})$.

For the case of classical domains of tubular type, this assertion is presented in [2].

Proof A subgroup of automorphisms leaving 0 fixed acts on S transitively. They are called unitary transformations, since they are linear, and given by unitary matrices for the case of domains consisting of square matrices. Since S is invariant with respect to unitary transformations (like the domain D), condition (3.4.2) also holds for arbitrary points $\lambda \in S$.

Let us parametrize the manifold S as follows:

$$\zeta = t\lambda, \quad t = e^{i\varphi}, \quad 0 \le \varphi \le 2\pi, \quad \lambda \in S',$$

if $\zeta \in S$. The manifold S' is defined differently for domains D of different types. For domains of the first type and $m = n$, this is a group $SU(n)$ of special unitary matrices, for domains of the fourth type, it is a sphere, and so on. The measure $d\mu$ can be written in the form

$$d\mu = \frac{d\varphi}{2\pi} \wedge d\mu_0(\lambda) = \frac{1}{2\pi i} \frac{dt}{t} \wedge d\mu_0(\lambda),$$

where $d\mu_0(\lambda)$ is a differential form defining a positive measure on S'.

Multiplying relation (3.4.2) by $d\mu_0$ and integrating over S', we obtain from (3.4.2) that

$$\int_S f(\psi(\zeta)) \, \zeta_k \, d\mu(\zeta) = 0, \tag{3.4.3}$$

where ζ_k are the components of the vector ζ, $k = 1, \ldots, N$.

Consider an automorphism ψ_A transforming the point A in D into 0. It is defined with accuracy up to a unitary transformation. Then substituting the automorphism ψ_A^{-1} instead of ψ in (3.4.3) and making the change of variables $W = \psi_A^{-1}(\zeta)$, we obtain

$$\int_S f(W) \, \psi_k^A(W) \, d\mu(\psi_A(W)) = 0, \tag{3.4.4}$$

where ψ_k^A are the components of the automorphism ψ_A. As was shown in [36, Lemma 3.4] (for the case of square matrices, see the proof of Theorem 4.6.3 in [32]),

$$d\mu(\psi_A(W)) = P(W, A) \, d\mu(W),$$

where $P(W, A)$ is the invariant Poisson kernel of the domain D. Therefore, we obtain from condition (3.4.4) that

$$\int_S f(W) \psi_k^A(W) \, P(W, A) \, d\mu(W) = 0 \tag{3.4.5}$$

for all points A from D and all $k = 1, \ldots, N$.

A further proof of Theorem 3.4.1 follows from Theorem 5.7.1 in [32] on the properties of the Poisson integral of continuous functions and analogue of the Hartogs–Bochner theorem. $\qquad\square$

3.4.3 Analogue of the Hartogs–Bochner Theorem in Classical Domains

Theorem 3.4.2 ([38]) *If a function* $f \in \mathscr{L}^1(S)$ *and condition (3.4.5) for this function holds for all automorphisms* ψ_A *of the domain D, that transform a point A from D into 0, and for all* $k = 1, \ldots, N$, *then the function f is the radial boundary value of a certain function* $F \in \mathscr{H}^1(D)$.

Proof

1. Let D be a domain of the first type and suppose for certainty that $m \le n$. The invariant Poisson kernel for the domain D has the form (see [32, p. 98])

$$P(W,A) = \frac{\left(\det(I^{(m)} - AA^*)\right)^n}{|\det(I^{(m)} - AW^*)|^{2n}} = \frac{\left(\det(I^{(m)} - AA^*)\right)^n}{\det(I^{(m)} - AW^*)^n \det(I^{(m)} - WA^*)^n}.$$

Let the matrices be $A = \|a_{sp}\|$ and $W = \|w_{sp}\|$ ($s = 1, \ldots, m; p = 1, \ldots, n$). Now we compute the expression

$$\sum_{s=1}^{m} \sum_{p=1}^{n} \bar{a}_{sp} \frac{\partial P(W,A)}{\partial \bar{a}_{sp}}. \tag{3.4.6}$$

We denote $I^{(m)} - WA^* = \|\alpha_{qj}\|$ ($q,j = 1, \ldots, m$), where

$$\alpha_{qj} = \delta_{qj} - \sum_{k=1}^{n} w_{qk} \bar{a}_{jk}, \quad q,j = 1, \ldots, m,$$

and δ_{qj} is the Kronecker symbol ($\delta_{qj} = 0$ for $q \ne j$, and $\delta_{qq} = 1$).

Using the usual rule for differentiating a determinant, it is easy to check that for any $s = 1, \ldots, m$

$$\sum_{p=1}^{n} \bar{a}_{sp} \frac{\partial \det(I^{(m)} - WA^*)}{\partial \bar{a}_{sp}} = \begin{vmatrix} \alpha_{11} & \cdots & \alpha_{1s} & \cdots & \alpha_{1m} \\ \cdots & \cdots & \cdots & \cdots & \cdots \\ \alpha_{s1} & \cdots & \alpha_{ss} - 1 & \cdots & \alpha_{sm} \\ \cdots & \cdots & \cdots & \cdots & \cdots \\ \alpha_{m1} & \cdots & \alpha_{ms} & \cdots & \alpha_{mm} \end{vmatrix}$$

$$= \det(I^{(m)} - WA^*) - \det(I^{(m)} - WA^*)[s,s],$$

where $\det(I^{(m)} - WA^*)[s,s]$ means the cofactor to the element α_{ss} in the matrix $(I^{(m)} - WA^*)$. Then

$$\sum_{s=1}^{m}\sum_{p=1}^{n}\bar{a}_{sp}\frac{\partial \det(I^{(m)} - WA^*)}{\partial \bar{a}_{sp}} = m\det(I^{(m)} - WA^*) - \sum_{s=1}^{m}\det(I^{(m)} - WA^*)[s,s].$$

Similarly,

$$\sum_{s=1}^{m}\sum_{p=1}^{n}\bar{a}_{sp}\frac{\partial \det(I^{(m)} - AA^*)}{\partial \bar{a}_{sp}} = m\det(I^{(m)} - AA^*) - \sum_{s=1}^{m}\det(I^{(m)} - AA^*)[s,s].$$

Hence we have that expression (3.4.6) equals

$$mP(W,A)\left[\frac{\sum_{s=1}^{m}\det(I^{(m)} - WA^*)[s,s]}{\det(I^{(m)} - WA^*)} - \frac{\sum_{s=1}^{m}\det(I^{(m)} - AA^*)[s,s]}{\det(I^{(m)} - AA^*)}\right]$$

$$= mP(W,A)\left[\operatorname{tr}(I^{(m)} - WA^*)^{-1} - \operatorname{tr}(I^{(m)} - AA^*)^{-1}\right], \tag{3.4.7}$$

where tr W denotes the trace of the matrix W.

As is known, the conditions

$$I^{(m)} - ZZ^* > 0 \quad \text{and} \quad I^{(n)} - Z^*Z = I^{(n)} - Z'(Z')^* > 0$$

are equivalent (see [32, p. 37]). Therefore, the map $Z \to Z'$ transforms the domain D to a domain D' also of a first type.

Consider the automorphism $\psi_{A'}$ transforming a point A' in D' into $0'$ of the following form:

$$\psi_{A'}(W') = Q'(W' - A')(I^{(m)} - (A')^*W')^{-1}(R')^{-1}$$

(see [32, p. 85]), where non-degenerate $n \times n$-matrix Q' and non-degenerate $m \times m$-matrix R' are chosen so that

$$R'(I^{(m)} - A'(A')^*)(R')^* = I^{(m)}, \quad Q'(I^{(n)} - (A')^*A')(Q')^* = I^{(n)}.$$

Then the automorphism $\psi_A(W)$ of the domain D, transforming the point A into 0 will have the form

$$\psi_A(W) = R^{-1}(I^{(m)} - WA^*)^{-1}(W - A)Q.$$

If condition (3.4.5) holds for components of the map ψ_A, then the same condition holds for components of the map

$$\varphi_A(W) = (I^{(m)} - AA^*)^{-1}(I^{(m)} - WA^*)^{-1}(W - A),$$

since the matrices P, Q, $(I^{(m)} - AA^*)$ are non-degenerate and depend only on A.

Denoting components $\varphi_A(W)$ by $\varphi_{sp}^A(W)$, $s = 1, \ldots, m$, $p = 1, \ldots, n$, from (3.4.5) we obtain

$$\int_S f(W)\varphi_{sp}^A(W)\, P(W, A)\, d\mu(W) = 0. \tag{3.4.8}$$

We find the sum

$$\sum_{s=1}^{m}\sum_{p=1}^{n} \bar{a}_{sp}\varphi_{sp}^A.$$

Obviously, the desired expression equals

$$\text{tr}(\varphi_A(W)A^*).$$

We need the following property of the matrix trace

$$\text{tr}(AB) = \text{tr}(BA), \tag{3.4.9}$$

where the rectangular matrices A and B are such that the products AB and BA are defined. Indeed, if $m \times n$-matrix A consists of elements a_{ks}, $k = 1, \ldots, m$, $s = 1, \ldots, n$, and $n \times m$-matrix B consists of elements b_{lq}, $l = 1, \ldots, n$; $q = 1, \ldots, m$ then

$$\text{tr}(AB) = \sum_{k=1}^{m}\sum_{s=1}^{n} a_{ks}b_{sk} = \text{tr}(BA).$$

Using the form of the map φ_A and property (3.4.9) of the matrix trace, we obtain

$$\text{tr}(\varphi_A(W)A^*) = \text{tr}\left[(I^{(m)} - AA^*)^{-1}(I^{(m)} - WA^*)^{-1}(W - A)A^*\right]$$

$$= \text{tr}\left[(I^{(m)} - AA^*)^{-1}(I^{(m)} - WA^*)^{-1}((WA^* - I^{(m)}) + (I^{(m)} - AA^*))\right]$$

$$= \text{tr}\left[(I^{(m)} - WA^*)^{-1} - (I^{(m)} - AA^*)^{-1}\right]. \tag{3.4.10}$$

Comparing formulas (3.4.7) and (3.4.10), from condition (3.4.8) we obtain

$$\sum_{s=1}^{m}\sum_{p=1}^{n} \bar{a}_{sp} \frac{\partial F(A)}{\partial \bar{a}_{sp}} = 0, \qquad (3.4.11)$$

where

$$F(A) = \int_{S} f(W) P(W,A) \, d\mu(W)$$

is the Poisson integral of the function f.

Function $F(A)$ is real-analytic in D so expanding it in a Taylor series in a neighborhood of 0, we obtain

$$F(A) = \sum_{|\alpha|,|\beta|\geq 0} c_{\alpha,\beta} a^{\alpha} \bar{a}^{\beta},$$

where $\alpha = \|\alpha_{qj}\|$ and $\beta = \|\beta_{qj}\|$ $(q = 1, \ldots, m, j = 1, \ldots, n)$ are the matrices with non-negative integer entries,

$$|\alpha| = \sum_{q=1}^{m}\sum_{j=1}^{n} \alpha_{qj} \quad \text{and} \quad a^{\alpha} = \prod_{q=1}^{m}\prod_{j=1}^{n} a_{qj}^{\alpha_{qj}}.$$

Then from condition (3.4.11) we obtain

$$\sum_{s=1}^{m}\sum_{p=1}^{n} \bar{a}_{sp} \frac{\partial F(A)}{\partial \bar{a}_{sp}} = \sum_{|\alpha|\geq 0, |\beta|>0} |\beta| \, c_{\alpha,\beta} \, a^{\alpha} \bar{a}^{\beta} = 0,$$

hence all the coefficients $c_{\alpha,\beta}$ with $|\beta| > 0$ are equal to 0. Therefore, the function $F(A)$ is holomorphic in D and belongs to the class $\mathscr{H}^{1}(D)$.

2. For domains of type D_{II} and D_{III} the proof repeats the proof for domains D_{I}. In this case it is however less involved since for a domain of the second type $Z' = Z$ is true, and for a domain of the third type $Z' = -Z$ is true.

3. Let $D = D_{IV}$. We will denote points on S by w, and points in D by a. So, we have

$$\int_{S} f(w) \psi_{k}^{a}(w) \, P(w,a) \, d\mu(w) = 0 \qquad (3.4.12)$$

for all points $a \in D$ and all components $\psi_{k}^{a}(w)$ of the automorphism $\psi_{a}(w)$.

The Poisson kernel for a domain of the fourth type has the form (see [32, p. 99])

$$P(w, a) = \frac{(1 + |\langle a, a \rangle|^2 - 2|a|^2)^{\frac{n}{2}}}{|\langle w - a, w - a \rangle|^n}.$$

Here and below

$$\langle w, z \rangle = wz' = w_1 z_1 + \ldots + w_n z_n.$$

Calculating

$$\sum_{k=1}^{n} \bar{a}_k \frac{\partial P(w, a)}{\partial \bar{a}_k},$$

we obtain

$$\sum_{k=1}^{n} \bar{a}_k \frac{\partial P(w, a)}{\partial \bar{a}_k} = nP(w, a) \left[\frac{|a|^2 - 1}{1 + |\langle a, a \rangle|^2 - 2|a|^2} - \frac{\overline{\langle w, a \rangle} - \overline{\langle w, w \rangle}}{\langle a - w, a - w \rangle} \right].$$

Using the representation $w = e^{i\varphi} x$, where x is the real vector such that $\langle x, x \rangle = 1$, we have

$$\sum_{k=1}^{n} \bar{a}_k \frac{\partial P(w, a)}{\partial \bar{a}_k} = nP(w, a) \left[\frac{|a|^2 - 1}{1 + |\langle a, a \rangle|^2 - 2|a|^2} \right.$$

$$\left. - \frac{\langle w, \bar{a} \rangle - 1}{1 + \langle w, w \rangle \overline{\langle a, a \rangle} - 2 \langle w, \bar{a} \rangle} \right]. \tag{3.4.13}$$

An automorphism ψ_a has the form (see [32, p. 88])

$$\psi_a(w) = \left\{ \left[\left(\frac{1}{2}(ww' + 1), \frac{i}{2}(ww' - 1) \right) A' - w X_0' A' \right] \binom{1}{i} \right\}^{-1} \times$$

$$\times \left\{ wQ' - \left(\frac{1}{2}(ww' + 1), \frac{i}{2}(ww' - 1) \right) X_0 Q' \right\},$$

where

$$A = \frac{1}{2}(1 + |aa'|^2 - 2\bar{a}a')^{\frac{1}{2}} \begin{pmatrix} -i(aa' - \overline{aa'}) & aa' + \overline{aa'} - 2 \\ aa' + \overline{aa'} + 2 & i(aa' - \overline{aa'}) \end{pmatrix},$$

$$X_0 = \frac{1}{1 - |aa'|^2} \begin{pmatrix} a + \bar{a} - (a(\overline{aa'}) + \bar{a}(aa')) \\ i(a - \bar{a}) + i(a(\overline{aa'}) - \bar{a}(aa')) \end{pmatrix},$$

and the non-degenerate matrix Q is chosen so that

$$Q(I^{(n)} - X_0'X_0)Q' = I^{(n)}.$$

Let us write this automorphism ψ_a in our notation:

$$\psi_a(w) = \frac{(1 + |\langle a, a\rangle|^2 - 2|a|^2)^{\frac{1}{2}}}{1 + \langle w, w\rangle\overline{\langle a, a\rangle} - 2\langle w, \bar{a}\rangle} \left(w - a + \frac{\langle a, a\rangle - \langle w, w\rangle}{1 - |\langle a, a\rangle|^2}(\bar{a} - a\overline{\langle a, a\rangle}) \right) Q'.$$

Since the non-degenerate matrix Q depends only on a, then condition (3.4.12) will be satisfied for components of the map as well

$$\varphi_a(w) = \frac{1}{1 + \langle w, w\rangle\overline{\langle a, a\rangle} - 2\langle w, \bar{a}\rangle} \left(w - a + \frac{\langle a, a\rangle - \langle w, w\rangle}{1 - |\langle a, a\rangle|^2}(\bar{a} - a\overline{\langle a, a\rangle}) \right).$$

Next we find the sum $\langle \varphi_a, \bar{a}\rangle$, which appears to be equal to the expression

$$\frac{1}{1 + \langle w, w\rangle\overline{\langle a, a\rangle} - 2\langle w, \bar{a}\rangle} \times \left(\langle w, \bar{a}\rangle - |a|^2 + \frac{\langle a, a\rangle - \langle w, w\rangle}{1 - |\langle a, a\rangle|^2}(\overline{\langle a, a\rangle}(1 - |a|^2)) \right).$$

Obviously, this expression differs from (4.2.4) only in the factor $P(w, a)$ and the factor that depends only on a. Thus, from (3.4.12) we have

$$\sum_{k=1}^{n} \bar{a}_k \frac{\partial F(a)}{\partial \bar{a}_k} = 0,$$

where $F(a)$ (as above) is the Poisson integral of the function f and further proof is the same as in Item 1. □

Let the classical domain D be such that the dimension of S is strictly greater than N. Recall that the function $f \in \mathscr{L}^1(S)$ is a CR-function if

$$\int_S f(\zeta)\bar{\partial}(\omega \wedge d\zeta) = 0$$

for all exterior differential forms ω with coefficients of class $\mathscr{C}^\infty$ in the neighborhood of S, of the corresponding dimension, where $d\zeta$ is the exterior product of all differentials $d\zeta_k$.

The Rossi–Vergne Theorem [70] says that any CR-function $f \in \mathscr{L}^1(S)$ is the radial boundary value on S of a function $F \in \mathscr{H}^1(D)$.

The Poisson kernel in the classical domain D is expressed through the Cauchy–Szegö kernel $C(W, A)$ as follows:

$$P(W, A) = \frac{C(W, A)C(A, W)}{C(A, A)}.$$

As was shown in [7], the expression $C(W, A)d\mu(W)$ is a restriction on S of the Cauchy–Fantappié kernel $\Omega(W, A)$, which is a $\bar{\partial}$-closed differential form. Therefore, in condition (3.4.5) of Theorem 3.4.2, we obtain a product of the Cauchy–Fantappié kernel Ω and the holomorphic function that vanishes at the point A. In this case, this product is a certain $\bar{\partial}$-exact form with coefficients of class $\mathscr{C}^\infty$ in a neighborhood of S. Therefore, Theorem 3.4.2 is a generalization of the Rossi–Vergne theorem from [70].

The proof of Theorem 3.4.2 shows that it remains true if condition (3.4.5) holds only for those automorphisms ψ_A for which point A belongs to a certain open set $V \subset D$. It suffices to apply the uniqueness theorem for real-analytic functions. Therefore, the following generalization of Theorem 3.4.2 holds.

Theorem 3.4.3 ([38]) *If a function $f \in \mathscr{L}^1(S)$ satisfies condition (3.4.5) for all points A belonging to a certain open $V \subset D$ and all components of the automorphism ψ_A, then f is the radial boundary value on S of a certain function $F \in \mathscr{H}^1(D)$.*

Therefore, Theorem 3.4.1 also admits a generalization.

Theorem 3.4.4 ([38]) *Let a function $f \in \mathscr{C}(S)$, and let condition (3.4.2) hold for all automorphisms ψ, transforming the point 0 to points of a certain open set $V \subset D$, then f is holomorphically extended into D up to a certain function $F \in \mathscr{C}(\overline{D})$.*

In the case of the ball D, this theorem generalizes the Nagel–Rudin theorem from [66]. Denote by $\triangle_\psi$ the analytic disk of the form

$$\triangle_\psi = \{\zeta : \zeta = \psi(t\lambda_0), \; |t| < 1\},$$

where λ_0 is a fixed point from the skeleton S and ψ is the automorphism of the domain D. Then the boundary T_ψ of this analytic disk lies on S.

Corollary 3.4.1 ([1, 38]) *If a function $f \in \mathscr{C}(S)$ is holomorphically extended (with respect to t) to analytic disks $\triangle_\psi$ for all automorphisms ψ (or for all automorphisms ψ transforming the point 0 into points of a certain fixed open set $V \subset D$), then the function f is holomorphically extended into D.*

In [1] Agranovskii proved this assertion for the case of all automorphisms of the domain D. In the same paper, M.L. Agranovskii described $\mathscr{U}$-invariant subspaces of the space $\mathscr{C}(S)$ in the classical domain and formulated integral conditions for the holomorphic extension of a continuous function from the skeleton S to the domain D, which are similar to but do not coincide with conditions (3.4.3). His proof was based on the description of the boundary values of holomorphic functions from [8].

Corollary 3.4.1 is an analogue of the Stout theorem [78] on functions with the one-dimensional holomorphic extension property for classical domains (see also [4, 26, 48]). Moreover, it generalizes (for the given class of domains S) Theorem 5.5 of Tumanov, which says that if a smooth function f is holomorphically extended to all analytic disks with a boundary on S, then f is a CR-function on S [79].

3.5 Multidimensional Analogue of the Morera Theorem for Real-Analytic Functions

This section contains some results related to the analytic continuation of real-analytic functions given on the boundary of a bounded domain to this domain. We consider functions that satisfy the Morera property (Definition 3.1.1). So let us consider a set of complex lines intersecting the germ of a real-analytic manifold of real dimension $(2n - 2)$ to be a sufficient set.

Let $D \subset \mathbb{C}^n$ $(n > 1)$ be a bounded domain with a connected real-analytic boundary of the form

$$D = \{ z \in \mathbb{C}^n : \rho(z) < 0 \},$$

where $\rho(z)$ is a real-analytic real-valued function in a neighborhood of the set $\overline{D}$ such that $d\rho|_{\partial D} \neq 0$. We identify $\mathbb{C}^n$ with $\mathbb{R}^{2n}$ in the following way: $z = (z_1, \ldots, z_n)$, where $z_j = x_j + i y_j$, $x_j, y_j \in \mathbb{R}, j = 1, \ldots, n$.

Consider complex lines $l_{z,b}$ of the form (3.2.1)

$$l_{z,b} = \{ \zeta \in \mathbb{C}^n : \zeta_j = z_j + b_j t, \ j = 1, \ldots, n, \ t \in \mathbb{C} \},$$

passing through the point $z \in \mathbb{C}^n$ in the direction of the vector $b = \{ b_1, \ldots, b_n \} \in \mathbb{CP}^{n-1}$ (the direction b is determined up to multiplication by a complex number $\lambda \neq 0$).

Let Γ be the germ of a real-analytic manifold of real dimension $(2n - 2)$. We assume that $0 \in \Gamma$ and the manifold Γ has the form

$$\Gamma = \{ \zeta \in \mathbb{C}^n : \Phi(\zeta) + i\Psi(\zeta) = 0 \},$$

in some neighborhood of zero, where Φ, Ψ are the real-analytic real-valued functions in the neighborhood of the point 0. Here $\zeta = (\zeta_1, \ldots, \zeta_n)$ and $\zeta_j = \xi_j + i\eta_j$, $\xi_j, \eta_j \in \mathbb{R}, j = 1, \ldots, n$. The smoothness condition of the manifold Γ can be written down as

$$\text{rang} A = \text{rang} \begin{pmatrix} \dfrac{\partial \Phi}{\partial \xi_1} & \cdots & \dfrac{\partial \Phi}{\partial \xi_n} & \dfrac{\partial \Phi}{\partial \eta_1} & \cdots & \dfrac{\partial \Phi}{\partial \eta_n} \\ \dfrac{\partial \Psi}{\partial \xi_1} & \cdots & \dfrac{\partial \Psi}{\partial \xi_n} & \dfrac{\partial \Psi}{\partial \eta_1} & \cdots & \dfrac{\partial \Psi}{\partial \eta_n} \end{pmatrix} = 2$$

at every point $\zeta \in \Gamma$.

Consider complex lines of the form (3.2.1), and let $b_j = c_j + id_j$, $c_j, d_j \in \mathbb{R}$, $j = 1, \ldots, n$ and $t = u + iv$, $u, v \in \mathbb{R}$. Then lines $l_{z,b}$ will be defined as

$$l_{z,b} = \{\xi, \eta \in \mathbb{R}^n : \xi_j = x_j + c_j u - d_j v, \ \eta_j = y_j + d_j u + c_j v, \ j = 1, \ldots, n\} \quad (3.5.1)$$

in real coordinates.

Lemma 3.5.1 *Let the vector $b^0 = (b_1^0, \ldots, b_n^0) \in \mathbb{CP}^{n-1}$ be such that $D \cap l_{0,b^0} \neq \varnothing$. Then there exists $\varepsilon > 0$ such that for any z such that $|z| < \varepsilon$, and any b such that $|b - b^0| < \varepsilon$, the intersection $D \cap l_{z,b} \neq \varnothing$ and $\Gamma \cap l_{z,b} \neq \varnothing$.*

Proof The intersection $\Gamma \cap l_{z,b}$ is given by the system of equations

$$\begin{cases} \varphi_{z,b}(u, v) = \Phi(\xi_1, \ldots, \xi_n \eta_1, \ldots, \eta_n), \\ \psi_{z,b}(u, v) = \Psi(\xi_1, \ldots, \xi_n \eta_1, \ldots, \eta_n), \end{cases}$$

where ξ_j and η_j are given by Eq. (3.5.1).

Choose the vector b^0 such that the determinant

$$|J| = \begin{vmatrix} \dfrac{\partial \varphi_{0,b^0}}{\partial u} & \dfrac{\partial \varphi_{0,b^0}}{\partial v} \\[2mm] \dfrac{\partial \psi_{0,b^0}}{\partial u} & \dfrac{\partial \psi_{0,b^0}}{\partial u} \end{vmatrix} (0,0) \neq 0. \quad (3.5.2)$$

Indeed, since

$$\frac{\partial \varphi_{0,b^0}}{\partial u} = \sum_{j=1}^{n} \frac{\partial \Phi}{\partial \xi_j} c_j + \sum_{j=1}^{n} \frac{\partial \Phi}{\partial \eta_j} d_j,$$

$$\frac{\partial \varphi_{0,b^0}}{\partial v} = -\sum_{j=1}^{n} \frac{\partial \Phi}{\partial \xi_j} d_j + \sum_{j=1}^{n} \frac{\partial \Phi}{\partial \eta_j} c_j,$$

$$\frac{\partial \psi_{0,b^0}}{\partial u} = \sum_{j=1}^{n} \frac{\partial \Psi}{\partial \xi_j} c_j + \sum_{j=1}^{n} \frac{\partial \Psi}{\partial \eta_j} d_j,$$

$$\frac{\partial \psi_{0,b^0}}{\partial v} = -\sum_{j=1}^{n} \frac{\partial \Psi}{\partial \xi_j} d_j + \sum_{j=1}^{n} \frac{\partial \Psi}{\partial \eta_j} c_j,$$

then the determinant (3.5.2)

$$
|J| =
\begin{vmatrix}
\sum\limits_{j=1}^{n} \dfrac{\partial \Phi}{\partial \xi_j} c_j + \sum\limits_{j=1}^{n} \dfrac{\partial \Phi}{\partial \eta_j} d_j - \sum\limits_{j=1}^{n} \dfrac{\partial \Phi}{\partial \xi_j} d_j + \sum\limits_{j=1}^{n} \dfrac{\partial \Phi}{\partial \eta_j} c_j \\[2mm]
\sum\limits_{j=1}^{n} \dfrac{\partial \Psi}{\partial \xi_j} c_j + \sum\limits_{j=1}^{n} \dfrac{\partial \Psi}{\partial \eta_j} d_j - \sum\limits_{j=1}^{n} \dfrac{\partial \Psi}{\partial \xi_j} d_j + \sum\limits_{j=1}^{n} \dfrac{\partial \Psi}{\partial \eta_j} c_j
\end{vmatrix}
$$

$$
= \sum_{j,k} \left(-c_j d_j
\begin{vmatrix}
\dfrac{\partial \Phi}{\partial \xi_j} & \dfrac{\partial \Phi}{\partial \xi_k} \\[2mm]
\dfrac{\partial \Psi}{\partial \xi_j} & \dfrac{\partial \Phi}{\partial \xi_k}
\end{vmatrix}
+ c_j c_k
\begin{vmatrix}
\dfrac{\partial \Phi}{\partial \xi_j} & \dfrac{\partial \Phi}{\partial \eta_k} \\[2mm]
\dfrac{\partial \Psi}{\partial \xi_j} & \dfrac{\partial \Phi}{\partial \eta_k}
\end{vmatrix}
\right.
$$

$$
\left.
- d_j d_k
\begin{vmatrix}
\dfrac{\partial \Phi}{\partial \eta_j} & \dfrac{\partial \Phi}{\partial \xi_k} \\[2mm]
\dfrac{\partial \Psi}{\partial \eta_j} & \dfrac{\partial \Phi}{\partial \xi_k}
\end{vmatrix}
+ d_j c_k
\begin{vmatrix}
\dfrac{\partial \Phi}{\partial \eta_j} & \dfrac{\partial \Phi}{\partial \eta_k} \\[2mm]
\dfrac{\partial \Psi}{\partial \eta_j} & \dfrac{\partial \Phi}{\partial \eta_k}
\end{vmatrix}
\right). \qquad (3.5.3)
$$

Suppose expression (3.5.3) is equal to zero for all b such that $\Gamma \cap l_{0,b} \neq \varnothing$ and b is an open set in $\mathbb{CP}^{n-1}$. Then expression (3.5.3) is identically equal to zero since it is a real-analytic function with respect to $b_j = c_j + id_j$. Without generality restriction we may assume $b_1 = 1 + i0$. Then expression (3.5.3) acquires the form

$$
|J| =
\begin{vmatrix}
\dfrac{\partial \Phi}{\partial \xi_1} & \dfrac{\partial \Phi}{\partial \eta_1} \\[2mm]
\dfrac{\partial \Psi}{\partial \xi_1} & \dfrac{\partial \Psi}{\partial \eta_1}
\end{vmatrix}
+ \sum_{k=2}^{n} \left(-d_k
\begin{vmatrix}
\dfrac{\partial \Phi}{\partial \xi_1} & \dfrac{\partial \Phi}{\partial \xi_k} \\[2mm]
\dfrac{\partial \Psi}{\partial \xi_1} & \dfrac{\partial \Psi}{\partial \xi_k}
\end{vmatrix}
+ c_k
\begin{vmatrix}
\dfrac{\partial \Phi}{\partial \xi_1} & \dfrac{\partial \Phi}{\partial \eta_k} \\[2mm]
\dfrac{\partial \Psi}{\partial \xi_1} & \dfrac{\partial \Psi}{\partial \eta_k}
\end{vmatrix}
\right)
$$

$$
+ \sum_{j,k=2}^{n} \left(-c_j d_k
\begin{vmatrix}
\dfrac{\partial \Phi}{\partial \xi_j} & \dfrac{\partial \Phi}{\partial \xi_k} \\[2mm]
\dfrac{\partial \Psi}{\partial \xi_j} & \dfrac{\partial \Psi}{\partial \xi_k}
\end{vmatrix}
+ c_j c_k
\begin{vmatrix}
\dfrac{\partial \Phi}{\partial \xi_j} & \dfrac{\partial \Phi}{\partial \eta_k} \\[2mm]
\dfrac{\partial \Psi}{\partial \xi_j} & \dfrac{\partial \Psi}{\partial \eta_k}
\end{vmatrix}
\right.
$$

$$
\left.
- d_j d_k
\begin{vmatrix}
\dfrac{\partial \Phi}{\partial \eta_j} & \dfrac{\partial \Phi}{\partial \xi_k} \\[2mm]
\dfrac{\partial \Psi}{\partial \eta_j} & \dfrac{\partial \Psi}{\partial \xi_k}
\end{vmatrix}
+ d_j c_k
\begin{vmatrix}
\dfrac{\partial \Phi}{\partial \eta_j} & \dfrac{\partial \Phi}{\partial \eta_k} \\[2mm]
\dfrac{\partial \Psi}{\partial \eta_j} & \dfrac{\partial \Psi}{\partial \eta_k}
\end{vmatrix}
\right) \equiv 0.
$$

Then the determinants of all second-order minors of the matrix A vanish, which contradicts the smoothness of Γ at 0. Therefore there exists a vector b^0 such that $|J| \neq 0$ and $l_{0,b^0} \cap \Gamma \neq \varnothing$. $\qquad \square$

Lemma 3.5.2 *For $\zeta \in \partial D \cap l_{z,b}$ let the function ρ, defining the domain D, satisfy the condition*

$$\sum_{j=1}^{n} \frac{\partial \rho}{\partial \zeta_j} b_j \neq 0, \qquad (3.5.4)$$

for some z and all ζ, b such that $D \cap l_{z,b} \neq \varnothing$, then the curves $\partial D \cap l_{z,b}$ are smooth and analytically dependent on the parameter b.

Proof Consider the function

$$\varphi_{z,b}(t) = \rho(z_1 + b_1 t, \ldots, z_n + b_n t),$$

then

$$\operatorname{grad} \varphi_{z,b}(t) = \left(\frac{\partial \varphi}{\partial t}, \frac{\partial \varphi}{\partial \bar{t}} \right) = \left(\frac{\partial \varphi}{\partial t}, \overline{\frac{\partial \varphi}{\partial t}} \right).$$

Therefore $\operatorname{grad} \varphi_{z,b}(t) \neq 0$ if and only if $\dfrac{\partial \varphi}{\partial t} \neq 0$. Hence the smoothness condition for the curve $D \cap l_{z,b}$ is equivalent the condition $\dfrac{\partial \varphi}{\partial t} \neq 0$. Finally $\dfrac{\partial \varphi}{\partial t} = \sum_{j=1}^{n} \dfrac{\partial \rho}{\partial t} \neq 0$, implies the statement of the lemma.   □

For example, strongly convex or strongly linearly convex domains in $\mathbb{C}^n$ satisfy the conditions of Lemma 3.5.2.

Let $\mathscr{C}^{\omega}(\partial D)$ denote the space of real-analytic functions on the boundary of the domain D.

Theorem 3.5.1 ([51]) *Let a domain $D \subset \mathbb{C}^n$ satisfy conditions (3.5.4) for the points z, lying in the neighbourhood of a manifold Γ such that $\partial D \cap \Gamma = \varnothing$. Let a function $f \in \mathscr{C}^{\omega}(\partial D)$ satisfy the generalized Morera property, i.e.,*

$$\int_{\partial D \cap l_{z,b}} f(z_1 + b_1 t, \ldots, z_n + b_n t) t^m dt = 0 \qquad (3.5.5)$$

for all $z \in \Gamma$, $b \in \mathbb{CP}^{n-1}$ and for a fixed integral non-negative number m. Then the function f has the holomorphic extension into the domain D.

Proof Consider the Bochner–Martinelli kernel of the form (1.1.1):

$$U(\zeta, z) = \frac{(n-1)!}{(2\pi i)^n} \sum_{k=1}^{n} (-1)^{k-1} \frac{\bar{\zeta}_k - \bar{z}_k}{|\zeta - z|^{2n}} d\bar{\zeta}[k] \wedge d\zeta.$$

By Lemma 3.2.1 the kernel $U(\zeta, z)$ in terms of the coordinates b and t has the form

$$U(\zeta, z) = \lambda(b) \wedge \frac{dt}{t},$$

where $\lambda(b)$ is a differential form of type $(n-1, n-1)$ in $\mathbb{CP}^{n-1}$ such that it does not depend on t, and the point $z \notin \partial D$.

Consider the integral

$$M_\alpha f(z) = \int_{\partial D_\zeta} (\zeta - z)^\alpha f(\zeta) U(\zeta, z),$$

where $\alpha = (\alpha_1, \dots, \alpha_n)$ is an arbitrary multi-index such that

$$\|\alpha\| = \alpha_1 + \dots + \alpha_n = m + 1$$

and

$$(\zeta - z)^\alpha = (\zeta_1 - z_1)^{\alpha_1} \cdots (\zeta_n - z_n)^{\alpha_n}.$$

By the Fubini theorem and Lemma 3.2.1 we obtain

$$M_\alpha f(z) = \int_{\mathbb{CP}^{n-1}} b^\alpha \lambda(b) \int_{\partial D \cap l_{z,b}} f(z_1 + b_1 t, \dots, z_n + b_n t) t^m dt.$$

Then by the condition of Theorem 3.5.1 and Lemma 3.5.1 the integrals

$$\int_{\partial D \cap l_{z,b}} f(z_1 + b_1 t, \dots, z_n + b_n t) t^m dt = 0$$

for any z with a sufficiently small $|z|$ and b close to b^0. By the condition of Theorem 3.5.1 and Lemma 3.5.2 this integral is a real-analytic function with respect to b, hence it is identically equal to zero, then

$$M_\alpha f(z) = \int_{\partial D_\zeta} (\zeta - z)^\alpha f(\zeta) U(\zeta, z) \equiv 0 \qquad (3.5.6)$$

for all z such that $|z| < \varepsilon$.

We rewrite the function $M_\alpha f(z)$ in a different form. Consider the differential forms $U_s(\zeta, z)$ of the form (3.2.5):

$$U_s(\zeta, z) = \frac{(-1)^s (n-2)!}{(2\pi i)^n} \left(\sum_{j=1}^{s-1} (-1)^j \frac{\bar{\xi}_j - \bar{z}_j}{|\zeta - z|^{2n-2}} \, d\bar{\xi}[j, s] \right.$$

$$\left. + \sum_{j=s+1}^{n} (-1)^{j-1} \frac{\bar{\xi}_j - \bar{z}_j}{|\zeta - z|^{2n-2}} \, d\bar{\xi}[s, j] \right) \wedge d\zeta.$$

It is easy to verify that

$$\bar{\partial} \left(\frac{1}{\xi_s - z_s} U_s(\zeta, z) \right) = U(\zeta, z)$$

at $\zeta_s \neq z_s$, $s = 1, \ldots, n$. Then condition (3.5.6) can be rewritten as

$$\int_{\partial D_\zeta} f(\zeta) \bar{\partial} \left((\zeta - z)^\beta U_s(\zeta, z) \right) \equiv 0 \tag{3.5.7}$$

for z such that $|z| < \varepsilon$ and for all monomials $(\zeta - z)^\beta$ with $\|\beta\| = m$.

Let us show that condition (3.5.7) holds for monomials $(\zeta - z)^\gamma$ with $\|\gamma\| < m$. Indeed, consider the monomial $(\zeta - z)^\gamma$ with $\|\gamma\| = m - 1$. Then condition (3.5.7) holds for monomials

$$(\zeta - z)^\beta (\zeta_k - z_k), \qquad k = 1, \ldots, n,$$

since the degree of these monomials is m. The equality

$$\frac{\partial}{\partial \zeta_k} ((\zeta - z)^\gamma (\zeta_k - z_k) U_s(\zeta, z)) = (\gamma_k + 1)(\zeta - z)^\gamma U_s(\zeta, z)$$

$$- (n-1)(\zeta - z)^\gamma \frac{(\zeta_k - z_k)(\bar{\xi}_k - \bar{z}_k)}{|\zeta - z|^2} U_s(\zeta, z) \tag{3.5.8}$$

holds. Summing equalities (3.5.8) by k, we obtain

$$\sum_{k=1}^{n} \frac{\partial}{\partial \zeta_k} ((\zeta - z)^\gamma (\zeta_k - z_k) U_s(\zeta, z)) = (\|\gamma\| + 1)(\zeta - z)^\gamma U_s(\zeta, z). \tag{3.5.9}$$

Since condition (3.5.7) can be differentiated in z for $|z| < \varepsilon$, and the derivatives with respect to z and ζ in expression (3.5.9) differ only in sign, then from (3.5.9) it follows that the degree of the monomial in (3.5.7) can be reduced by one.

Consequentially reducing this degree we obtain the conditions

$$\int_{\partial D_\zeta} f(\zeta)\bar{\partial}U_s(\zeta, z) \equiv 0$$

for $|z| < \varepsilon$ and $s = 1, \ldots, n$, i.e.,

$$\int_{\partial D_\zeta} (\zeta_s - z_s)f(\zeta)U(\zeta, z) \equiv 0 \qquad (3.5.10)$$

for $|z| < \varepsilon$ and $s = 1, \ldots, n$.

Applying the Laplace operator

$$\Delta = \frac{\partial^2}{\partial z_1 \partial \bar{z}_1} + \ldots + \frac{\partial^2}{\partial z_n \partial \bar{z}_n},$$

to the left-hand side of (3.5.10) we obtain

$$\frac{\partial}{\partial \bar{z}_s} \int_{\partial D_\zeta} f(\zeta)U(\zeta, z) \equiv 0$$

for $|z| < \varepsilon$ and $s = 1, \ldots, n$. Here we have used the harmonicity of the kernel $U(\zeta, z)$ and the identity

$$\Delta(gh) = h\Delta g + g\Delta h + \sum_{j=1}^{n} \frac{\partial g}{\partial \bar{z}_j}\frac{\partial h}{\partial z_j} + \sum_{j=1}^{n} \frac{\partial g}{\partial z_j}\frac{\partial h}{\partial \bar{z}_j}.$$

Consequently, the Bochner–Martinelli integral of the function f

$$Mf(z) = \int_{\partial D_\zeta} f(\zeta)U(\zeta, z)$$

is a function holomorphic in the neighborhood of zero.

If $\Gamma \subset \mathbb{C}^n \setminus \overline{D}$, then $Mf(z) \equiv 0$ outside $\overline{D}$, and then by Corollary 2.5.3 the function f is holomorphically extended to the domain D. If $\Gamma \subset D$, then by Corollary 2.5.2 the function Mf is holomorphic in D and the boundary values of Mf coincide with f. $\square$

For $m = 0$ condition (3.5.5) takes us to the boundary Morera property [26]

$$\int_{\partial D \cap l_{z,b}} f(z_1 + b_1 t, \ldots, z_n + b_n t)\, dt = 0. \qquad (3.5.11)$$

Corollary 3.5.1 *Let a domain D satisfy the conditions of Theorem 3.5.1, and a function $f \in \mathscr{C}^\omega(\partial D)$ satisfy condition (3.5.11) for all $z \in \Gamma$ and $b \in \mathbb{CP}^{n-1}$, then f is holomorphically extended into the domain D.*

Chapter 4
Functions with the One-Dimensional Holomorphic Extension Property

Abstract The first result related to our subject was obtained by Agranovskii and Val'sky in (Sib. Math. J. **12**, 1–7, 1971), who studied functions with the one-dimensional holomorphic extension property in a ball. Their proof was based on the properties of the automorphism group of the ball. Stout (Duke Math. J. **44**, 105–108, 1977) used the complex Radon transform to extend the Agranovskii–Val'sky theorem to arbitrary bounded domains with smooth boundaries. An alternative proof of the Stout theorem was suggested in Integral Representations and Residues in (Multidimensional Complex Analysis. AMS, Providence, 1983) by Kytmanov, who applied the Bochner–Martinelli integral. The idea of using integral representations (those of Bochner–Martinelli, Cauchy–Fantappié, and the logarithmic residue) turns out to be useful in studying functions with a one-dimensional holomorphic extension property along complex curves (Kytmanov and Myslivets, Sib. Math. J. **38**, 302–311, 1997; Kytmanov and Myslivets, J. Math. Sci. **120**, 1842–1867, 2004).

4.1 Sufficient Families of Complex Lines Intersecting a Generic Manifold Lying Outside the Domain

This section contains some results related to the sufficiency of a family of complex lines intersecting a generating manifold. We will be talking about functions with the one-dimensional holomorphic extension property along families of such complex lines.

Let D be a bounded domain in $\mathbb{C}^n$ ($n > 1$) with a connected smooth boundary ∂D of class $\mathscr{C}^2$. Consider complex one-dimensional lines l of the form (3.2.1)

$$l = \{\zeta \in \mathbb{C}^n : \zeta_j = z_j + b_j t, \ j = 1, \ldots, n, \ t \in \mathbb{C}\}, \qquad (4.1.1)$$

that pass through a point $z \in \mathbb{C}^n$ in the direction of the vector $b \in \mathbb{CP}^{n-1}$ (the direction of b is determined up to multiplication by a complex number $\lambda \neq 0$). Consider functions f with the one-dimensional holomorphic extension property (Definition 3.2.1) along families of complex lines of the form (4.1.1).

In [26], the problem of finding *sufficient* sets of complex lines $\mathfrak{L} = \{l\}$ for which conditions of Definition 3.2.1 for $l \in \mathfrak{L}$ implies a holomorphic extension of the function f to D was posed.

© Springer International Publishing Switzerland 2015

A.M. Kytmanov, S.G. Myslivets, *Multidimensional Integral Representations*,
DOI 10.1007/978-3-319-21659-1_4

4.1.1 Examples of Families of Complex Lines that are Not Sufficient

It is clear that any family of complex lines passing through one point is not sufficient. Let us show that, generally speaking, a family of all complex lines passing through finitely many points is not sufficient either.

Example 4.1.1 ([50]) Suppose that the domain D is a unit ball in $\mathbb{C}^n$:

$$D = \left\{ z \in \mathbb{C}^n : \sum_{j=1}^{n} |z_j|^2 < 1 \right\}.$$

Consider a set of parallel complex lines of the form

$$l_{c,b} = \{ z \in \mathbb{C}^n : z_1 = t, \ z_j = c_j + b_j t, \ j = 2, \ldots, n, \ t \in \mathbb{C} \}, \tag{4.1.2}$$

where $b = (b_2, \ldots, b_n) \in \mathbb{C}^{n-1}$ is the fixed vector and $c = (c_2, \ldots, c_n) \in \mathbb{C}^{n-1}$ is the current vector. On the sphere ∂D, we have

$$|t|^2 + \sum_{j=2}^{n} |c_j + b_j t|^2 = 1,$$

or, equivalently,

$$|t|^2 + |t|^2 \sum_{j=2}^{n} |b_j|^2 + \sum_{j=2}^{n} |c_j|^2 + t \sum_{j=2}^{n} b_j \bar{c}_j + \bar{t} \sum_{j=2}^{n} \bar{b}_j c_j = 1.$$

Therefore,

$$\bar{t} = \frac{1 - \sum_{j=2}^{n} |c_j|^2 - t \sum_{j=2}^{n} b_j \bar{c}_j}{t \left(1 + \sum_{j=2}^{n} |b_j|^2 \right) + \sum_{j=2}^{n} \bar{b}_j c_j}$$

on ∂D. Consider the function $f = |z_1|^2 P(z)$ on ∂D, where $P(z)$ is the polynomial

$$P(z) = z_1 \left(1 + \sum_{j=2}^{n} |b_j|^2 \right) + \sum_{j=2}^{n} \bar{b}_j (z_j - b_j z_1) = z_1 + \sum_{j=2}^{n} \bar{b}_j z_j.$$

On family (4.1.2) (i.e., on the sets $\partial D \cap l_{c,b}$, where $c \in \mathbb{C}^{n-1}$) the function f has the form

$$f = |t|^2 \left(t\left(1 + \sum_{j=2}^{n} |b_j|^2\right) + \sum_{j=2}^{n} \bar{b}_j c_j \right) = t\left(1 - \sum_{j=2}^{n} |c_j|^2 - t\sum_{j=2}^{n} b_j \bar{c}_j \right).$$

Therefore, it extends holomorphically along the given family of complex lines from the curves $\partial D \cap l_{c,b}$ to the sets $D \cap l_{c,b}$.

On the other hand, f is not a CR-function on ∂D, because is does not satisfy the tangent Cauchy–Riemann equations. For a finite set of points $b^k = (b_2^k, \ldots, b_n^k)$, where $k = 1, \ldots, m$, we define f on ∂D as

$$f = |z_1|^2 \prod_{k=1}^{m} \left(z_1\left(1 + \sum_{j=2}^{n} |b_j^k|^2\right) + \sum_{j=2}^{n} \bar{b}_j^k (z_j - b_j^k z_1) \right).$$

This function f is not a CR-function on ∂D either, but it extends holomorphically to all intersections $D \cap l_{c,b^k}$, where $c \in \mathbb{C}^{n-1}$ and $k = 1, \ldots, m$. A set of lines of the form (4.1.2) determines the point in the infinite complex hyperplane $\Pi = \mathbb{CP}^n \setminus \mathbb{C}^n$.

To obtain sets passing through finite points in $\mathbb{C}^n$, we transform the plane Π into a complex hyperplane L_0 in $\mathbb{C}^n$ by a linear-fractional map. This map turns the domain D into some bounded domain D^* and transforms the function f into a function f^*. Since any linear-fractional transformation turns complex lines into complex lines, we obtain the required example of a domain, a function, and a finite set of points for which the holomorphic extendability of the function f^* along all complex lines passing through the given point does not imply that f^* holomorphically extends to D^*. Note that these points will lie on a complex hyperplane not intersecting the closure of the domain D^*.

Example 4.1.2 Consider a unit ball B in $\mathbb{C}^2$:

$$B = \{(z, w) \in \mathbb{C}^2 : |z|^2 + |w|^2 < 1\}$$

and the complex manifold $\Gamma = \{(z, w) \in \mathbb{C}^2 : w = 0\}$, which coincides with its complex tangent space at each point and intersects $\bar{B}$.

Consider the complex lines intersecting Γ:

$$l_a = \{(z, w) \in \mathbb{C}^2 : z = a + bt, \ w = ct, \ t \in \mathbb{C}\}. \tag{4.1.3}$$

These lines pass through the point $(a, 0) \in \Gamma$. The point $(a, 0)$ lies on B for $|a| < 1$ and it does not for $|a| > 1$. Without loss of generality, we can assume that $|b|^2 + |c|^2 = 1$. The intersection $l_a \cap \partial B$ forms a circle

$$|t|^2 + a\bar{b}\bar{t} + \bar{a}bt = 1 - |a|^2 \quad \text{or} \quad |t + a\bar{b}|^2 = 1 - |c|^2|a|^2. \tag{4.1.4}$$

This set $l_a \cap \partial B$ is not empty, if $|a|^2 |c|^2 < 1$. Thus the condition

$$\bar{t} = \frac{1 - |a|^2 - \bar{a}bt}{t + a\bar{b}} \tag{4.1.5}$$

holds for $l_a \cap \partial B$.

Consider the function

$$f_a(z, w) = (1 - \bar{a}z)\frac{w^{k+2}}{\bar{w}}, \quad k \in \mathbb{Z}, \quad k \geq 0.$$

This function is a smooth function of class $\mathscr{C}^k$ on ∂B, since the ratio of $\dfrac{w}{\bar{w}}$ is bounded, then the function $\dfrac{w^2}{\bar{w}}$ is continuous and $\dfrac{w^{k+2}}{\bar{w}}$ is the $\mathscr{C}^k$-smoothness function. On the set $l_a \cap \partial B$ the function f_a is equal to

$$\frac{1 - \bar{a}(a + bt)}{1 - |a|^2 - \bar{a}bt} (t + a\bar{b}) (ct)^{k+2} = (t + a\bar{b}) (ct)^{k+2}.$$

Thus the restriction of f_a is holomorphically extended to the set $l_a \cap B$ for all complex lines l_a, passing through $(a, 0)$ and intersecting $\bar{B}$.

Considering an arbitrary finite set of points $(a_m, 0)$ with $|a_m| > 1, m = 1, \ldots, N$, and the function

$$f(z, w) = \frac{w^{k+2}}{\bar{w}} \prod_{m=1}^{N} (1 - \bar{a}_m z),$$

we obtain that f has the one-dimensional holomorphic extension property along all complex lines l_{a_m}, intersecting $\bar{B}$. Nevertheless, f can not be extended holomorphically into a ball B from the boundary ∂B, since it is obvious that f is not a CR-function on ∂B.

Example 4.1.3 In a ball B, we consider a part of the complex manifold $\Gamma_1 = \{(z, w) \in B : w = 0\}$. As shown by Globevnik [25], the function $f_1 = \dfrac{w^{k+2}}{\bar{w}}$ ($k \in \mathbb{Z}, k \geq 0$) has the one-dimensional holomorphic extension property from ∂B along the complex lines of the family $\mathfrak{L}_{\Gamma_1}$, this is a smooth function on ∂B, which, however, does not extend holomorphically into B.

Indeed, since equality (4.1.5) holds for complex lines of the form (4.1.3) on ∂B, the function f_1 on ∂B takes the form

$$f_1 = \frac{t + a\bar{b}}{1 - |a|^2 - \bar{a}bt} (ct)^{k+2}.$$

The denominator of this fraction vanishes at the point $t_0 = \dfrac{1 - |a|^2}{\bar{a}b}$. Substituting this point into expression (4.1.4), we obtain

$$\frac{(1 - |a|^2)^2}{|a|^2|b|^2} + 1 - |a|^2 > 0 \quad \text{for } |a| < 1.$$

Therefore, the point of the line l_a, corresponding to t_0, lies outside the ball B. So, the function f_1 extends holomorphically into $l_a \cap B$.

This motivates the problem of finding families of manifolds (desirably, of minimal dimension) for which the sets of complex lines intersecting these manifolds are sufficient for holomorphic extension.

4.1.2 Sufficient Families of Complex Lines

In this section, we consider families of complex lines passing through a generic manifold. The real dimension of such a manifold is at least n.

Recall that a smooth manifold Γ of class $\mathscr{C}^\infty$ is said to be *generic* if the complex linear span of the tangent space $T_z(\Gamma)$ coincides with $\mathbb{C}^n$ for each point $z \in \Gamma$. We denote the family of all complex lines intersecting Γ by $\mathfrak{L}_\Gamma$.

Theorem 4.1.1 ([50]) *If Γ is a germ of a generic manifold in $\mathbb{C}^n \setminus \overline{D}$ and a function $f \in \mathscr{C}(\partial D)$ has the one-dimensional holomorphic extension property along all complex lines from $\mathfrak{L}_\Gamma$, then the function f is holomorphically extended in D.*

Proof Consider the Bochner–Martinelli integral of f of the form (1.2.4)

$$Mf(z) = \int_{\partial D} f(\zeta)U(\zeta, z), \quad z \notin \partial D,$$

where $U(\zeta, z)$ is the Bochner–Martinelli kernel of the form (1.1.1).

Lemma 4.1.1 *If $z \in \mathbb{C}^n \setminus \overline{D}$ and a function f has the one-dimensional holomorphic extension property along almost all complex lines passing through z, then $Mf(z) = 0$, and all derivatives (by z) of Mf of order $\alpha = (\alpha_1, \ldots, \alpha_n)$ vanish as well:*

$$\frac{\partial^\alpha Mf}{\partial z^\alpha}(z) = \frac{\partial^{\|\alpha\|} Mf}{\partial z_1^{\alpha_1} \cdots \partial z_n^{\alpha_n}}(z) = 0,$$

where $\|\alpha\| = \alpha_1 + \ldots + \alpha_n$.

Proof Consider complex lines l of the form (4.1.1) passing through the point z in the direction of the vector $b \in \mathbb{CP}^{n-1}$.

By Lemma 3.2.1 in the coordinates t and b, the Bochner–Martinelli kernel is written as

$$U(\zeta, z) = \lambda(b) \wedge \frac{dt}{t},$$

where $\lambda(b)$ is a differential form of type $(n-1, n-1)$ on $\mathbb{CP}^{n-1}$ not depending on t. We need a similar representation for the derivatives of this integral. We have

$$\frac{\partial^\alpha Mf}{\partial z^\alpha}(z) = \int_{\partial D} f(\zeta) \frac{\partial^\alpha U(\zeta, z)}{\partial z^\alpha} = (-1)^{\|\alpha\|} \int_{\partial D} f(\zeta) \frac{\partial^\alpha U(\zeta, z)}{\partial \zeta^\alpha}.$$

We assume that $z = 0$. Expressing the derivatives of the Bochner–Martinelli integral in terms of the variables t and b, we obtain

$$\frac{\partial^\alpha U(\zeta, 0)}{\partial \zeta^\alpha} = \frac{(n + \|\alpha\| - 1)!}{(2\pi i)^n} \frac{(-1)^{\|\alpha\|} \bar{\zeta}^\alpha \sum_{k=1}^n (-1)^{k-1} \bar{\zeta}_k d\bar{\zeta}[k] \wedge d\zeta}{|\zeta|^{2n+2\|\alpha\|}}$$

$$= \frac{(n + \|\alpha\| - 1)!}{(n-1)!} \frac{(-1)^{\|\alpha\|} \bar{\zeta}^\alpha}{|\zeta|^{2\|\alpha\|}} U(\zeta, z) = \frac{(-1)^{\|\alpha\|+n-1}(n + \|\alpha\| - 1)!}{(2\pi i)^n} \times$$

$$\times \frac{\bar{b}^\alpha \sum_{k=1}^n (-1)^{k-1} \bar{b}_k d\bar{b}[k] \wedge \sum_{k=1}^n (-1)^{k-1} b_k db[k]}{|b|^{2n+2\|\alpha\|}} \wedge \frac{dt}{t^{\|\alpha\|+1}}$$

$$= \lambda_\alpha(b) \wedge \frac{dt}{t^{\|\alpha\|+1}},$$

where $\bar{\zeta}^\alpha = \bar{\zeta}_1^{\alpha_1} \cdots \bar{\zeta}_n^{\alpha_n}$.

These calculations are similar to those performed in Lemma 3.2.1. Since the point $z = 0$ is outside $\overline{D}$, it follows that

$$\int_{\partial D \cap l} f(b_1 t, \ldots, b_n t) \frac{dt}{t^{\|\alpha\|+1}} = 0,$$

therefore, by Fubini's theorem, we have

$$\frac{\partial^\alpha Mf}{\partial z^\alpha}(0) = \int_{\mathbb{CP}^{n-1}} \lambda_\alpha(b) \int_{\partial D \cap l} f(b_1 t, \ldots, b_n t) \frac{dt}{t^{\|\alpha\|+1}} = 0.$$

$\square$

Let us go back to the proof of the theorem. Suppose that Γ is a germ of a generic manifold in $\mathbb{C}^n \setminus \overline{D}$, i.e., there exists an open set W on which Γ is a smooth generic manifold of class $\mathscr{C}^\infty$. If a function f has the one-dimensional property of holomorphic extension along complex lines from $\mathfrak{L}_\Gamma$, then, by Lemma 4.1.1, the Bochner–Martinelli integral and all of its derivatives with respect to z vanish on Γ:

$$Mf|_\Gamma = 0, \qquad \frac{\partial^\alpha Mf}{\partial z^\alpha}\bigg|_\Gamma = 0 \quad \text{for all multi-indices} \quad \alpha. \tag{4.1.6}$$

The generic manifold Γ can be reduced by a locally biholomorphic transformation to the form (see [11])

$$\Gamma : \begin{cases} v_1 = h_1(z_1, \ldots, z_k, u_1, \ldots, u_m), \\ \ldots\ldots\ldots\ldots \\ v_m = h_m(z_1, \ldots, z_k, u_1, \ldots, u_m), \end{cases}$$

where $k + m = n$, $z_j = x_j + iy_j$, $j = 1, \ldots, k$, $w_s = u_s + iv_s$, and $s = 1, \ldots, m$. Moreover, the real-valued vector-function $h = (h_1, \ldots, h_m)$ is of class $\mathscr{C}^\infty$ in a neighborhood W of 0, and

$$h_p(0) = 0, \quad \frac{\partial h_p}{\partial x_j}(0) = \frac{\partial h_p}{\partial y_j}(0) = \frac{\partial h_p}{\partial u_s}(0) = 0$$

for $j, p = 1, \ldots, m$, $s = 1, \ldots, k$. Any biholomorphic transformation turns derivatives with respect to holomorphic variables into similar derivatives; thus, condition (4.1.6) can be rewritten as

$$Mf|_\Gamma = 0, \qquad \frac{\partial^{\alpha+\beta} Mf}{\partial z^\alpha \partial w^\beta}\bigg|_\Gamma = 0 \quad \text{for all multi-indices} \quad \alpha, \beta. \tag{4.1.7}$$

Lemma 4.1.2 *If a real-analytic function Mf defined on the neighborhood W of a set Γ satisfies conditions (4.1.7), then it vanishes on W.*

Proof Let us show that all Taylor coefficients of the function Mf in a neighborhood of zero vanish. For $j = 1, \ldots, k$ and $s = 1, \ldots, m$ we denote the partial derivatives of the function $Mf(z, u + ih(z, u))$ with respect to the variables x_j, y_j, and u_s by D_{x_j}, D_{y_j}, and D_{u_s}, respectively. Since

$$0 = D_{x_j} Mf = \frac{\partial Mf}{\partial x_j} + \sum_{l=1}^m \frac{\partial Mf}{\partial v_l} \frac{\partial h_l}{\partial x_j} \quad \text{and} \quad \frac{\partial h_l}{\partial x_j}(0) = 0,$$

it follows that $\dfrac{\partial Mf}{\partial x_j}(0) = 0$. Similarly, $\dfrac{\partial Mf}{\partial y_j}(0) = 0$ and $\dfrac{\partial Mf}{\partial u_s}(0) = 0$ for $j = 1, \ldots, k$ and $s = 1, \ldots, m$. Since

$$0 = \frac{\partial Mf}{\partial w_s}\bigg|_\Gamma = \frac{1}{2}\left(\frac{\partial Mf}{\partial u_s} - i\frac{\partial Mf}{\partial v_s}\right)\bigg|_\Gamma \quad \text{and} \quad \frac{\partial Mf}{\partial u_s}(0) = 0,$$

it follows that $\dfrac{\partial Mf}{\partial v_s}(0) = 0$ for $s = 1, \ldots, m$. Thus, all of the first derivatives of Mf vanish at 0.

Let us show that the second derivatives vanish at zero as well. We have

$$0 = D^2_{x_j x_l} Mf = \frac{\partial^2 Mf}{\partial x_j \partial x_l} + \sum_{p=1}^{m} \frac{\partial^2 Mf}{\partial x_j \partial v_p} \frac{\partial h_p}{\partial x_l}$$

$$+ \sum_{p=1}^{m} \left(\frac{\partial^2 Mf}{\partial v_p \partial x_l} \frac{\partial h_p}{\partial x_j} + \sum_{q=1}^{m} \left(\frac{\partial^2 Mf}{\partial v_p \partial v_q} \frac{\partial h_p}{\partial x_j} \frac{\partial h_q}{\partial x_l} \right) + \frac{\partial Mf}{\partial v_p} \frac{\partial^2 h_p}{\partial x_j \partial x_l} \right).$$

The relations $\dfrac{\partial h_p}{\partial x_l}(0) = 0$ and $\dfrac{\partial Mf}{\partial v_p}(0) = 0$ imply $\dfrac{\partial^2 Mf}{\partial x_j \partial x_l}(0) = 0$. Similar arguments show that the second derivatives of Mf with respect to the variables x_j, y_j, and u_s vanish as well at 0 .

Consider

$$0 = D_{x_j} \left(\frac{\partial Mf}{\partial w_s} \right) = \frac{\partial^2 Mf}{\partial w_s \partial x_j} + \sum_{p=1}^{m} \frac{\partial^2 Mf}{\partial w_s \partial v_p} \frac{\partial h_p}{\partial x_j}.$$

Since $\dfrac{\partial h_p}{\partial x_j}(0) = 0$, it follows that $\dfrac{\partial^2 Mf}{\partial w_s \partial x_j}(0) = 0$. Thus, we have

$$0 = \frac{\partial^2 Mf}{\partial w_s \partial x_j}(0) = \frac{1}{2} \left(\frac{\partial^2 Mf}{\partial u_s \partial x_j}(0) - i \frac{\partial^2 Mf}{\partial v_s \partial x_j}(0) \right) \quad \text{and} \quad \frac{\partial^2 Mf}{\partial u_s \partial x_j}(0) = 0,$$

therefore, $\dfrac{\partial^2 Mf}{\partial v_s \partial x_j}(0) = 0$. Similarly, $\dfrac{\partial^2 Mf}{\partial v_s \partial y_j}(0) = 0$ and $\dfrac{\partial^2 Mf}{\partial v_s \partial u_l}(0) = 0$. Further,

$$0 = \frac{\partial^2 Mf}{\partial w_l \partial w_s}\bigg|_\Gamma = \frac{1}{4} \left(\frac{\partial^2 Mf}{\partial u_l \partial u_s} - i \frac{\partial^2 Mf}{\partial u_l \partial v_s} - i \frac{\partial^2 Mf}{\partial v_l \partial u_s} - \frac{\partial^2 Mf}{\partial v_l \partial v_s} \right),$$

hence $\dfrac{\partial^2 Mf}{\partial v_l \partial v_s}(0) = 0$.

Applying induction, we can show in a similar way that all higher-order derivatives of the function Mf vanish at the point 0. Thus, the Taylor expansion of Mf vanishes at 0, and the function Mf itself vanishes on W.   □

Let us complete the proof of the theorem. We have shown that the Bochner–Martinelli integral vanishes on W. Since this integral is a real-analytic function and the complement $\mathbb{C}^n \setminus \overline{D}$ is connected, it follows that $Mf(z) \equiv 0$ on $\mathbb{C}^n \setminus \overline{D}$. Applying the assertion that functions representable by a Bochner–Martinelli integral are holomorphic (Corollary 2.5.2), we see that the function Mf is holomorphic on D, and its boundary values coincide with those of f on ∂D. $\square$

4.2 Sufficiency of a Family of Lines Intersecting a Generic Manifold Lying Inside the Domain

Let D be a bounded domain in $\mathbb{C}^n$ ($n > 1$) with a connected smooth boundary ∂D of class $\mathscr{C}^2$. In the previous section we considered a family of complex lines passing through the germ of a generic manifold lying outside the domain D. Here we will deal with the case of a generic manifold lying in the domain.

Let $\mathfrak{L}_\Gamma$ denote a set of complex lines of the form

$$l = \{\zeta \in \mathbb{C}^n : \zeta_j = z_j + b_j t, \ j = 1, \ldots, n, \ t \in \mathbb{C}\}, \tag{4.2.1}$$

passing through the point $z \in \Gamma$ in the direction of the vector $b \in \mathbb{CP}^{n-1}$.

In this section we consider the case where the germ of a generic manifold lies in the domain D, so it will require additional conditions on the domain D. To do this, we first prove some lemmas. Consider the Bochner–Martinelli integral of the form (1.2.4) for the function f:

$$Mf(z) = \int_{\partial D} f(\zeta) U(\zeta, z), \quad z \notin \partial D,$$

where $U(\zeta, z)$ is the Bochner–Martinelli kernel of the form (1.1.1). We assume that $0 \in D$ and the generic manifold Γ is in some neighborhood of $W \subset D$, and $0 \in \Gamma$.

The generic manifold Γ can be reduced by local biholomorphic transformation to the form (see [11])

$$\begin{cases} v_1 = h_1(z_1, \ldots, z_k, u_1, \ldots, u_m), \\ \ldots\ldots\ldots\ldots\ldots \\ v_m = h_m(z_1, \ldots, z_k, u_1, \ldots, u_m), \end{cases} \tag{4.2.2}$$

where $k + m = n$, $z_j = x_j + iy_j$, $j = 1, \ldots, k$, $w_s = u_s + iv_s$, $s = 1, \ldots, m$, $h = (h_1, \ldots, h_m)$ is a real-valued vector-function of class $\mathscr{C}^\infty$ in the neighborhood W of 0 and the conditions

$$h_p(0) = 0, \quad \frac{\partial h_p}{\partial x_j}(0) = \frac{\partial h_p}{\partial y_j}(0) = \frac{\partial h_p}{\partial u_s}(0), \quad j, p = 1, \ldots, m, \ s = 1, \ldots k$$

hold.

Lemma 4.2.1 *If a real-analytic function Mf, defined in W, satisfies the conditions*

$$Mf\big|_\Gamma = 0, \qquad \frac{\partial^{\alpha+\beta} Mf}{\partial \bar{z}^\alpha \partial \bar{w}^\beta}\bigg|_\Gamma = 0 \quad \text{for all multi-indices} \quad \alpha, \beta, \tag{4.2.3}$$

then it is equal to zero on W.

Proof We show that all coefficients of the expansion of F in a Taylor series vanish in the neighborhood of zero.

Denote full partial derivatives along the manifold Γ with respect to x_j, y_j, u_s by $D_{x_j}, D_{y_j}, D_{u_s}$, and $j = 1, \ldots, k, s = 1, \ldots, m$. Since

$$0 = D_{x_j} Mf = \frac{\partial Mf}{\partial x_j} + \sum_{l=1}^{m} \frac{\partial Mf}{\partial v_l} \frac{\partial h_l}{\partial x_j} \quad \text{and} \quad \frac{\partial h_l}{\partial x_j}(0) = 0,$$

then $\dfrac{\partial Mf}{\partial x_j}(0) = 0$. Similarly $\dfrac{\partial Mf}{\partial y_j}(0) = 0, \dfrac{\partial Mf}{\partial u_s}(0) = 0, j = 1, \ldots, k, s = 1, \ldots, m$. Since

$$0 = \frac{\partial Mf}{\partial \bar{w}_s}\bigg|_\Gamma = \frac{1}{2}\left(\frac{\partial Mf}{\partial u_s} + i\frac{\partial Mf}{\partial v_s}\right) \quad \text{and} \quad \frac{\partial Mf}{\partial u_s}(0) = 0,$$

then $\dfrac{\partial Mf}{\partial v_s}(0) = 0, s = 1, \ldots, m$. So, all of the first derivatives of the function Mf vanish at 0.

We now show that all second derivatives also vanish at 0. We have

$$0 = D^2_{x_j x_l} Mf = \frac{\partial^2 Mf}{\partial x_j \partial x_l} + \sum_{p=1}^{m} \frac{\partial^2 Mf}{\partial x_j \partial v_p} \frac{\partial h_p}{\partial x_l}$$

$$+ \sum_{p=1}^{m} \left(\frac{\partial^2 Mf}{\partial v_p \partial x_l} \frac{\partial h_p}{\partial x_j} + \sum_{q=1}^{m} \left(\frac{\partial^2 Mf}{\partial v_p \partial v_q} \frac{\partial h_p}{\partial x_j} \frac{\partial h_q}{\partial x_l} \right) + \frac{\partial Mf}{\partial v_p} \frac{\partial^2 h_p}{\partial x_j \partial x_l} \right).$$

Since $\dfrac{\partial h_p}{\partial x_l}(0) = 0, \dfrac{\partial Mf}{\partial v_p}(0) = 0$, then $\dfrac{\partial^2 Mf}{\partial x_j \partial x_l}(0) = 0$. Similarly, all of the second derivatives of F with respect to y_j, u_s vanish at 0.

Consider

$$0 = D_{x_j}\left(\frac{\partial Mf}{\partial \bar{w}_s}\right) = \frac{\partial^2 Mf}{\partial \bar{w}_s \partial x_j} + \sum_{p=1}^{m} \frac{\partial^2 Mf}{\partial \bar{w}_s \partial v_p} \frac{\partial h_p}{\partial x_j}.$$

Since $\dfrac{\partial h_p}{\partial x_j}(0) = 0$, then $\dfrac{\partial^2 Mf}{\partial \bar{w}_s \partial x_j}(0) = 0$. Since

$$0 = \frac{\partial^2 Mf}{\partial \bar{w}_s \partial x_j}(0) = \frac{1}{2}\left(\frac{\partial^2 Mf}{\partial u_s \partial x_j}(0) + i\frac{\partial^2 Mf}{\partial v_s \partial x_j}(0)\right) \quad \text{and} \quad \frac{\partial^2 Mf}{\partial u_s \partial x_j}(0) = 0,$$

then $\dfrac{\partial^2 Mf}{\partial v_s \partial x_j}(0) = 0$. Similarly, $\dfrac{\partial^2 Mf}{\partial v_s \partial y_j}(0) = 0$, $\dfrac{\partial^2 Mf}{\partial v_s \partial u_l}(0) = 0$. Further

$$0 = \left.\frac{\partial^2 Mf}{\partial \bar{w}_l \partial \bar{w}_s}\right|_\Gamma = \frac{1}{4}\left(\frac{\partial^2 Mf}{\partial u_l \partial u_s} + i\frac{\partial^2 Mf}{\partial u_l \partial v_s} + i\frac{\partial^2 Mf}{\partial v_l \partial u_s} - \frac{\partial^2 Mf}{\partial v_l \partial v_s}\right),$$

so $\dfrac{\partial^2 Mf}{\partial v_l \partial v_s}(0) = 0$.

Applying induction, it can be shown just as above, that all higher derivatives of the function Mf vanish at 0. Thus, the Taylor series at 0 of the function is zero, so the function itself is zero in W. □

Clearly, Lemma 4.2.1 is also true for the original variables z (to bring the manifold Γ to the form (4.2.2)).

We define the functions

$$M_j f(z) = \int_{\partial D} f(\zeta)(\zeta_j - z_j) U(\zeta, z), \quad z \in D, \quad j = 1, \ldots, n.$$

They are real-analytic in D.

Lemma 4.2.2 *If for real-analytic functions $M_j f$ the conditions*

$$\left.\frac{\partial^{\alpha+\beta} M_j f}{\partial \bar{z}^\alpha \partial \bar{w}^\beta}\right|_\Gamma = 0 \tag{4.2.4}$$

hold for all multi-indices α, β at $\|\alpha\| + \|\beta\| > 0$, then $M_j f$ are holomorphic on D.

Proof We apply Lemma 4.2.1 to the functions $\dfrac{\partial M_j f}{\partial \bar{z}_p}$, $\dfrac{\partial M_j f}{\partial \bar{w}_s}$, $p = 1, \ldots, k$, $s = 1, \ldots, m$, to show that these functions are equal to zero in W, i.e., the functions F_j are holomorphic in W, and therefore in D. □

In what follows we will need the definition of a domain with the Nevanlinna property (see [17]). Let $G \subset \mathbb{C}$ be a simply connected domain and $t = k(\tau)$ be a conformal mapping of the unit circle $\Delta = \{\tau : |\tau| < 1\}$ on G.

Domain G is *a domain with the Nevanlinna property*, if there are two bounded holomorphic functions u and v in G such that almost everywhere on $S = \partial\Delta$, the equality

$$\bar{k}(\tau) = \frac{u(k(\tau))}{v(k(\tau))}$$

holds in terms of the angular boundary values. Essentially this means

$$\bar{t} = \frac{u(t)}{v(t)} \quad \text{on} \quad \partial G.$$

Give a characterization of a domains with the Nevanlinna property (Proposition 3.1 in [17]). Domain G is a domain with the Nevanlinna property if and only if $k(\tau)$ admits a holomorphic pseudocontinuation through S in $\overline{\mathbb{C}} \setminus \overline{\Delta}$, i.e., there are bounded holomorphic functions u_1 and v_1 in $\overline{\mathbb{C}} \setminus \overline{\Delta}$ such that the function $\tilde{k}(\tau) = \dfrac{u_1(\tau)}{v_1(\tau)}$ coincides almost everywhere with the function $k(\tau)$ on S.

The above definition and statement will be applied to bounded domains G with a boundary of class $\mathscr{C}^2$, therefore (due to the principle of correspondence of boundaries) the function $k(\tau)$ extends to $\overline{\Delta}$ as a function of class $\mathscr{C}^1(\overline{\Delta})$. The same can be said about the function $\tilde{k}$.

Various examples of domains with the Nevanlinna property are given in [17]. For example, if ∂G is real-analytic, then $k(\tau)$ is a rational function with no poles on the closure Δ. In our further consideration we will need the domain G to possess *a strengthened Nevanlinna property*, i.e., the function $u_1(\tau) \neq 0$ in $\mathbb{C} \setminus \Delta$ and $\tilde{k}(\infty) \neq 0$. For example, such domains will include domains for which $k(\tau)$ is a rational function with no poles on $\overline{\Delta}$ and no zeros in $\mathbb{C} \setminus \Delta$.

Lemma 4.2.3 *If the domain G has a strengthened Nevanlinna property, then the function $\dfrac{1}{\bar{t}}$ extends holomorphically from ∂G into G.*

Proof Consider the function $\dfrac{1}{\bar{t}}$ on ∂G and $\tau \in S$

$$\frac{1}{\bar{t}} = \frac{1}{\overline{k(\tau)}} = \frac{1}{\tilde{\bar{k}}(\tau)} = \frac{\bar{v}_1(\tau)}{\bar{u}_1(\tau)} = \frac{\bar{v}_1\left(\frac{1}{\bar{\tau}}\right)}{\bar{u}_1\left(\frac{1}{\bar{\tau}}\right)}.$$

Then the function $h(\tau) = \dfrac{\bar{v}_1\left(\frac{1}{\bar{\tau}}\right)}{\bar{u}_1\left(\frac{1}{\bar{\tau}}\right)}$ is holomorphic in the circle Δ, since the denominator $\bar{u}_1\left(\frac{1}{\bar{\tau}}\right) \neq 0$ at $|\tau| > 1$ in $h(0) = \dfrac{1}{\tilde{k}(\infty)} \neq \infty$. Therefore, the function $h(\tau)$ gives a holomorphic extension of the function $\dfrac{1}{\tilde{k}(\tau)}$ in the circle Δ hence, the function $\dfrac{1}{\bar{t}}$ extends holomorphically into G. $\qquad\square$

Consider complex lines l of the form (4.2.1) passing through z in the direction of the vector $b \in \mathbb{CP}^{n-1}$. Consider also the following representation of the Bochner–Martinelli kernel in the variables t and b):

$$U(\zeta, z) = \lambda(b) \wedge \frac{dt}{t},$$

where $\lambda(b)$ is a differential form of type $(n-1, n-1)$ in $\mathbb{CP}^{n-1}$ independent of t.

Lemma 4.2.4 *If a function $f \in \mathscr{C}(\partial D)$ has the one-dimensional holomorphic extension property along almost all complex lines $l \in \mathfrak{L}_\Gamma$ and the connected components of the intersection $D \cap l$ are domains with the strengthened Nevanlinna property, then equalities (4.2.4) hold for all multi-indices α with $\|\alpha\| > 0$ and for all $j = 1, \ldots, n$.*

Proof Since

$$U(\zeta, z) = \sum_{j=1}^{n} (-1)^{j-1} \frac{\partial g}{\partial \zeta_j} d\bar{\zeta}[j] \wedge d\zeta,$$

where $g(\zeta, z) = -\dfrac{(n-2)!}{(2\pi i)^n} \dfrac{1}{|\zeta - z|^{2n-2}}$ is the fundamental solution of the Laplace equation, then

$$\frac{\partial^\alpha U}{\partial \bar{z}^\alpha} = (-1)^{\|\alpha\|} \sum_{j=1}^{n} (-1)^{j-1} \frac{\partial}{\partial \zeta_j} \left(\frac{\partial^\alpha g}{\partial \bar{\zeta}^\alpha} \right) d\bar{\zeta}[j] \wedge d\zeta.$$

Since

$$\frac{\partial^\alpha g}{\partial \bar{\zeta}^\alpha} = \frac{(-1)^{\|\alpha\|+1}}{(2\pi i)^n} \frac{(n + \|\alpha\| - 2)!(\zeta - z)^\alpha}{|\zeta - z|^{2n+2\|\alpha\|-2}},$$

then

$$\frac{\partial^\alpha U}{\partial \bar{z}^\alpha} = \frac{(n + \|\alpha\| - 2)!}{(2\pi i)^n} \sum_{j=1}^{n} (-1)^j \frac{\partial}{\partial \zeta_j} \left(\frac{(\zeta - z)^\alpha}{|\zeta - z|^{2n+\|\alpha\|-2}} \right) d\bar{\zeta}[j] \wedge d\zeta$$

$$= \frac{(n + \|\alpha\| - 2)!}{(2\pi i)^n} \times$$

$$\times \sum_{j=1}^{n} (-1)^j \left[\frac{\alpha_j(\zeta - z)^{\alpha - e_j}}{|\zeta - z|^{2n+\|\alpha\|-2}} - \frac{(n + \|\alpha\| - 1)(\zeta - z)^\alpha}{|\zeta - z|^{2n+\|\alpha\|}} \right] d\bar{\zeta}[j] \wedge d\zeta$$

$$= \frac{(n + \|\alpha\| - 2)!}{(2\pi i)^n} \sum_{j=1}^{n} (-1)^j \frac{\alpha_j (\zeta - z)^{\alpha - e_j}}{|\zeta - z|^{2n + 2\|\alpha\| - 2}} d\bar{\zeta}[j] \wedge d\zeta$$

$$+ \frac{(n + \|\alpha\| - 1)!}{(n-1)!} \frac{(\zeta - z)^\alpha}{|\zeta - z|^{2\|\alpha\|}} U(\zeta, z).$$

Let us calculate this form in the variables b and t, i.e., $\zeta_j - z_j = b_j t, j = 1, \ldots, n$. In the calculation we assume that $d\bar{t} \wedge dt = 0$ on $\partial D \cap l$ and that $b \in \mathbb{CP}^{n-1}$. We obtain

$$\frac{\partial^\alpha U}{\partial \bar{z}^\alpha} = \frac{(n + \|\alpha\| - 2)!}{(2\pi i)^n} \sum_{j=1}^{n} (-1)^j \frac{\alpha_j b^{\alpha - e_j}}{t\bar{t}^{\|\alpha\|} |b|^{2\|\alpha\|}} d\bar{b}[j] \wedge \sum_{s=1}^{n} (-1)^{s-1} b_s db[s] \wedge dt$$

$$+ \frac{(n + \|\alpha\| - 1)!}{(n-1)!} \frac{b^\alpha}{t\bar{t}^{\|\alpha\|} |b|^{2\|\alpha\|}} \lambda(b) \wedge dt.$$

Thus we see that

$$(\zeta_j - z_j) \frac{\partial^\alpha U}{\partial \bar{z}^\alpha} = \mu(b) \wedge \frac{dt}{\bar{t}^{\|\alpha\|}}.$$

It remains to show that

$$\int_{\partial D \cap l} f(z + bt) \frac{dt}{\bar{t}^{\|\alpha\|}} = 0,$$

but this follows from Lemma 4.2.3.   □

Theorem 4.2.1 ([52]) *Let Γ be the germ of a generic manifold in D, the function $f \in \mathscr{C}(\partial D)$ has the one-dimensional holomorphic extension property along almost all complex lines $l \in \mathfrak{L}_\Gamma$ and the connected components of the intersection $D \cap l$ be domains with the strengthened Nevanlinna property, then there exists a holomorphic function $Mf \in \mathscr{C}(\overline{D})$ in D that coincides with the function f on the boundary ∂D.*

Proof From Lemmas 4.2.1–4.2.4 it follows that $M_j f(z)$ is holomorphic in D. Since

$$\Delta M_j f = -\frac{\partial Mf}{\partial \bar{z}_j} = 0$$

in D, then the function Mf is holomorphic in D. Therefore, its boundary values coincide with f (see Corollary 2.5.3).   □

Consider examples of the domains for which Theorem 4.2.1 is true.

Example 4.2.1 Let $D = B$ be a ball of radius R with the center at the origin, i.e., $D = \{\zeta : |\zeta| < R\}$. Then for $z \in D$ and $|b| = 1$, the intersection of this domain with the complex line

$$l = \{\zeta : \zeta_j = z_j + b_j t, \; j = 1, \ldots, n\}$$

is a circle

$$G = \{t : |t + \langle z, \bar{b}\rangle|^2 < R^2 - |z|^2 + |\langle z, \bar{b}\rangle|^2\}.$$

On the boundary of this circle the following holds

$$\frac{1}{\bar{t}} = \frac{t + \langle z, \bar{b}\rangle}{R^2 - |z|^2 - t\langle \bar{z}, b\rangle}.$$

It is easy to verify that the denominator of this function does not vanish at 0 in G, so the function $\dfrac{1}{\bar{t}}$ satisfies Lemma 4.2.3 for all points $z \in B$ and for all complex lines l.

Example 4.2.2 Let $w_j = \dfrac{L_j(z)}{L(z)}, \; j = 1, \ldots, n$, where $L_j(z)$, $L(z)$ are the linear functions, and zeros of $L(z)$ do not intersect the closure of the ball, then, with this mapping, the image of the ball B (provided it is biholomorphic on the closure of the ball B) is a bounded domain, for which Theorem 4.2.1 is valid. Indeed, it is easy to verify that all intersections of this domain with complex lines are circles.

Example 4.2.3 Let D be a complete circular domain with respect to all points $z \in \Gamma$, then Theorem 4.2.1 is true for it.

Lemma 4.2.3 shows that for Theorem 4.2.1 to be true it is sufficient that the function $\dfrac{1}{\bar{t}}$ in the sections be holomorphically continued from the boundary of the section into the cross section itself.

Theorem 4.2.2 ([52]) *Let Γ be the germ of a generic manifold in D, and the function $f \in \mathscr{C}(\partial D)$ has the one-dimensional holomorphic extension property along almost all complex lines $l \in \mathfrak{L}_\Gamma$ and the function $\dfrac{1}{\bar{t}}$ holomorphically extends from $\partial D \cap l$ into $D \cap l$ on the connected components of the intersection $D \cap l$, then there exists a function $Mf \in \mathscr{C}(\overline{D})$ is holomorphic in D and coincides with the function f on the boundary ∂D.*

We seek to answer the question: What domains (other than circles) in the complex plane have such a property.

Example 4.2.4 In the complex plane $\mathbb{C}$ consider an open set

$$\{t \in \mathbb{C} : R|t|^{k+1} < |P(t)|\}, \quad 0 < R < \infty, \tag{4.2.5}$$

where $P(t)$ is the polynomial of degree k and $P(0) \neq 0$.

Obviously, this set is bounded and contains a neighborhood of zero. Denote the connected component containing 0 by G. By Sard's theorem, for almost all R, $0 < R < \infty$, the boundary of G consists of a finite number of smooth curves.

For sufficiently small R the domain G is obtained from some domain after the neighborhoods of zero of the polynomial P have been deleted. If R is sufficiently large then G is a simply connected neighborhood of zero. Consider the boundary of G:

$$S = \{t \in \mathbb{C} : R|t|^{k+1} = |P(t)|\} = \{t \in \mathbb{C} : R^2 t^{k+1} \bar{t}^{k+1} = P(t)\overline{P(t)}\}.$$

Denote $w = \dfrac{1}{t}$. Then on S we have the equality

$$\frac{R^2 t^{k+1}}{w^{k+1}} = P(t)\tilde{P}\left(\frac{1}{w}\right),$$

where $\tilde{P}(t) = \displaystyle\sum_{j=0}^{k} \bar{a}_j t^j$, if $P(t) = \displaystyle\sum_{j=0}^{k} a_j t^j$. Then

$$\overline{P(t)} = \tilde{P}(\bar{t}) = \tilde{P}\left(\frac{1}{w}\right) = \frac{1}{w^k} P^s(w),$$

where $P^s(w) = \displaystyle\sum_{j=0}^{k} \bar{a}_j w^{k-j}$. Then on S we have the equality

$$\frac{R^2 t^{k+1}}{w^{k+1}} = P(t)P^s(w)\frac{1}{w^k}.$$

Hence

$$R^2 t^{k+1} = P(t)wP^s(w),$$

i.e.,

$$wP^s(w) = \frac{R^2 t^{k+1}}{P(t)}. \tag{4.2.6}$$

From the form of G we also find that in G

$$\frac{R|t|^{k+1}}{|P(t)|} < 1. \tag{4.2.7}$$

Consider the function $\zeta = \varphi(w) = wP^s(w)$, then $\varphi'(0) = P^s(0) \neq 0$, since the polynomial P has the degree k. Therefore, the function φ is a conformal map of some neighborhood of zero U_w on the neighborhood of zero V_ζ. Therefore, there is an inverse function $w = \varphi^{-1}(\zeta) : V_\zeta \to U_w$. From equality (4.2.6) we have

$$w = \varphi^{-1}\left(\frac{R^2 t^{k+1}}{P(t)}\right) \quad \text{on} \quad S.$$

Since, by virtue of inequality (4.2.7)

$$R\left|\frac{R t^{k+1}}{P(t)}\right| < R \quad \text{in} \quad G,$$

then, for sufficiently small R, the point $\dfrac{R^2 t^{k+1}}{P(t)}$ is in U_w and the denominator $P(t) \neq$

0 in G. Therefore, the function $w = \varphi^{-1}\left(\dfrac{R^2 t^{k+1}}{P(t)}\right)$ is holomorphic in G. Thus

$w = \dfrac{1}{t}$ extends holomorphically to G. Hence, the following assertion is true.

Lemma 4.2.5 *For a domain G of the form (4.2.5) for sufficiently small R the function $\dfrac{1}{t}$ extends holomorphically from the boundary ∂G into G.*

Consider a moved domain

$$G_a = \{t \in \mathbb{C} : R|t + a|^{k+1} < |P_1(t)|\}, \tag{4.2.8}$$

where $P_1(t) = P(t + a)$. Then on $S_a = \partial G_a$ we have the equality

$$R^2 (t + a)^{k+1}(\bar{t} + \bar{a})^{k+1} = P_1(t)\overline{P_1(t)}$$

and if $\bar{t} = \dfrac{1}{w}$, then

$$R^2 (t + a)^{k+1}(1 + \bar{a}w)^{k+1} = P_1(t)w P_1^s(w).$$

The function $\zeta = \psi(w) = \dfrac{w P_1^s(w)}{(1 + \bar{a}w)^{k+1}}$ maps conformally the neighborhood of zero U_w onto the neighborhood of zero V_ζ, therefore, as above, the function

$$w = \psi^{-1}(\zeta) = \psi^{-1}\left(\frac{R^2 (t + a)^{k+1}}{P_1(t)}\right)$$

extends holomorphically to G_a for sufficiently small R.

Lemma 4.2.6 *For a domain G_a of the form (4.2.8) for sufficiently small R the function $\dfrac{1}{t}$ extends holomorphically from the boundary ∂G into G.*

It is easy to give examples of domains in the complex plane, for which Lemma 4.2.3 is not satisfied. An ordinary ellipse

$$G = \left\{ t = x + iy \in \mathbb{C} : \frac{x^2}{a^2} + \frac{y^2}{b^2} < 1 \right\}, \quad a, b > 0, \quad a \neq b$$

is one of such domains. It is not difficult to show that from the ellipse boundary equation we obtain

$$w = \frac{1}{\bar{t}} = \frac{t(a^2 + b^2) \pm 2ab\sqrt{z^2 - a^2 + b^2}}{4a^2 b^2 - z^2(b^2 - a^2)}.$$

This function has two singular points (branch points) in the ellipse G and two poles outside $\overline{G}$. Furthermore

$$\frac{1}{2\pi i} \int_{\partial G} \frac{dt}{\bar{t}} = - \left| \frac{a - b}{a + b} \right| \neq 0.$$

Consider a domain in $\mathbb{C}^n$ of the form

$$D = \{ z \in \mathbb{C}^n : R|z|^{k+1} < |P(z)| \}, \quad 0 < R < \infty, \qquad (4.2.9)$$

where $P(z)$ is the k-degree polynomial and $P(0) \neq 0$.

It is clear that this domain is bounded and contains a neighborhood of zero and the boundary D is smooth for almost all R. Intersections of this domain with the complex lines $l = \{ \zeta : \zeta_1 = b_1 t, \ldots, \zeta_n = b_n t \}$ form domains of the form (4.2.5), while intersections of the domain as in (4.2.9) with the lines

$$l = \{ \zeta : \zeta_1 = z_1 + b_1 t, \ldots, \zeta_n = z_n + b_n t \}$$

can be reduced to domains of the form (4.2.8) so by Theorem 4.2.2 and Lemmas 4.2.5 and 4.2.6 we obtain the assertion.

Corollary 4.2.1 *Let Γ be the germ of a generic manifold in D of the form (4.2.9) (R is sufficiently small), $0 \in \Gamma$ and the function $f \in (\partial D)$ has the one-dimensional holomorphic extension property along almost all complex lines $l \in \mathfrak{L}_\Gamma$, then there exists a function $Mf \in \mathscr{C}(\overline{D})$, that is holomorphic in D and coincides with the function f on the boundary ∂D.*

4.3 Sufficient Families of Complex Lines of Minimal Dimension

4.3.1 Preliminary Results

Consider a bounded simply connected domain $D \subset \mathbb{C}^n$ with a connected boundary ∂D of class $\mathscr{C}^2$. Let Γ be the germ of a complex manifold of dimension $(n-1)$ in $\mathbb{C}^n$, which lies outside $\overline{D}$. Having done the shift and unitary transformation, we can assume that $0 \in \Gamma$, $0 \notin \overline{D}$ and that the complex hypersurface Γ in some neighborhood U of 0 has the form

$$\Gamma = \{z \in U : z_n = \varphi(z'), \; z' = (z_1, \dots, z_{n-1})\},$$

where φ is the holomorphic function in a neighborhood of zero in $\mathbb{C}^{n-1}$ and $\varphi(0) = 0$, $\dfrac{\partial \varphi}{\partial z_k}(0) = 0$, $k = 1, \dots, n-1$.

Further on we will assume that there is a direction $b^0 \neq 0$ such that

$$\langle b^0, \bar{\zeta} \rangle \neq 0 \quad \text{for all} \quad \zeta \in \overline{D}. \tag{4.3.1}$$

Let $\mathfrak{L}_\Gamma$ denote a set of complex lines of the form

$$l = \{\zeta \in \mathbb{C}^n : \zeta_j = z_j + b_j t, \; j = 1, \dots, n, \; t \in \mathbb{C}\}, \tag{4.3.2}$$

passing through the point $z \in \Gamma$ in the direction of $b \in \mathbb{CP}^{n-1}$.

By Sard's theorem, the intersection $\partial D \cap l$ is a finite set of piecewise smooth curves for almost all $z \in \mathbb{C}^n$ and almost all $b \in \mathbb{CP}^{n-1}$ (excluding the degenerate case when $\partial D \cap l = \varnothing$). Consider the Bochner–Martinelli integral of the form (1.2.4) of the function f:

$$Mf(z) = \int_{\partial D} f(\zeta) U(\zeta, z), \quad z \notin \partial D,$$

where $U(\zeta, z)$ is the Bochner–Martinelli kernel of the form (1.1.1).

From Lemma 4.1.1 we obtain the assertion.

Proposition 4.3.1 *If a function $f \in \mathscr{C}(\partial D)$ has the one-dimensional holomorphic extension property along the complex lines of the family $\mathfrak{L}_\Gamma$, then for any multi-indices α the following holds*

$$Mf\big|_\Gamma = 0, \qquad \frac{\partial^\alpha Mf}{\partial z^\alpha}\bigg|_\Gamma = 0 \tag{4.3.3}$$

Consider a kernel of the form

$$U_{\mathbb{C}}(\zeta, z, w) = \frac{(n-1)!}{(2\pi i)^n} \sum_{k=1}^{n} (-1)^{k-1} \frac{\bar{\zeta}_k - w_k}{\left(\sum_{j=1}^{n} (\zeta_j - z_j)(\bar{\zeta}_j - w_j)\right)^n} d\bar{\zeta}[k] \wedge d\zeta$$

and the integral

$$\Phi f(z, w) = \int_{\partial D_\zeta} f(\zeta) U_{\mathbb{C}}(\zeta, z, w). \qquad (4.3.4)$$

It is clear, that the function $\Phi f(z, w)$ is holomorphic in a neighborhood of $(0, 0) \in \mathbb{C}^{2n}$ because $0 \notin \overline{D}$.

Lemma 4.3.1 *Suppose D is simply connected and satisfies (4.3.1), then there is an unbounded open and connected set $\Omega \subset \mathbb{C}^{2n}$, $(0, 0) \in \Omega$, in which the function $\Phi f(z, w)$ is defined and holomorphic, and there are $\varepsilon > 0$, $R > 0$ such that points of the form (tb, w) belong to Ω at $|w| < \varepsilon$, $|b - b^0| < \varepsilon$ and $|t| > R$ $(t \in \mathbb{C})$.*

Proof Consider the denominator of the kernel $U_{\mathbb{C}}(\zeta, tb^0, w)$:

$$\psi = \langle \zeta - tb^0, \bar{\zeta} - w \rangle = |\zeta|^2 - \langle \zeta, w \rangle - t\langle b^0, \bar{\zeta} \rangle + t\langle b^0, w \rangle.$$

Find conditions under which $\psi \neq 0$. Let

$$\max_{\overline{D}} |\zeta| = M, \quad \min_{\overline{D}} |\zeta| = m > 0, \quad \min_{\overline{D}} |\langle b^0, \bar{\zeta} \rangle| = c > 0.$$

Then for $|w| < \varepsilon$ we obtain that

$$|\langle b^0, w \rangle| \leq |b^0| \, |w| \leq \varepsilon |b^0|, \quad |\langle \zeta, w \rangle| \leq M\varepsilon.$$

Equating ψ to zero, we obtain

$$t = \frac{|\zeta|^2 - \langle \zeta, w \rangle}{\langle b^0, \bar{\zeta} \rangle - \langle b^0, w \rangle}. \qquad (4.3.5)$$

We estimate the numerator in (4.3.5) for $\zeta \in \overline{D}$

$$||\zeta|^2 - \langle \zeta, w \rangle| \geq |\zeta|^2 - |\langle \zeta, w \rangle| \geq m^2 - M\varepsilon > 0$$

at $\varepsilon < \dfrac{m^2}{M}$. Now estimate the denominator in (4.3.5) for $\zeta \in \overline{D}$

$$|\langle b^0, \bar{\zeta} \rangle - \langle b^0, w \rangle| \geq |\langle b^0, \bar{\zeta} \rangle| - |\langle b^0, w \rangle| \geq c - |b^0|\varepsilon > 0$$

at $\varepsilon < \dfrac{c}{|b^0|}$. Thus, if $|w| \le \varepsilon$ (where ε satisfies both inequalities) the image $\overline{D}$ under the mapping

$$\frac{|\zeta|^2 - \langle \zeta, w \rangle}{\langle b^0, \bar{\zeta} \rangle - \langle b^0, w \rangle}$$

is some compact $K_{b^0,\varepsilon}$ in the complex plane $\mathbb{C}$, not containing the point 0. Since D is simply connected, then the complement $K_{b^0,\varepsilon}$ is connected, i.e., 0 lies in the unbounded component of the complement.

Therefore, when $t \notin K_{b^0,\varepsilon}$ the function $\psi \ne 0$. In particular, this is true for $t = 0$, $|w| \le \varepsilon$ and $|t| > R$, where R is sufficiently large. It is clear that for all b with $|b - b^0| < \varepsilon$ this argument remains valid, reducing, if necessary, ε. So, the existence of the domain Ω is ensured.   □

From Lemma 4.3.1 and the kernel form $U_{\mathbb{C}}(\zeta, z, w)$ it follows that the function $\Phi f(z, w)$ and all its derivatives tend to zero as $|z| \to \infty$, $|w| \to \infty$ and $(z, w) \in \Omega$. Note that $\Phi f(z, \bar{z}) = Mf(z)$ and $\dfrac{\partial^{\alpha} \Phi f}{\partial z^{\alpha}}\Big|_{w=\bar{z}} = \dfrac{\partial^{\alpha} Mf}{\partial z^{\alpha}}$. We introduce the differential operator in $\mathbb{C}^{2n}$

$$\Delta_{\mathbb{C}} = \Delta_{\mathbb{C}}(z, w) = \sum_{k=1}^{n} \frac{\partial^2}{\partial z_k \partial w_k}.$$

When $w = \bar{z}$ we obtain the Laplace operator $\Delta = \sum_{k=1}^{n} \dfrac{\partial^2}{\partial z_k \partial \bar{z}_k}$. Let $\Gamma_{\mathbb{C}}$ denote a complex manifold in $\mathbb{C}^{2n}$ of the form

$$\Gamma_{\mathbb{C}} = \{(z, w) \in U \times U : z_n = \varphi(z'), \ w_n = \overline{\varphi(\bar{w}')}\}.$$

Choosing U sufficiently small, we can assume that the function $\Phi f(z, w)$ is defined and holomorphic in $U \times U$. When $w = \bar{z}$ we get $\Gamma_{\mathbb{C}} = \Gamma$ or $\Gamma_{\mathbb{C}}|_{w=\bar{z}} = \Gamma$.

Proposition 4.3.2 *If equalities (4.3.3) are satisfies for the function $Mf(z)$, so are the equalities*

$$\Phi f\big|_{\Gamma_{\mathbb{C}}} = 0, \qquad \frac{\partial^{\alpha} \Phi f}{\partial z^{\alpha}}\bigg|_{\Gamma_{\mathbb{C}}} = 0 \qquad (4.3.6)$$

for the function $\Phi f(z, w)$ and for all multi-indices α.

Proof The manifold $G = \{(z, w) \in \mathbb{C}^{2n} : w = \bar{z}\}$ is a generic in $\mathbb{C}^{2n}$, i.e., the complex linear span of the tangent space $T_z(G)$ coincides with the tangent space $T_z(\mathbb{C}^{2n})$ for each point $(z, w) \in G$. Indeed, writing G in the form

$$\{(z, w) : \mathrm{Re}(z_j - w_j) = 0, \ \mathrm{Re}(i(z_j + w_j)) = 0, \ j = 1, \ldots, n\}$$

and making a nonsingular complex linear transformation $\tilde{z}_j = z_j - w_j$, $\tilde{w}_j = i(z_j + w_j)$, $j = 1, \ldots, n$, the manifold G can be rewritten as

$$G = \{(\tilde{z}, \tilde{w}) : \operatorname{Re} \tilde{z}_j = 0, \ \operatorname{Re} \tilde{w}_j = 0, \ j = 1, \ldots, n\}.$$

So G is, obviously, a generic in $\mathbb{C}^{2n}$ in the new coordinates. Since the generatedness property is not affected by holomorphic transformations, then G is a generic in the old coordinates as well.

Then, obviously, the submanifold $\{(z, w) \in M : z_n = 0, \ w_n = 0\}$ is a generic in the manifold $\{(z, w) \in \mathbb{C}^{2n} : z_n = 0, \ w_n = 0\}$. Hence the manifold Γ, written as

$$\{(z, w) \in M \cap (U \times U) : z_n = \varphi(z'), \ w_n = \overline{\varphi(\bar{w}')}\}$$

is a generic in $\Gamma_{\mathbb{C}}$. Here again we use the statement that the generatedness property does not change under holomorphic transformations.

To prove the proposition, it remains to use the statement that the generic manifold is a uniqueness set for holomorphic functions [67]. $\qquad\square$

Lemma 4.3.2 *The kernel $U_{\mathbb{C}}(\zeta, z, w)$ satisfies the condition*

$$\Delta_{\mathbb{C}}(z, w) U_{\mathbb{C}}(\zeta, z, w) = 0$$

outside the zeros of the denominator of this kernel.

Proof It suffices to verify this equality for functions of the form

$$\frac{1}{\left(\sum_{j=1}^{n} (\zeta_j - z_j)(\bar{\zeta}_j - w_j)\right)^{n-1}}.$$

Since

$$\frac{\partial}{\partial w_k}\left(\frac{1}{\left(\sum_{j=1}^{n} (\zeta_j - z_j)(\bar{\zeta}_j - w_j)\right)^{n-1}}\right) = \frac{(n-1)(\zeta_k - z_k)}{\left(\sum_{j=1}^{n} (\zeta_j - z_j)(\bar{\zeta}_j - w_j)\right)^{n}},$$

then

$$\frac{\partial^2}{\partial z_k \partial w_k}\left(\frac{1}{\left(\sum_{j=1}^{n} (\zeta_j - z_j)(\bar{\zeta}_j - w_j)\right)^{n-1}}\right)$$

$$= \frac{\partial}{\partial z_k}\left(\frac{(n-1)(\zeta_k - z_k)}{\left(\sum_{j=1}^{n} (\zeta_j - z_j)(\bar{\zeta}_j - w_j)\right)^{n}}\right) = (1-n)\times$$

$$\times \frac{\left(\sum_{j=1}^{n}(\zeta_j - z_j)(\bar{\zeta}_j - w_j)\right)^n - n(\zeta_k - z_k)(\bar{\zeta}_k - w_k)\left(\sum_{j=1}^{n}(\zeta_j - z_j)(\bar{\zeta}_j - w_j)\right)^{n-1}}{\left(\sum_{j=1}^{n}(\zeta_j - z_j)(\bar{\zeta}_j - w_j)\right)^{2n}}$$

$$= \frac{(1-n)\left(\sum_{j=1}^{n}(\zeta_j - z_j)(\bar{\zeta}_j - w_j) - n(\zeta_k - z_k)(\bar{\zeta}_k - w_k)\right)}{\left(\sum_{j=1}^{n}(\zeta_j - z_j)(\bar{\zeta}_j - w_j)\right)^{n+1}}.$$

And hence,

$$\Delta_{\mathbb{C}}\left(\frac{1}{\left(\sum_{j=1}^{n}(\zeta_j - z_j)(\bar{\zeta}_j - w_j)\right)^{n-1}}\right)$$

$$= \frac{(1-n)n\left(\sum_{j=1}^{n}(\zeta_j - z_j)(\bar{\zeta}_j - w_j) - \sum_{k=1}^{n}(\zeta_k - z_k)(\bar{\zeta}_k - w_k)\right)}{\left(\sum_{j=1}^{n}(\zeta_j - z_j)(\bar{\zeta}_j - w_j)\right)^{n+1}} = 0.$$

□

Lemma 4.3.3 *The function* $\Phi f(z,w)$ *satisfies* $\Delta_{\mathbb{C}}\Phi f(z,w) = 0$ *in its domain of definition.*

Lemma 4.3.4 *The relation*

$$\Delta_{\mathbb{C}}(hg) = h\Delta_{\mathbb{C}}g + g\Delta_{\mathbb{C}}h + \sum_{k=1}^{n}\frac{\partial h}{\partial z_k}\frac{\partial g}{\partial w_k} + \sum_{k=1}^{n}\frac{\partial h}{\partial w_k}\frac{\partial g}{\partial z_k}$$

holds true for holomorphic functions h *and* g *in* $\mathbb{C}^{2n}$.

We make a holomorphic change of the variables in the neighborhood of the point $(0,0) \in \mathbb{C}^{2n}$:

$$\begin{cases} z_1 = u_1 \\ \cdots\cdots \\ z_{n-1} = u_{n-1} \\ z_n = u_n + \varphi(u') \end{cases} \qquad \begin{cases} w_1 = v_1 \\ \cdots\cdots \\ w_{n-1} = v_{n-1} \\ w_n = v_n + \overline{\varphi(\bar{v}')}. \end{cases}$$

Let U^* be the image of the neighborhood U under this change. A reverse change of the variables looks as follows:

$$\begin{cases} u_1 = z_1 \\ \cdots\cdots \\ u_{n-1} = z_{n-1} \\ u_n = z_n - \varphi(z') \end{cases} \qquad \begin{cases} v_1 = w_1 \\ \cdots\cdots \\ v_{n-1} = w_{n-1} \\ v_n = w_n - \overline{\varphi(\bar{w}')}. \end{cases}$$

Then under this change, $\Gamma_{\mathbb{C}}$ will become part of the plane

$$\Gamma_{\mathbb{C}}^* = \{(u, v) \in U^* \times U^* : u_n = 0, \ v_n = 0\}.$$

And the plane $\Gamma_{\mathbb{C}}^*$ will become part of the hyperplane

$$\Gamma^* = \{u \in U^* : u_n = 0\}$$

at $v = \bar{u}$.

Lemma 4.3.5 *Let* $\Phi^*(u, v) = \Phi f(z(u), w(v))$. *Equality (4.3.6) can be rewritten as*

$$\Phi^* \big|_{\Gamma_{\mathbb{C}}^*} = 0, \tag{4.3.7}$$

$$\frac{\partial^\alpha \Phi^*}{\partial u^\alpha} \bigg|_{\Gamma_{\mathbb{C}}^*} = 0, \tag{4.3.8}$$

$$\frac{\partial^{\alpha+\beta'} \Phi^*}{\partial u^\alpha \partial v^{\beta'}} \bigg|_{\Gamma_{\mathbb{C}}^*} = 0 \tag{4.3.9}$$

for all multi-indices α and multi-indices β' of the form $\beta' = (\beta_1, \ldots, \beta_{n-1}, 0)$.

Proof Equality (4.3.7) is obvious. Since the derivatives of Φ^* in the variables u_j, $j = 1, \ldots, n$, are expressed only in terms of the derivatives of the function Φf by z_k, $k = 1, \ldots, n$, equality (4.3.8) follows from (4.3.6). We obtain (4.3.9) from Eqs. (4.3.7), (4.3.8) and the type of the plane $\Gamma_{\mathbb{C}}^*$. $\qquad \square$

Consider expansion of the function $\Phi^*(u, v)$ in a Taylor series in the variable v_n at $v_n = 0$

$$\Phi^*(u, v) = \sum_{k=0}^{\infty} \frac{1}{k!} \frac{\partial^k \Phi^*(u, v', 0)}{\partial v_n^k} v_n^k. \tag{4.3.10}$$

Lemma 4.3.6 *Let conditions (4.3.7)–(4.3.9) be satisfied for the function $\Phi^*(u, v)$, then the coefficient $\Phi^*(u, v', 0) = 0$ in the series (4.3.10) and, therefore,*

$$\Phi^*(u, v) = v_n \Psi(u, v).$$

Proof We expand the function $\Phi^*(u, v)$ in a Taylor series in the neighborhood of $(0, 0)$ in the variables u and v:

$$\Phi^*(u, v) = \sum_{\|\alpha\| \geq 0, \|\beta\| \geq 0} c_{\alpha,\beta} u^\alpha v^\beta,$$

where $u^\alpha = u_1^{\alpha_1} \cdots u_n^{\alpha_n}$, $v^\beta = v_1^{\beta_1} \cdots v_n^{\beta_n}$.

We show that in this series there are no monomials of the form $c_{\alpha,\beta'} u^\alpha v^{\beta'}$, where $\beta' = (\beta_1, \ldots, \beta_{n-1}, 0)$. Indeed, if $c_{\alpha,\beta'} \neq 0$, then applying the differential operator $\dfrac{\partial^{\alpha+\beta'}}{\partial u^\alpha \partial v^{\beta'}}$ to the function $\Phi^*(u, v)$ and substituting $u_n = 0$, $v_n = 0$, we obtain a power series in the variables u', v' with a non-zero free term. This contradicts equalities (4.3.7)–(4.3.9). □

4.3.2 Sufficient Families Associated with a Complex Hypersurface

Theorem 4.3.1 ([55]) *If the function $\Phi f(z, w)$ satisfies the conditions (4.3.6), then $\Phi f(z, w) \equiv 0$ in the neighborhood of $(0, 0)$.*

Proof We go back from (u, v) to the old variables z and w. By Lemma 4.3.6 we obtain the expansion

$$\Phi f(z, w) = \sum_{k=1}^{n} \frac{1}{k!} \frac{\partial^k \Phi^*}{\partial v_n^k}(u, v', 0) v_n^k$$

$$= \sum_{k=1}^{n} \frac{1}{k!} \left(w_n - \overline{\varphi(\overline{w}')}\right)^k \frac{\partial^k \Phi f}{\partial w_n^k}(z, w', \overline{\varphi(\overline{w}')}), \qquad (4.3.11)$$

since $\dfrac{\partial \Phi^*}{\partial v_n} = \dfrac{\partial \Phi f}{\partial w_n}$. We apply the operator $\Delta_{\mathbb{C}}$ to equality (4.3.11), and obtain

$$0 = \sum_{k=1}^{\infty} \frac{1}{k!} \Delta_{\mathbb{C}} \left[\left(w_n - \overline{\varphi(\overline{w}')}\right)^k \frac{\partial^k \Phi}{\partial w_n^k}(z, w', \overline{\varphi(\overline{w}')})\right].$$

Next we regroup the resulting series with respect to the powers $\left(w_n - \overline{\varphi(\overline{w}')}\right)^k$, and obtain

$$0 = \sum_{k=0}^{\infty} \left(w_n - \overline{\varphi(\overline{w}')}\right)^k c_k(z, w').$$

All the coefficients $c_k(z, w') \equiv 0$ by the uniqueness of decomposition in this series, which follows from the uniqueness property of decomposition of the power series in v_n obtained in the new variables u, v.

We compute successively $\Delta_{\mathbb{C}}$ from $\left(w_n - \overline{\varphi(\overline{w'})}\right)^k \dfrac{\partial^k \Phi f}{\partial w_n^k}$, for $k = 1$. By Lemma 4.3.4 we have

$$
\Delta_{\mathbb{C}}\left[\left(w_n - \overline{\varphi(\overline{w'})}\right) \frac{\partial \Phi f}{\partial w_n}\left(z, w', \overline{\varphi(\overline{w'})}\right) \right]
$$

$$
= \left(w_n - \overline{\varphi(\overline{w'})}\right) \Delta_{\mathbb{C}}\left(\frac{\partial \Phi f}{\partial w_n}\right) + \frac{\partial}{\partial z_n}\left(\frac{\partial \Phi f}{\partial w_n}\right) - \sum_{k=1}^{n-1} \frac{\partial \overline{\varphi}}{\partial w_k} \frac{\partial}{\partial z_k}\left(\frac{\partial \Phi f}{\partial w_n}\right).
$$

Hence

$$
c_0(z, w') = \left(\frac{\partial}{\partial z_n} - \sum_{k=1}^{n-1} \frac{\partial \overline{\varphi}}{\partial w_k} \frac{\partial}{\partial z_k} \right)\left(\frac{\partial \Phi f}{\partial w_n} \right) \equiv 0. \tag{4.3.12}
$$

Thus, for fixed w' the derivatives of the function $\dfrac{\partial \Phi f}{\partial w_n}\left(z, w', \overline{\varphi(\overline{w'})}\right)$ in the direction of the vector $s = \left(-\dfrac{\partial \overline{\varphi}}{\partial w_1}, \dots, -\dfrac{\partial \overline{\varphi}}{\partial w_{n-1}}, 1 \right)$ are identically equal to zero. We fix a point $\left(z^0, w'^0, \overline{\varphi(\overline{w'^0})}\right)$ in the domain Ω from Lemma 4.3.1 such that the complex line

$$
\left\{ z : z_j = z_j^0 - \frac{\partial \overline{\varphi}}{\partial w_j} t, \, j = 1, \dots, n-1, \, z_n = z_n^0 + t, \, t \in \mathbb{C} \right\}
$$

does not intersect $\overline{D}$ for sufficiently small $|w|$. This can be achieved by taking $|z^0|$ large enough (see Lemma 4.3.1).

By Eq. (4.3.12) on the complex line

$$
l_{z^0, s} = \left\{ \left(z, w'^0, \overline{\varphi(\overline{w'^0})}\right) \in \mathbb{C}^n \times U : z_j = z_j^0 - \frac{\partial \overline{\varphi}}{\partial w_j} t, \, j = 1, \dots, n-1, \right.
$$

$$
\left. z_n = z_n^0 + t, \, t \in \mathbb{C} \right\}
$$

the derivative $\dfrac{\partial}{\partial s}\left(\dfrac{\partial \Phi f}{\partial w_n}\right) = \dfrac{d}{dt}\left(\dfrac{\partial \Phi f}{\partial w_n}\right) = 0$ for sufficiently small $|t|$. The domain Ω was chosen in Lemma 4.3.1 so that the function $\Phi f(z, w)$ is holomorphic in Ω, i.e., the denominator of the kernel $U_{\mathbb{C}}(\zeta, z, w)$ remains nonzero for all $\zeta \in \overline{D}$ and all $(z, w) \in \Omega$.

Consider this denominator on the line $l_{z^0,s}$. We have

$$\sum_{j=1}^{n-1} (\zeta_j - z_j)(\bar{\zeta}_j - w_j^0) + (\zeta_n - z_n)(\bar{\zeta}_n - \overline{\varphi(\bar{w}'^0)})$$

$$= \sum_{j=1}^{n-1} (\zeta_j - z_j^0)(\bar{\zeta}_j - w_j^0) + (\zeta_n - z_n^0)(\bar{\zeta}_n - \overline{\varphi(\bar{w}'^0)})$$

$$+ t\left(\bar{\zeta}_n - \overline{\varphi(\bar{w}'^0)} - \sum_{j=1}^{n-1} \frac{\partial \bar{\varphi}}{\partial w_j}(\bar{\zeta}_j - w_j^0) \right).$$

The expression

$$\sum_{j=1}^{n-1} (\zeta_j - z_j^0)(\bar{\zeta}_j - w_j^0) + (\zeta_n - z_n^0)(\bar{\zeta}_n - \overline{\varphi(\bar{w}'^0)}) \neq 0$$

for all $\zeta \in \overline{D}$. So the values of the expression on the complex plane form a compact set not containing 0 at $\zeta \in \overline{D}$ and w' from a compact neighborhood of the point $0 \in \mathbb{C}^n$. We can assume (making shift in z), that $z^0 = 0$.

For $z^0 = 0$, $w'^0 = 0$ the expression

$$t\left(\bar{\zeta}_n - \overline{\varphi(\bar{w}'^0)} - \sum_{j=1}^{n-1} \frac{\partial \bar{\varphi}}{\partial w_j}(\bar{\zeta}_j - w_j^0) \right) = t\bar{\zeta}_n.$$

Since $\bar{\zeta}_n \neq 0$ on $\overline{D}$, then the values of the expression

$$\bar{\zeta}_n - \overline{\varphi(\bar{w}'^0)} - \sum_{j=1}^{n-1} \frac{\partial \bar{\varphi}}{\partial w_j}(\bar{\zeta}_j - w_j^0)$$

in the complex plane $\mathbb{C}$ also form a compact set not containing 0 for $\zeta \in \overline{D}$ and z, w' from a compact neighborhood of $(0,0)$. Therefore, the denominator of the kernel $U_\mathbb{C}(\zeta, z, w)$ on the line $l_{z^0,s}$ can only become zero for the t, lying on a compact of the complex plane that does not contain zero. Thus, the denominator is not zero outside this compact, and hence the function $\Phi f(z, w)$ is holomorphic in the complex line $l_{z^0,s}$ except for some compact set $K_{z^0,s}$ not containing zero. Since the addition of this compact is connected, then $(0,0)$ lies in the unbounded component of a holomorphy set $\Phi f(z, w', 0)$ for all z and w' in some neighborhood of $(0,0)$.

Hence $\dfrac{d}{dt}\left(\dfrac{\partial \Phi f}{\partial w_n}\right) = 0$ in $\mathbb{C} \setminus K_{z^0,s}$. So $\dfrac{\partial \Phi f}{\partial w_n}\Big|_{\mathbb{C}\setminus K_{z^0,s}} = \text{const.}$ From (4.3.4) of

the function $\Phi f(z,w)$ we get that $\Phi f\big|_{\mathbb{C}\setminus K_{z^0,s}} \to 0$ and $\dfrac{\partial \Phi f}{\partial w_n}\Big|_{\mathbb{C}\setminus K_{z^0,s}} \to 0$ as $|t| \to \infty$.

Therefore $\dfrac{\partial \Phi f}{\partial w_n}\Big|_{\mathbb{C}\setminus K_{z^0,s}} = 0$, and so we obtain that $\dfrac{\partial \Phi f}{\partial w_n}\Big|_{\mathbb{C}\setminus K_{z^0,s}} \equiv 0$ for all z^0 and w' in some neighborhood of 0. From Lemma 4.3.1 we find that the derivative $\dfrac{\partial \Phi f}{\partial w_n}(z,w',0) = 0$ in the unbounded component of its domain of definition.

Therefore, series (4.3.11) begins with $k = 2$. Applying the same argument to the expression $\Delta_{\mathbb{C}}\left[\left(w_n - \varphi(w')\right)^2 \dfrac{\partial^2 \Phi f}{\partial w_n^2}\right]$, we get that $\dfrac{\partial^2 \Phi f}{\partial w_n^2}\Big|_{\mathbb{C}\setminus K_{z^0,s}} \equiv 0$ etc. $\qquad\square$

Corollary 4.3.1 *Let the function $Mf(z)$ satisfy conditions (4.3.3), then $Mf(z) \equiv 0$ in a neighborhood of zero.*

Theorem 4.3.2 ([55]) *Let the function $f \in \mathscr{C}(\partial D)$ and conditions (4.3.3) be satisfied for its Bochner–Martinelli integral $Mf(z)$, then f extends holomorphically to the domain D.*

Proof follows from Corollary 4.3.1 and Corollary 15.5 from [45]. $\qquad\square$

Theorem 4.3.3 ([55]) *Let D be a simply connected bounded domain and condition (4.3.1) be fulfilled. If the function $f \in \mathscr{C}(\partial D)$ and has the one-dimensional holomorphic extension property along the complex lines of the family $\mathfrak{L}_\Gamma$, then f extends holomorphically into the domain D.*

Proof follows from Proposition 4.3.1 and Theorem 4.3.2. $\qquad\square$

If Γ is the germ of a complex hypersurface in $\mathbb{C}^n$, then condition (4.3.1) becomes superfluous. Indeed, let $\tilde{\Gamma}$ be a complex hypersurface (a complex manifold of dimension $(n-1)$) in $\mathbb{C}^n$ and $\Gamma = \tilde{\Gamma} \cap U$. The surface $\tilde{\Gamma}$ is a connected unbounded set in $\mathbb{C}^n$. Still, Γ does not intersect $\overline{D}$, whereas $\tilde{\Gamma}$ can intersect $\overline{D}$. Then $\tilde{\Gamma} \cap D$ is a relatively compact open set on $\tilde{\Gamma}$. Let $\tilde{\Gamma} \setminus (\tilde{\Gamma} \cap \overline{D})$ be connected.

We assume that $f \in \mathscr{C}(\partial D)$ has the one-dimensional property of holomorphic extension along the complex lines $l \in \mathfrak{L}_\Gamma$, then Proposition 4.3.1 is true for the Bochner–Martinelli integral F, i.e., equality (4.3.3). Due to the integral being real-analytic, this condition is fulfilled on the whole set $\tilde{\Gamma} \setminus (\tilde{\Gamma} \cap \overline{D})$. Since the set is unbounded, there exist a point $z^0 \in \tilde{\Gamma}$ and a direction b^0 such that $\langle b^0, \bar{\zeta} - \bar{z}^0 \rangle \neq 0$ for all $\zeta \in \overline{D}$. Thus we come to our original terms for the domain D and the germ $\tilde{\Gamma}$ already in the neighborhood of the point z^0. Thus, the assertion is true

Theorem 4.3.4 ([55]) *Let D be a simply connected bounded domain with a connected smooth boundary, and $\tilde{\Gamma}$ be a complex hypersurface in $\mathbb{C}^n$, provided the set $\tilde{\Gamma} \setminus (\tilde{\Gamma} \cap \overline{D})$ is connected and $\Gamma = \tilde{\Gamma} \cap U$ does not intersect $\overline{D}$. If the function $f \in \mathscr{C}(\partial D)$ and has the one-dimensional holomorphic extension property along the complex lines of the family $\mathfrak{L}_\Gamma$, then f extends holomorphically into the domain D.*

Theorems 4.3.3 and 4.3.4, generally speaking, are not true for complex hypersurfaces, lying in a domain. See Example 18.2.

4.3.3   Sufficient Families on a Generic Manifold, Laying on the Complex Hypersurface

Let $\gamma \subset \Gamma$, $0 \in \gamma$ and γ be a generic manifold of class $\mathscr{C}^\infty$ in Γ, i.e., the complex linear span of the tangent space $T_z(\gamma)$ coincides with the tangent space $T_z(\Gamma)$ for every point $z \in \Gamma$. Note that the real dimension of γ is at least $(n - 1)$. We use $\mathfrak{L}_\gamma$ to denote a set of complex lines intersecting γ.

Theorem 4.3.5 ([55]) *Let D and Γ satisfy the conditions of Theorem 4.3.3 or Theorem 4.3.4. If the function $f \in \mathscr{C}(\partial D)$ and has the one-dimensional holomorphic extension property along the complex lines of the family $\mathfrak{L}_\gamma$, then f extends holomorphically into the domain D.*

Proof Let $Mf(z)$ be an integral of the form (1.2.4), then, by Lemma 4.1.1 the equalities

$$Mf(z) = 0, \qquad \frac{\partial^\alpha Mf}{\partial z^\alpha}(z) = 0 \qquad (4.3.13)$$

hold for all $z \in \gamma$ and all multi-indices α. Again making the change of variables

$$\begin{cases} z_1 = u_1 \\ \cdots\cdots \\ z_{n-1} = u_{n-1} \\ z_n = u_n + \varphi(u') \end{cases}$$

we get that Γ goes into $\Gamma^* = \{u \in \mathbb{C}^n : u_n = 0\}$ and γ goes into the generic manifold $\gamma^* \subset \Gamma^*$. Since the u-derivatives are expressed in terms of the derivatives of the same order with respect to z, then from condition (4.3.13) we obtain

$$F^*(u) = 0, \qquad \frac{\partial^\alpha F^*}{\partial u^\alpha}(u) = 0, \quad u \in \gamma^*,$$

where $F^* = Mf(z(u))$.

We apply Lemma 4.2.1 to the functions $F^*(u', 0)$ and $\dfrac{\partial^{\alpha_n} F^*}{\partial u^{\alpha_n}}(u', 0)$. These are equal to zero on γ^*. We get $\dfrac{\partial^\alpha F^*}{\partial u^\alpha}(u) = 0$ on Γ^*. Making inverse change in these equations, we find that for the function Mf conditions (4.3.13) are fulfilled for $z \in \Gamma$.

To complete the proof of Theorem 4.3.5, it remains to apply Theorem 4.3.3 or Theorem 4.3.4.   □

Consider the germ γ of a real-analytic manifold of real dimension $(n - 1)$ in $\mathbb{C}^n \setminus \overline{D}$. We can assume that $0 \in \gamma$, and the manifold γ in some neighborhood U of the point 0 has the form

$$\gamma = \{z \in U : z_j = \psi_j(t_1, \ldots, t_{n-1}), \ t = (t_1, \ldots, t_{n-1}) \in V, \ j = 1, \ldots, n\}.$$
(4.3.14)

The functions $\psi_j(t)$ are real-analytic functions in the neighborhood V of the point $0 \in \mathbb{R}^{n-1}$, $\psi_j(0) = 0, j = 1, \ldots, n$, and the rank of the Jacobian matrix is equal to $(n - 1)$, i.e.,

$$\text{rang} \ \frac{\partial(\psi_1, \ldots, \psi_n)}{\partial(t_1, \ldots, t_{n-1})} = n - 1.$$
(4.3.15)

The functions $\psi_j(t)$ in the neighborhood V can be expanded in a Taylor series

$$\psi_j(t) = \sum_{\|\beta\| \geq 0} c_\beta t^\beta, \quad j = 1, \ldots, n,$$

where $\beta = (\beta_1, \ldots, \beta_{n-1})$, $t^\beta = t_1^{\beta_1} \cdots t_{n-1}^{\beta_{n-1}}$.
Consider the complexification Γ of the manifold γ, then Γ has the form

$$\Gamma = \{z \in U : z_j = \tilde{\psi}_j(t_1, \ldots, t_{n-1}), \ t = (t_1, \ldots, t_{n-1}) \in \mathbb{C}^{n-1}, \ j = 1, \ldots, n\},$$

where $\tilde{\psi}_j(t) = \sum_{\|\beta\| \geq 0} c_\beta t^\beta, t \in \mathbb{C}^{n-1}, j = 1, \ldots, n$. Condition (4.3.15) shows that Γ is a complex analytic manifold of dimension $(n - 1)$ in U. Let us show that γ is a generic manifold in Γ. Indeed, the tangent plane $T_{z^0}(\gamma)$ at the point $z^0 \in \gamma$ has the form

$$T_{z^0}(\gamma) = \left\{z \in \mathbb{C}^n : z_j = z_j^0 + \sum_{k=1}^{n-1} \frac{\partial \psi_j}{\partial t_k}(t^0)(t_k - t_k^0), \ t \in \mathbb{R}^{n-1}, \ j = 1, \ldots, n\right\},$$

where $\psi_j(t^0) = z_j^0, j = 1, \ldots, n$. Consider the complex linear span of $T_{z^0}(\gamma)$, obviously having the form

$$\left\{z \in \mathbb{C}^n : z_j = z_j^0 + \sum_{k=1}^{n-1} \frac{\partial \tilde{\psi}_j}{\partial t_k}(t^0)(t_k - t_k^0), \ t \in \mathbb{C}^{n-1}, \ j = 1, \ldots, n\right\}.$$

This set is exactly $T_{z^0}(\Gamma)$. Therefore, γ is a generic manifold in Γ.
We will also continue assuming that the domain D satisfies Condition (4.3.1).

Corollary 4.3.2 *Let the function* $f \in \mathscr{C}(\partial D)$ *and has the one-dimensional holomorphic extension property along the complex lines of the family* $\mathfrak{L}_\gamma$, *where* γ *is the germ of a real-analytic manifold of the form (4.3.14). Then the function* f *extends holomorphically to* D.

Consider a complex line l_0, passing through zero and intersecting the domain D. Assume that τ is a generic manifold in l_0.

Theorem 4.3.6 ([55]) *Let the domain* $D \subset \mathbb{C}^n$ *be a strictly convex with a boundary of class* $\mathscr{C}^\infty$ *and let the function* $f \in \mathscr{C}^\infty(\partial D)$ *has the one-dimensional holomorphic extension property along the complex lines of the family* $\mathfrak{L}_\tau$. *Then* f *extends holomorphically to the domain* D.

Proof Consider a complex two-dimensional plane Π_{ζ_0}, containing l_0 and passing through the point $\zeta^0 \in \partial D$. The intersection $D \cap \Pi_{\zeta^0}$ is a strictly convex domain in $\mathbb{C}^2$ with a boundary of class $\mathscr{C}^\infty$. The conditions of Theorem 4.3.5 are fulfilled for the domain $D \cap \Pi_{\zeta^0}$, so the function f extends holomorphically to $\partial D \cap \Pi_{\zeta^0}$ in $D \cap \Pi_{\zeta^0}$ up to the function $Mf(z)$. This function is uniquely defined in D, since the intersection of two different planes Π_{ζ^0} and Π_{w^0} coincides with l_0. And a continuation from $\partial D \cap l_0$ to $D \cap l_0$ is given by the Cauchy integral. Moreover, the function $Mf(z)$ is a function of class $\mathscr{C}^\infty$ in the domain D, since its holomorphic extension from $\partial D \cap \Pi_{\zeta^0}$ is given by the Bochner–Martinelli integral infinitely smoothly dependent on the parameter.

We choose a point $z^0 \in D \cap l_0$, then the function $Mf(z)$ is holomorphic in $D \cap l$, where l is an arbitrary complex line passing through the point z^0. Because the lines l and l_0 define a two-dimensional plane Π, the function $Mf(z)$ is holomorphic in $D \cap \Pi$. By the Forelli Theorem [71, Theorem 4.4.5] the function $Mf(z)$ is holomorphic in some neighborhood z^0. And, therefore, by Hartogs' extension theorem [73, Item 26] the function $Mf(z)$ is holomorphic in D. □

4.4 Functions with the One-Dimensional Holomorphic Extension Property in a Ball

Historically, the first statements about the functions with the one-dimensional holomorphic extension property along the complex lines were obtained in a ball by Agranovskii and Val'sky [4]. In the proof of their assertion, they used only the Morera property along complex lines intersecting the ball. So, in fact, they obtained a boundary Morera theorem. The ball thus became a model example, to obtain a series of statements, which were then extended to the case of domains of a more general form.

A number of papers dealt with classes of complex lines (or curves), sufficient for holomorphic extension into a ball. Thus, in the monograph by Rudin [71, Theorem 12.3.11] it is shown that if a function $f \in \mathscr{C}(\partial B)$ (B is a unit ball in $\mathbb{C}^n$ centered at the origin) has the one-dimensional holomorphic extension property along all complex lines that are lying at a distance r from the center of the ball for $0 < r < 1$, then this is a CR-function on ∂B. The proof is based on the description of $\mathscr{U}$-invariant subspaces of functions in the ball. This statement was generalized to strictly convex domains with a real-analytic boundary by Agranovskii [5].

Finer families of complex lines sufficient for holomorphic continuation, were studied in [22, 72]. In [23] Globevnik shows that a two-dimensional compact manifold of complex lines is a sufficient family for holomorphic extension into $\mathbb{C}^2$.

Agranovskii and Semenov in [3] prove the following result. Let R be a smooth analytic disk in $\mathbb{C}^n$, i.e., $R = \varphi(\triangle)$, where $\triangle$ is an open unit disk in the complex plane $\mathbb{C}$, and $\varphi : \triangle \to \mathbb{C}^n$ is a holomorphic map of class $\mathscr{C}^1(\overline{\triangle})$. Denote the Shilov boundary of R by γ, i.e., $\gamma = \varphi(\partial\triangle)$. We put

$$\Omega = \bigcup_{u \in \mathscr{U}(n)} u(\gamma),$$

where $\mathscr{U}(n)$ is the group of unitary transformations in $\mathbb{C}^n$. The set Ω is a spherical layer

$$\Omega = \left\{ \zeta : \min_{z \in \gamma} |z| \le |\zeta| \le \max_{z \in \gamma} |z| \right\}.$$

Theorem 4.4.1 (Agranovskii and Semenov [3]) *Assume the following conditions to be fulfilled:*

1. $0 \notin R \cup \gamma$;
2. γ is not contained in any complex line in $\mathbb{C}^n$, passing through 0.

Let $f \in \mathscr{C}^1(\Omega)$ and for any $u \in \mathscr{U}(n)$ the restriction f on $u(\gamma)$ admits a holomorphic extension to $u(R)$, which is smooth on $\overline{u(R)}$. Then f is holomorphic into Ω (and therefore extends holomorphically in the corresponding ball).

As noted in [3], if we require that the set R be symmetric relative to the mapping $z \to \bar{z}$, then Theorem 4.4.1 remains valid for continuous functions f. Thus it generalizes the already-mentioned Rudin theorem from [71]. The theorem fails without the above condition (2).

As already noted, Grinberg [29] was the first to formulate the boundary Morera theorem for a ball (in the case of complex lines), although one of the assertions by Nagel and Rudin [66] (see [71, Item 13.4]) can also be treated as a boundary Morera theorem.

We present one of the theorems of [26], in which the class of complex lines is significantly narrowed.

Theorem 4.4.2 (Globevnik and Stout [26]) *Consider a unit ball $B \subset \mathbb{C}^2$. Suppose that the given number r, $0 < r < 1$, is such that the expression $r^{-1}(1 - r^2)^{1/2}$ is not a root of any polynomial with integer coefficients. Assume that the function*

$f \in \mathscr{C}(\partial B)$ *and satisfies the Morera property along all complex lines lying at a distance r from the center of the ball, then f extends to B as a function from $\mathscr{A}(B)$.*

Proof By $\mathscr{P}_{p,q}$ we denote the space of all homogeneous harmonic polynomials in z, $\bar{z}$, having the degree of homogeneity of p for z and q for $\bar{z}$.

Let $\mathscr{X} \subset \mathscr{C}(\partial B)$ be a subspace of all continuous functions g such that the condition

$$\int_{-\pi}^{\pi} g(x + e^{i\theta}y)e^{i\theta}d\theta = 0$$

is fulfilled for each $x \in \mathbb{C}^2$, $|x| = r$ and each $y \in \mathbb{C}^2$ such that $(x, \bar{y}) = 0$, $|x|^2 + |y|^2 = 1$. This means that the function g satisfies the Morera property along complex lines lying at a distance r from the center of the ball.

Then $\mathscr{X}$ is a closed unitary-invariant subspace in $\mathscr{C}(\partial B)$ (i.e., invariant with respect to unitary transformations.) Such a subspace was described by Nagel and Rudin in [66] (see also [71, Chap. 12]).

In order to prove that g extends holomorphically to B, it is enough by the theorem from [66], to prove that $\mathscr{X}$ does not contain any spaces $\mathscr{P}_{p,q}$ for $q \geq 1$. As the function $z^p\bar{z}^q$ is in $\mathscr{P}_{p,q}$, it suffices to show that $z^p\bar{z}^q$ does not belong to $\mathscr{X}$ for $p \geq 0$, $q \geq 1$.

Any $x \in \mathbb{C}^2$, provided $|x| = r$ can be represented as

$$x = e^{i\alpha}(\rho, (r^2 - \rho^2)^{1/2}e^{i\omega}),$$

where $0 \leq \rho \leq r$ and $\alpha, \omega \in \mathbb{R}$. Then the vector y can be represented as

$$y = t((r^2 - \rho^2)^{1/2}e^{-i\omega}, -\rho)$$

for $t = r^{-1}(1 - r^2)^{1/2}$. Thus $(x, \bar{y}) = 0$ and $|x|^2 + |y|^2 = 1$. Let $g(z, \bar{z}) = z_1^m \bar{z}_2^n$, where $n \geq 1$, and $g \in \mathscr{X}$. This implies

$$\int_{-\pi}^{\pi} (e^{i\alpha}\rho + e^{i\theta}t(r^2 - \rho^2)^{1/2}e^{-i\omega})^m \times$$

$$\times (e^{-i\alpha}(r^2 - \rho^2)^{1/2}e^{-i\omega} - e^{-i\theta}t\rho)^n e^{i\theta}d\theta = 0 \qquad (4.4.1)$$

for each ρ, $0 \leq \rho \leq r$ and each $\alpha, \omega \in \mathbb{R}$.

Since

$$\int_{-\pi}^{\pi} e^{ik\theta}d\theta = 0$$

for $k \neq 0$, then (4.4.1) yields

$$0 = \binom{m}{0}(e^{i\alpha}\rho)^m \binom{n}{1}(e^{-i\alpha}(r^2 - \rho^2)^{1/2}e^{-i\omega})^{n-1}(t\rho)^1(-1)^1$$

$$+ \binom{m}{1}(e^{i\alpha}\rho)^{m-1}[t(r^2 - \rho^2)^{1/2}e^{-i\omega}]^1\binom{n}{2} \times$$

$$\times (e^{-i\alpha}(r^2 - \rho^2)^{1/2}e^{-i\omega})^{n-2}(t\rho)^2(-1)^2$$

$$+ \binom{m}{2}(e^{i\alpha}\rho)^{m-2}[t(r^2 - \rho^2)^{1/2}e^{-i\omega}]^2\binom{n}{3} \times$$

$$\times (e^{-i\alpha}(r^2 - \rho^2)^{1/2}e^{-i\omega})^{n-3}(t\rho)^3(-1)^3 + \cdots$$

$$+ \begin{cases} \binom{m}{m}(e^{i\alpha}\rho)^{m-m}[t(r^2 - \rho^2)^{1/2}e^{-i\omega}]^m\binom{n}{m+1} \times \\ \quad \times (e^{-i\alpha}(r^2 - \rho^2)^{1/2}e^{-i\omega})^{n-(m+1)}(t\rho)^{m+1}(-1)^{m+1}, \quad \text{if } n \geq m+1, \\ \binom{m}{n-1}(e^{i\alpha}\rho)^{m-(n-1)}[t(r^2 - \rho^2)^{1/2}e^{-i\omega}]^{n-1}\binom{n}{n} \times \\ \quad \times (e^{-i\alpha}(r^2 - \rho^2)^{1/2}e^{-i\omega})^{n-n}(t\rho)^n(-1)^n, \qquad \text{if } n < m+1. \end{cases}$$

After transformations have been performed, this relation becomes

$$0 = (r^2 - \rho^2)^{(n-1)/2}(e^{i\alpha})^{m-n+1}(e^{-i\omega})^{n-1}\rho^{m+1} \times$$

$$\times \left[\binom{m}{0}\binom{n}{1}t(-1)^1 + \binom{m}{1}\binom{n}{2}t^3(-1)^2 + \cdots \right.$$

$$+ \left. \begin{cases} \binom{m}{m}\binom{n}{m+1}t^{2m+1}(-1)^{m+1} \right], & \text{if } n \geq m+1, \\ \binom{m}{n-1}\binom{n}{n}t^{2n-1}(-1)^n \right], & \text{if } n < m+1. \end{cases}$$

The latter relation is impossible, since $t = r^{-1}(1 - r^2)^{1/2}$ is not the root of any polynomial with integer coefficients. □

Example 4.4.1 Consider the example from [26], showing that the condition in Theorem 4.4.2 is essential. Let $r = \left(\dfrac{3}{5}\right)^{1/2}$. Then the function $g(z, \bar{z}) = z_1^3\bar{z}_2^2$ has the Morera property along any complex line lying at a distance r from the center of the ball, but obviously, g does not extend holomorphically in B from the boundary ∂B.

Indeed, the calculations in Theorem 4.4.2, show that

$$\int_{-\pi}^{\pi} (e^{i\alpha}\rho + e^{i\theta}t(r^2-\rho^2)^{1/2}e^{-i\omega})^3(e^{-i\alpha}(r^2-\rho^2)^{1/2}e^{-i\omega} - e^{-i\theta}t\rho)^2 e^{i\theta} \, d\theta$$

$$= 2\pi(r^2-\rho^2)^{1/2}(e^{i\alpha})^{1/2}(e^{-i\omega})\rho^3 \left[\binom{3}{0}\binom{2}{1}t(-1)^1 + \binom{3}{1}\binom{2}{2}t^3(-1)^2 \right]$$

$$= 2\pi(r^2-\rho^2)^{1/2}e^{2i\alpha}e^{-i\omega}\rho^3(-2t+3t^3) = 0,$$

since $-2t + 3t^3 = t\left(3\left(1-\dfrac{3}{5}\right)\dfrac{5}{3} - 2\right) = 0.$

Recently Agranovskii [6] and Globevnik [25] have shown that a family of complex lines passing through two fixed points in $\overline{D}$ is sufficient for holomorphic extension for real-analytic functions on the boundary of a ball. A family of complex lines passing through one point on the boundary of a ball was proved to be sufficient for holomorphic extension by Baracco in [12].

Theorem 4.4.3 (Baracco) *Let the point $z_0 \in \partial B$, and the function f be of class $\mathscr{C}^\omega(\partial B)$, and suppose that f extends holomorphically from ∂B along each line passing through z_0. Then f extends holomorphically to B.*

Proof

(a) We first prove the result for a ball B in $\mathbb{C}^2$. Without loss of generality, we can assume that z_0 is the point $(0, 1)$. Disks passing through the point $(0, 1)$, can be parameterized by parameter $a \in \mathbb{C}$ as a set of the form

$$D_a(\tau) = \left(\frac{\tau-1}{1+|a|^2}a, \frac{\tau-1}{1+|a|^2}+1 \right), \qquad \tau \in \overline{\Delta},$$

where Δ is a unit disk in $\mathbb{C}$. We note that when $|a| \gg 1$ the disks D_a become close to the complex tangent line to the sphere at z_0 and, moreover, D_a lie in the neighborhood of z_0.

Since $f \in \mathscr{C}^\omega(\partial B)$ and $\dfrac{\partial}{\partial \bar{z}_2}$ is transversal to ∂B at z_0, then f can be extended holomorphically with respect to z_2 $\left(\bar{z}_2 = \dfrac{z_2}{1-|z_1|^2} \right)$ to a neighborhood of z_0. We denote this extension again by f. Consider the expansion of f in a power series at z_0:

$$f(z_1, \bar{z}_1, z_2) = \sum_{l=0}^{+\infty} \sum_{h+k+2m=l} b_{h,k,m} z_1^h \bar{z}_1^k (z_2-1)^m.$$

Then we regroup the terms of the series on weighted degree (by assigning a weight of 2 to the variable z_2). Taking $|a|$ large enough, we consider the N-moment conditions for the disk D_a:

$$0 = G(a, N) = \int_{\partial \Delta} \tau^N f(D_a(\tau)) \, d\tau$$

$$= \int_{\partial \Delta} \tau^N \sum_{l=0}^{+\infty} \sum_{h+k+2m=l} b_{h,k,m} \left(\frac{\tau - 1}{1 + |a|^2} a \right)^h \left(\frac{\tau - 1}{1 + |a|^2} a \right)^k \left(\frac{\tau - 1}{1 + |a|^2} \right)^m d\tau.$$

We want to prove that the coefficients $b_{h,k,m} = 0$ for $k > 0$. For this we take the smallest weighted degree l_0 such that $b_{h,k,m} \neq 0$ for some $k > 0$, and let k_0 be the highest degree in $\bar{z}_1$ in this case. We get $G(a, N) = 0$ for any N and any a, in particular, for ta at $|a| = 1$ and $t \to +\infty$.

Using the fact that for $k > N$

$$\int_{\partial \Delta} \tau^N (\tau - 1)^h (\bar{\tau} - 1)^k (\tau - 1)^m d\tau = \int_{\partial \Delta} (-1)^k \frac{(\tau - 1)^{h+k+m}}{\tau^{k-N}} d\tau$$

$$= (-1)^{h+m+k+N-1} \binom{h+k+m}{k-1-N},$$

we consider the limit

$$\lim_{t \to +\infty} G(ta, N) t^{l_0} = \lim_{t \to +\infty} \sum_{l=l_0}^{+\infty} \sum_{\substack{h+k+2m=l \\ k>N}} 2\pi i (-1)^{h+m+k+N-1} \times$$

$$\times \binom{h+k+m}{k-N-1} t^{h+k+l_0} a^h \bar{a}^k \frac{b_{h,k,m}}{(1 + t^2|a|^2)^{h+k+m}}$$

$$= \lim_{t \to +\infty} \sum_{l=l_0}^{+\infty} \sum_{\substack{h+k+2m=l \\ k>N}} 2\pi i (-1)^{h+m+k+N-1} \times$$

$$\times \binom{h+k+m}{k-N-1} a^h \bar{a}^k \left(\frac{1}{t^2} + |a|^2 \right)^m t^{l_0-l} \frac{b_{h,k,m}}{\left(\frac{1}{t^2} + |a|^2 \right)^l}$$

$$= \sum_{\substack{h+k+2m=l_0 \\ k>N}} 2\pi i (-1)^{h+m+k+N-1} \binom{h+k+m}{k-N-1} b_{h,k,m} \frac{a^h \bar{a}^k |a|^{2m}}{|a|^{2l_0}} = 0.$$

Now, choosing $N = k_0 - 1$, we obtain the following relations for the coefficients $b_{h,k,m}$:

$$\sum_{h+k_0+2m=l_0} (-1)^{h+m} \binom{h+k_0+m}{1} b_{h,k_0,m} a^{h+m} \bar{a}^{k_0+m} = 0.$$

Putting $a = e^{i\theta}$, we obtain

$$\sum_{h+k_0+2m=l_0} (-1)^{h+m} \binom{h+k_0+m}{1} b_{h,k_0,m} e^{i\theta(h-k_0)} = 0,$$

which implies that $b_{h,k_0,m} = 0$ for $h + k_0 + 2m = l_0$. Therefore, we find that $b_{h,k,m} = 0$ for any weighted degree l when $k \geq 1$.

Thus, the series for f converges in the neighborhood of z_0 and is a holomorphic function in this neighborhood. By Hartogs' extension theorem (see [81, Sect. 15.6]) the function f is a holomorphic function in the ball B. This completes the proof for dimension 2.

(b) Consider a ball $B \subset \mathbb{C}^n$, $n > 2$. We assume that the point $z_0 = (0, \ldots, 0, 1)$. According to (a) above f is holomorphic in the section of the ball B of the two-dimensional plane passing through 0 and z_0. The various extensions glue together to form a single function $F(z)$, defined in B because by Cauchy's formula for $L_0 \cap \partial B$ these coincide in the complex line L_0, passing through 0 and z_0, where there is an overlap.

The function F is real-analytic in a ball since its extension into the ball is given by, for example, the Bochner–Martinelli integral with the set of integration at the intersection of a two-dimensional complex plane with the ball B. This integral is a real-analytic function of the parameters. Making the complex lines pass through the center of the ball we see that F is a holomorphic function thereon (since we can construct a complex two-dimensional plane through these lines and L_0). Therefore, by Forelli's theorem [71, Sect. 4.4] it is holomorphic in B.   □

4.5   Boundary Analogue of the Forelli Theorem in a Strictly Convex Domain

Here we generalize Theorem 4.4.3 for the case of a strictly convex domain.

In this section we prove the boundary analogue of the Forelli theorem for real-analytic functions, that is, we show that any real-analytic function f defined on the boundary of a bounded strictly convex domain D in a multidimensional complex space and having the one-dimensional holomorphic extension property along a family of complex lines passing through the boundary point and intersecting the domain D, extends holomorphically to D as a function of several complex variables.

4.5.1   The Multidimensional Case

Consider a strictly convex domain $D \subset \mathbb{C}^n$. We recall that the domain D is called strictly convex, if the function $\rho\,(w_1, \ldots, w_n)$, which defines the domain D, i.e., $D = \{w : \rho\,(w) < 0\}$ and grad $\rho = \left(\dfrac{\partial \rho}{\partial w_1}, \ldots, \dfrac{\partial \rho}{\partial w_n} \right) \neq 0$ on ∂D, satisfies the condition

$$\sum_{p,j=1}^{n} \frac{\partial^2 \rho}{\partial w_p \partial w_j} \left(w^0\right) \xi_p \xi_j + \sum_{p,j=1}^{n} \frac{\partial^2 \rho}{\partial \bar{w}_p \partial \bar{w}_j} \left(w^0\right) \bar{\xi}_p \bar{\xi}_j$$

$$+ 2 \sum_{p,j=1}^{n} \frac{\partial^2 \rho}{\partial w_p \partial \bar{w}_j} \left(w^0\right) \xi_p \bar{\xi}_j > 0$$

for all $\xi \neq 0$ and $w^0 \in \overline{D}$.

In what follows D is a bounded strictly convex domain in $\mathbb{C}^n$ ($n > 1$) with a real-analytic boundary, that is, the defining function ρ is real-analytic in a neighborhood of the closure of $\overline{D}$.

We denote a family of complex lines passing through w_0, $w_0 \in \partial D$, by $\mathfrak{L}_{w_0}$.

Theorem 4.5.1 ([41]) *Let a function $f \in \mathscr{C}^w(\partial D)$ have the one-dimensional holomorphic extension property along all complex lines from $\mathfrak{L}_{w_0}$, intersecting D, then the function f extends holomorphically into D.*

Proof The proof of this result in the two-dimensional case will be given later in the following subsections. Here we prove the result for the case $n > 2$, assuming the statement to be true in the two-dimensional case.

Let $0 \in D$. We will take two-dimensional sections of D passing through a boundary point w_0 and the point 0 lying in D. The function defining the boundary of the two-dimensional section will satisfy the conditions of the theorem in the two-dimensional case. Therefore, f will continue holomorphically in the interior of these two-dimensional sections and this function will define the function F in D therein, since by the assumption of the theorem, these functions coincide at the intersection of these two-dimensional sections (i.e., on the complex lines). The union of these two-dimensional sections coincides with the whole domain D. Thus, the function F is uniquely defined in the whole domain D.

Since the holomorphic extension of f in two-dimensional sections is given by the two-dimensional Bochner–Martinelli integral or Khenkin–Ramirez integral, real-analytically dependent on the parameters, then the holomorphic extension of the function f is a real-analytic function. Thus, the function F belongs to the class $\mathscr{C}^w$ in the domain D.

Since the two-dimensional section is defined by two complex lines, then the function F, being holomorphic throughout the two-dimensional section, is also holomorphic on complex lines lying in this section. Thus, the function F is holomorphic at the intersection of D with each complex line passing through the point 0.

We adopt the conditions of Forelli's theorem [71, Theorem 4.4.5], and applying this theorem, we obtain that the function F is holomorphic in a neighborhood of 0. Since the function F is holomorphic in a neighborhood of 0 and real-analytic in D, then it is holomorphic in the whole domain D. $\qquad\qquad\square$

4.5.2 The Form of Sections of the Complex Line

In this section, we will describe the first step to the proof of Theorem 4.5.1 in the two-dimensional case and will prove the assertion about the form of the section of $D \subset \mathbb{C}^2$ of the complex line.

We consider a two-dimensional complex space $\mathbb{C}^2$, whose points will be denoted by $w = (w_1, w_2)$, $z = (z_1, z_2)$, etc. We make a shift to take the point $w_0 \in \partial D$ to 0 and perform a unitary transformation of coordinates $w = w(z)$ so that in the neighborhood of the boundary point 0 after switching from complex coordinates to real ones, i.e., representing $z_1 = x_1 + ix_2$, $z_2 = x_3 + ix_4$, the boundary defining function by the implicit function theorem, takes the form

$$x_4 = \varphi(x_1, x_2, x_3), \qquad (4.5.1)$$

where the function φ is real-analytic in a neighborhood of zero and satisfies $\varphi(0) = 0$, $\dfrac{\partial \varphi}{\partial x_k}(0) = 0, k = 1, 2, 3$.

Expanding the function $\varphi(x_1, x_2, x_3)$ in (4.5.1) in a Taylor series in the neighborhood of the boundary point 0, by virtue of the conditions on φ, we have

$$x_4 = T(x_1, x_2, x_3) + o(|x'|^2), \quad |x'| \to 0, \quad x' = (x_1, x_2, x_3), \qquad (4.5.2)$$

where

$$T(x_1, x_2, x_3) = c_{11}x_1^2 + c_{22}x_2^2 + c_{33}x_3^2 + c_{12}x_1x_2 + c_{13}x_1x_3 + c_{23}x_2x_3$$

is a positive definite quadratic form (due to the strict convexity of ρ).

Next, we consider the section $D_a(\tau)$ of D

$$D_a(\tau) = \left(\frac{\tau}{1 + |a|^2}a, \frac{\tau}{1 + |a|^2} \right), \qquad \tau \in \overline{\Delta}_a,$$

passing in the direction of the vector $(a, 1) \in \mathbb{C}^2$. The domain Δ_a of the change in the parameter τ is a domain in the complex plane with a real-analytic boundary (in the neighborhood of the boundary point 0).

Let $\tau = u + iv$, $a = a_1 + ia_2$. Then

$$\frac{\tau}{1 + |a|^2} a = \frac{(ua_1 - va_2) + i(ua_2 + va_1)}{1 + |a|^2},$$

$$\frac{\tau}{1 + |a|^2} = \frac{u + iv}{1 + |a|^2}.$$

Thus,

$$x_1 = \frac{ua_1 - va_2}{1 + |a|^2}, \quad x_2 = \frac{ua_2 + va_1}{1 + |a|^2}, \quad x_3 = \frac{u}{1 + |a|^2}, \quad x_4 = \frac{v}{1 + |a|^2}.$$

We write an expression for the quadratic form $T(x_1, x_2, x_3)$:

$$\begin{aligned}
T(x_1, x_2, x_3) &= c_{11}x_1^2 + c_{22}x_2^2 + c_{33}x_3^2 + c_{12}x_1x_2 + c_{13}x_1x_3 + c_{23}x_2x_3 \\
&= \frac{1}{(1 + |a|^2)^2}\Big[c_{11}\left(u^2a_1^2 - 2uva_1a_2 + v^2a_2^2\right) + c_{22}\left(u^2a_2^2 + 2uva_1a_2 + v^2a_1^2\right) \\
&\quad + c_{33}u^2 + c_{12}\left(u^2a_1a_2 + uva_1^2 - uva_2^2 - v^2a_1a_2\right) + c_{13}\left(u^2a_1 - uva_2\right) \\
&\quad + c_{23}\left(u^2a_2 + uva_1\right)\Big] = \frac{1}{(1 + |a|^2)^2}\Big[v^2\left(c_{11}a_2^2 + c_{22}a_1^2 - c_{12}a_1a_2\right) \\
&\quad + v\left(-2c_{11}ua_1a_2 + 2c_{22}ua_1a_2 + c_{12}ua_1^2 - c_{12}ua_2^2 - c_{13}ua_2 + c_{23}ua_1\right) \\
&\quad + \left(c_{11}u^2a_1^2 + c_{22}u^2a_2^2 + c_{33}u^2 + c_{12}u^2a_1a_2 + c_{13}u^2a_1 + c_{23}u^2a_2\right)\Big].
\end{aligned}$$

By substituting the x_4 from (4.5.1) and $T(x_1, x_2, x_3)$ into Eq. (4.5.2) and reducing the similar terms, we obtain

$$\begin{aligned}
&v^2\left(c_{11}a_2^2 + c_{22}a_1^2 - c_{12}a_1a_2\right) + v\left(-2c_{11}ua_1a_2 + 2c_{22}ua_1a_2 + c_{12}ua_1^2 - c_{12}ua_2^2\right. \\
&\left. - c_{13}ua_2 + c_{23}ua_1 - 1 - |a|^2\right) + \left(c_{11}u^2a_1^2 + c_{22}u^2a_2^2 + c_{33}u^2 + c_{12}u^2a_1a_2\right. \\
&\left. + c_{13}u^2a_1 + c_{23}u^2a_2\right) + o\left(|a|^2\right) = 0, \qquad |a| \to +\infty.
\end{aligned}$$

Choosing $|a|$ to be large enough, and replacing a by ta, where $|a| = 1$, $t \in \mathbb{R}$, we have

$$\begin{aligned}
&v^2\left(c_{11}a_2^2t^2 + c_{22}a_1^2t^2 - c_{12}a_1a_2t^2\right) + v\left(-2c_{11}ua_1a_2t^2 + 2c_{22}ua_1a_2t^2 + c_{12}ua_1^2t^2\right. \\
&\left. - c_{12}ua_2^2t^2 - c_{13}ua_2t + c_{23}ua_1t - 1 - |a|^2 t^2\right) + \left(c_{11}u^2a_1^2t^2 + c_{22}u^2a_2^2t^2 + c_{33}u^2\right. \\
&\left. + c_{12}u^2a_1a_2t^2 + c_{13}u^2a_1t + c_{23}u^2a_2t\right) + o\left(|t|^2\right) = 0, \qquad t \to +\infty.
\end{aligned}$$

Thus, dividing by t^2 and passing over to the limit as $t \to +\infty$ in this expression, we obtain

$$v^2 \left(c_{11} a_2^2 + c_{22} a_1^2 - c_{12} a_1 a_2 \right)$$
$$+ v \left(-2 c_{11} u a_1 a_2 + 2 c_{22} u a_1 a_2 + c_{12} u a_1^2 - c_{12} u a_2^2 - |a|^2 \right)$$
$$+ \left(c_{11} u^2 a_1^2 + c_{22} u^2 a_2^2 + c_{12} u^2 a_1 a_2 \right) = 0,$$

i.e.,

$$u^2 \left(c_{11} a_1^2 + c_{22} a_2^2 + c_{12} a_1 a_2 \right) + 2uv \left(-c_{11} a_1 a_2 + c_{22} a_1 a_2 + \frac{c_{12}}{2} a_1^2 - \frac{c_{12}}{2} a_2^2 \right)$$
$$+ v^2 \left(c_{11} a_2^2 + c_{22} a_1^2 - c_{12} a_1 a_2 \right) - v = 0. \qquad (4.5.3)$$

Proposition 4.5.1 *The domain Δ of the change in the parameter τ is the interior of the ellipse in the limiting case when $|a| \to +\infty$. Relation (4.5.3) defines the boundary $\partial \Delta$.*

Proof We write relations (4.5.3) as

$$b_{11} u^2 + 2 b_{12} uv + b_{22} v^2 + 2 b_1 u + 2 b_2 v + b_0 = 0 \qquad (4.5.4)$$

with the coefficients

$$b_{11} = c_{11} a_1^2 + c_{22} a_2^2 + c_{12} a_1 a_2,$$
$$b_{12} = -c_{11} a_1 a_2 + c_{22} a_1 a_2 + \frac{c_{12}}{2} a_1^2 - \frac{c_{12}}{2} a_2^2,$$
$$b_{22} = c_{11} a_2^2 + c_{22} a_1^2 - c_{12} a_1 a_2,$$
$$b_2 = -\frac{1}{2}, \quad b_1 = 0, \quad b_0 = 0.$$

Let λ_1, λ_2 be the roots of a characteristic equation of the quadratic form $b_{11} u^2 + 2 b_{12} uv + b_{22} v^2$. Then these satisfy the characteristic equation

$$(b_{11} - \lambda)(b_{22} - \lambda) - b_{12}^2 = 0,$$

or

$$\lambda^2 - \lambda (b_{11} + b_{22}) + \left(b_{11} b_{22} - b_{12}^2 \right) = 0. \qquad (4.5.5)$$

Let us find expressions belonging to quadratic equation (4.5.5)

$$b_{11} + b_{22} = c_{11}a_1^2 + c_{22}a_2^2 + c_{12}a_1a_2 + c_{11}a_2^2 + c_{22}a_1^2 - c_{12}a_1a_2$$

$$= c_{11}(a_1^2 + a_2^2) + c_{22}(a_1^2 + a_2^2) = c_{11} + c_{22},$$

$$b_{11}b_{22} - b_{12}^2 = (c_{11}a_1^2 + c_{22}a_2^2 + c_{12}a_1a_2)(c_{11}a_2^2 + c_{22}a_1^2 - c_{12}a_1a_2)$$

$$- \left(- c_{11}a_1a_2 + c_{22}a_1a_2 + \frac{c_{12}}{2}(a_1^2 - a_2^2) \right)^2$$

$$= c_{11}c_{22}(a_1^4 + a_2^4 + 2a_1^2a_2^2) - \frac{c_{12}^2}{4}(a_1^4 + 2a_1^2a_2^2 + a_2^4)$$

$$= c_{11}c_{22}(a_1^2 + a_2^2)^2 - \frac{c_{12}^2}{4}(a_1^2 + a_2^2)^2 = c_{11}c_{22} - \frac{c_{12}^2}{4}.$$

Substituting the values obtained for $b_{11} + b_{22}$, $b_{11}b_{22} - b_{12}^2$ into Eq. (4.5.5), we obtain the following characteristic equation

$$\lambda^2 - \lambda(c_{11} + c_{22}) + \left(c_{11}c_{22} - \frac{c_{12}^2}{4} \right) = 0.$$

The discriminant of this quadratic equation is

$$(c_{11} + c_{22})^2 - 4\left(c_{11}c_{22} - \frac{c_{12}^2}{4} \right) = c_{11}^2 - 2c_{11}c_{22} + c_{22}^2 + c_{12}^2 = (c_{11} - c_{22})^2 + c_{12}^2.$$

Then the roots λ_1, λ_2 of a characteristic equation of the quadratic form $b_{11}u^2 + 2b_{12}uv + b_{22}v^2$ will have the form

$$\lambda_1 = \frac{c_{11} + c_{22} - \sqrt{(c_{11} - c_{22})^2 + c_{12}^2}}{2},$$

$$\lambda_2 = \frac{c_{11} + c_{22} + \sqrt{(c_{11} - c_{22})^2 + c_{12}^2}}{2}.$$

(4.5.6)

Due to the positive definiteness of the quadratic form $T(x_1, x_2, x_3)$ and from the expression (4.5.6) for λ_2 we see that $\lambda_2 > 0$. In order to show that $\lambda_1 > 0$, it suffices to show that

$$c_{11} + c_{22} > \sqrt{(c_{11} - c_{22})^2 + c_{12}^2},$$

$$(c_{11} + c_{22})^2 > (c_{11} - c_{22})^2 + c_{12}^2,$$

$$2c_{11}c_{22} > -2c_{11}c_{22} + c_{12}^2,$$

$$c_{11}c_{22} - \frac{c_{12}^2}{4} > 0,$$

and this is true by virtue of the positive definiteness of the form $T(x_1, x_2, x_3)$. Thus, it is shown that the characteristic roots $\lambda_1 > 0$, $\lambda_2 > 0$. We make a coordinate transformation

$$\begin{cases} u = u' \cos \alpha - v' \sin \alpha, \\ v = u' \sin \alpha + v' \cos \alpha, \end{cases} \tag{4.5.7}$$

where the angle α is determined from the relation

$$\frac{\cos 2\alpha}{\sin 2\alpha} = \frac{b_{11} - b_{22}}{2b_{12}}. \tag{4.5.8}$$

After the coordinate transformation (4.5.7) equation (4.5.4) can be written as

$$\lambda_1 u'^2 + \lambda_2 v'^2 + 2b_1' u' + 2b_2' v' + b_0 = 0 \tag{4.5.9}$$

with the coefficients

$$b_1' = b_1 \cos \alpha + b_2 \sin \alpha = -\frac{1}{2} \sin \alpha,$$
$$b_2' = -b_1 \sin \alpha + b_2 \cos \alpha = -\frac{1}{2} \cos \alpha. \tag{4.5.10}$$

Let us transfer the origin to the point (u_0', v_0'), i.e., perform the transformation

$$\begin{cases} u' = u'' + u_0', \\ v' = v'' + v_0'. \end{cases} \tag{4.5.11}$$

After the coordinate transformation (4.5.11) equation (4.5.9) can be written as

$$\lambda_1 u''^2 + \lambda_2 v''^2 + 2\left(\lambda_1 u_0' + b_1'\right) u'' + 2\left(\lambda_2 v_0' + b_2'\right) v'' + b_0' = 0, \tag{4.5.12}$$

where the constant term b_0' is

$$b_0' = \lambda_1 u_0'^2 + \lambda_2 v_0'^2 + 2b_1' u_0' + 2b_2' v_0' + b_0.$$

The coordinates (u_0', v_0') are chosen to provide that the coefficients of u'' and v'' vanish in (4.5.12), i.e.,

$$\lambda_1 u_0' + b_1' = 0, \quad \lambda_2 v_0' + b_2' = 0. \tag{4.5.13}$$

From Eq. (4.5.13) we have

$$u_0' = -\frac{b_1'}{\lambda_1}, \quad v_0' = -\frac{b_2'}{\lambda_2}. \tag{4.5.14}$$

So, the initial equation (4.5.4) is transformed to

$$\lambda_1 u''^2 + \lambda_2 v''^2 + b_0' = 0 \tag{4.5.15}$$

in the new coordinate system.

We proceed to investigate Eq. (4.5.15). Since the coefficients λ_1 and λ_2 have the same sign, then Eq. (4.5.15) is of elliptic type. Using formula (4.5.14) to find an expression for b_0', we obtain

$$b_0' = -\frac{b_1'^2}{\lambda_1} - \frac{b_2'^2}{\lambda_2} = -\left(\frac{b_1'^2 \lambda_2 + b_2'^2 \lambda_1}{\lambda_1 \lambda_2}\right) < 0.$$

Since the common sign of λ_1 and λ_2 is opposite to b_0', then rewriting Eq. (4.5.15) in the form

$$\frac{u''^2}{-\dfrac{b_0'}{\lambda_1}} + \frac{v''^2}{-\dfrac{b_0'}{\lambda_2}} = 1,$$

shows that both denominators $-\dfrac{b_0'}{\lambda_1}$ and $-\dfrac{b_0'}{\lambda_2}$ are positive. We denote these by A^2 and B^2 and obtain the canonical ellipse equation

$$\frac{u''^2}{A^2} + \frac{v''^2}{B^2} = 1$$

with semi-axes A and B such that

$$A^2 = -\frac{b_0'}{\lambda_1} = \frac{b_1'^2 \lambda_2 + b_2'^2 \lambda_1}{\lambda_1^2 \lambda_2}, \qquad B^2 = -\frac{b_0'}{\lambda_2} = \frac{b_1'^2 \lambda_2 + b_2'^2 \lambda_1}{\lambda_1 \lambda_2^2}. \tag{4.5.16}$$

Let us find the relation between the "old" and "new" variables after coordinate transformations (4.5.7) and (4.5.11). We obtain

$$\tau = u + iv = \left(u' + iv'\right)\cos\alpha - v'\sin\alpha + iu'\sin\alpha$$

$$= \left(u' + iv'\right)\left(\cos\alpha + i\sin\alpha\right) = \tau' e^{i\alpha},$$

where $\tau' = u' + iv'$. Denoting

$$\tilde{\tau} = u'' + iv'', \quad \tau_0' = u_0' + iv_0' = -\left(\frac{b_1'}{\lambda_1} + i\frac{b_2'}{\lambda_2}\right),$$

we obtain

$$\tau' = u' + iv' = \left(u'' + iv''\right) + \left(u'_0 + iv'_0\right) = \tilde{\tau} + \tau'_0.$$

Thus, the coordinate transformation is performed according to the formula

$$\tau = \left(\tilde{\tau} + \tau'_0\right) e^{i\alpha}. \tag{4.5.17}$$

To conclude the proof, we must justify the fact that the angle α in (4.5.8) can take any value, i.e., the right-hand side is unbounded in expression (4.5.8). We are to show that the polynomials $b_{11} - b_{22}$ and b_{12} (with respect to the variables a_1, a_2 with $|a| = 1$) have no common roots, i.e., the system

$$\begin{cases} b_{11} - b_{22} = 0, \\ \qquad b_{12} = 0 \end{cases}$$

has no solution. Since

$$b_{11} - b_{22} = (c_{11} - c_{22})\, a_1^2 + 2c_{12}a_1a_2 + (c_{22} - c_{11})\, a_2^2,$$
$$b_{12} = \tfrac{c_{12}}{2} a_1^2 + (c_{22} - c_{11})\, a_1a_2 - \tfrac{c_{12}}{2} a_2^2,$$

we need to show that the system

$$\begin{cases} (c_{11} - c_{22})\, a_1^2 + 2c_{12}a_1a_2 + (c_{22} - c_{11})\, a_2^2 = 0, \\ \dfrac{c_{12}}{2} a_1^2 + (c_{22} - c_{11})\, a_1a_2 - \dfrac{c_{12}}{2} a_2^2 = 0 \end{cases}$$

will have no solutions. We assume that $a_2 \neq 0$. Dividing each equation by a_2^2 and denoting $\dfrac{a_1}{a_2} = y$, we obtain the system

$$\begin{cases} (c_{11} - c_{22})\, y^2 + 2c_{12}y + (c_{22} - c_{11}) = 0, \\ \dfrac{c_{12}}{2} y^2 + (c_{22} - c_{11})\, y - \dfrac{c_{12}}{2} = 0. \end{cases}$$

We recall that the resultant of the two quadratic polynomials

$$g_1(y) = A_0 y^2 + A_1 y + A_2, \qquad g_2(y) = B_0 y^2 + B_1 y + B_2$$

is an expression

$$R(g_1, g_2) = (A_0 B_2 - A_2 B_0)^2 - (A_0 B_1 - A_1 B_0)(A_1 B_2 - A_2 B_1).$$

It is well known that given polynomials with arbitrary leading coefficients, the resultant of these polynomials is zero if and only if these polynomials have a common root or if their leading coefficients are zero. In this case we have

$$A_0 = c_{11} - c_{22}, \quad A_1 = 2c_{12}, \qquad\qquad A_2 = c_{22} - c_{11} = -A_0,$$

$$B_0 = \frac{c_{12}}{2} = \frac{A_1}{4}, \quad B_1 = c_{22} - c_{11} = -A_0, \quad B_2 = -\frac{c_{12}}{2} = -B_0 = -\frac{A_1}{4}.$$

Then the resultant $R(g_1, g_2)$ has the form:

$$R(g_1, g_2) = -\left(-A_0^2 - A_1 \frac{A_1}{4}\right)\left(-A_1 \frac{A_1}{4} + A_0(-A_0)\right)$$

$$= -\left(A_0^2 + \frac{A_1^2}{4}\right)^2 = -\left((c_{11} - c_{22})^2 + c_{12}^2\right)^2 < 0.$$

Thus, we have shown that the polynomials $b_{11} - b_{22}$ and b_{12} (with respect to the variables a_1, a_2 with $|a| = 1$) have no common roots.   □

4.5.3 Proof of Theorem 4.5.1 in the Case of a Restriction Imposed on the Domain

In this section we present the proof of Theorem 4.5.1 for the two-dimensional case, when there are additional restrictions imposed on the domain D.

Consider the two-dimensional complex space $\mathbb{C}^2$. Let D be a bounded strongly convex domain in $\mathbb{C}^2$ with a real-analytic boundary ∂D. Suppose that for all points of the boundary of D the condition

$$\left(\frac{\partial \rho}{\partial w_2}(w)\right)^2 \frac{\partial^2 \rho}{\partial w_1^2}(w) - 2\frac{\partial \rho}{\partial w_1}(w)\frac{\partial \rho}{\partial w_2}(w)\frac{\partial^2 \rho}{\partial w_1 \partial w_2}(w)$$

$$+ \left(\frac{\partial \rho}{\partial w_1}(w)\right)^2 \frac{\partial^2 \rho}{\partial w_2^2}(w) = 0 \qquad (4.5.18)$$

is fulfilled. We recall that $\mathfrak{L}_{w_0}$ is a family of complex lines passing through the point w_0, $w_0 \in \partial D$.

Theorem 4.5.2 ([40]) *Let a function* $f \in \mathscr{C}^w(\partial D)$ *have the one-dimensional holomorphic extension property along all complex lines from* $\mathfrak{L}_{w_0}$, *intersecting* D, *then the function f extends holomorphically to* D.

Proof We make a shift to carry the point $w_0 \in \partial D$ to 0 and perform an orthogonal transformation

$$w = Bz,$$

given by the matrix

$$
B = \begin{pmatrix}
\dfrac{\partial \rho}{\partial w_2}(0) & i\dfrac{\partial \rho}{\partial \bar{w}_1}(0) \\[3mm]
-\dfrac{\partial \rho}{\partial w_1}(0) & i\dfrac{\partial \rho}{\partial \bar{w}_2}(0)
\end{pmatrix}.
$$

This transformation is non-degenerate, since $|B| \neq 0$. Under this transformation the real-analyticity of the function $\rho(Bz) = \tilde{\rho}(z)$ is preserved. When exploded, this transformation looks as follows:

$$
\begin{cases}
\dfrac{\partial \rho}{\partial w_2}(0)z_1 + i\dfrac{\partial \rho}{\partial \bar{w}_1}(0)z_2 = w_1, \\[3mm]
-\dfrac{\partial \rho}{\partial w_1}(0)z_1 + i\dfrac{\partial \rho}{\partial \bar{w}_2}(0)z_2 = w_2.
\end{cases}
$$

Let $z_1 = x_1 + ix_2$, $z_2 = x_3 + ix_4$. For further proof of the theorem we prove the following lemma.

Lemma 4.5.1 *Under a complex linear coordinate transformation $w = Bz$, condition (4.5.18) on the function $\rho(w_1, w_2)$, considered at the boundary point $w_0 = 0$, can be written as*

$$
\frac{\partial^2 \varphi}{\partial x_1 \partial x_2}(0) = 0, \qquad \frac{\partial^2 \varphi}{\partial x_1^2}(0) = \frac{\partial^2 \varphi}{\partial x_2^2}(0), \tag{4.5.19}
$$

where the implicit function $x_4 = \varphi(x_1, x_2, x_3)$ is defined by the equation $\rho(x_1, x_2, x_3, x_4) = 0$ and satisfies $\varphi(0) = 0$, $\dfrac{\partial \varphi}{\partial x_k}(0) = 0$, $k = 1, 2, 3$.

Proof Let us find the relation between the partial derivatives $\tilde{\rho}(z)$ and $\rho(w)$, and also conditions on the function $\tilde{\rho}(z)$. We obtain

$$
\begin{aligned}
\frac{\partial \tilde{\rho}}{\partial z_1} &= \frac{\partial \rho}{\partial w_1}\frac{\partial w_1}{\partial z_1} + \frac{\partial \rho}{\partial \bar{w}_1}\frac{\partial \bar{w}_1}{\partial z_1} + \frac{\partial \rho}{\partial w_2}\frac{\partial w_2}{\partial z_1} + \frac{\partial \rho}{\partial \bar{w}_2}\frac{\partial \bar{w}_2}{\partial z_1} \\[2mm]
&= \frac{\partial \rho}{\partial w_2}(0)\frac{\partial \rho}{\partial w_1} - \frac{\partial \rho}{\partial w_1}(0)\frac{\partial \rho}{\partial w_2},
\end{aligned}
$$

$$
\begin{aligned}
\frac{\partial \tilde{\rho}}{\partial z_2} &= \frac{\partial \rho}{\partial w_1}\frac{\partial w_1}{\partial z_2} + \frac{\partial \rho}{\partial \bar{w}_1}\frac{\partial \bar{w}_1}{\partial z_2} + \frac{\partial \rho}{\partial w_2}\frac{\partial w_2}{\partial z_2} + \frac{\partial \rho}{\partial \bar{w}_2}\frac{\partial \bar{w}_2}{\partial z_2} \\[2mm]
&= i\frac{\partial \rho}{\partial \bar{w}_1}(0)\frac{\partial \rho}{\partial w_1} + i\frac{\partial \rho}{\partial \bar{w}_2}(0)\frac{\partial \rho}{\partial w_2}.
\end{aligned}
$$

From the above calculations, it is evident that

$$\frac{\partial \tilde{\rho}}{\partial z_1}(0) = 0,$$

and the value

$$\frac{\partial \tilde{\rho}}{\partial z_2}(0) = i\left(\left|\frac{\partial \rho}{\partial w_1}(0)\right|^2 + \left|\frac{\partial \rho}{\partial w_2}(0)\right|^2\right) \neq 0$$

is purely imaginary.

We consider the second-order partial derivatives of $\tilde{\rho}(z)$:

$$\frac{\partial^2 \tilde{\rho}}{\partial z_1^2} = \frac{\partial \rho}{\partial w_2}(0)\left(\frac{\partial^2 \rho}{\partial w_1 \partial w_1}\frac{\partial w_1}{\partial z_1} + \frac{\partial^2 \rho}{\partial w_1 \partial w_2}\frac{\partial w_2}{\partial z_1}\right)$$

$$- \frac{\partial \rho}{\partial w_1}(0)\left(\frac{\partial^2 \rho}{\partial w_2 \partial w_1}\frac{\partial w_1}{\partial z_1} + \frac{\partial^2 \rho}{\partial w_2 \partial w_2}\frac{\partial w_2}{\partial z_1}\right)$$

$$= \left(\frac{\partial \rho}{\partial w_2}(0)\right)^2 \frac{\partial^2 \rho}{\partial w_1^2} - 2\frac{\partial \rho}{\partial w_1}(0)\frac{\partial \rho}{\partial w_2}(0)\frac{\partial^2 \rho}{\partial w_1 \partial w_2} + \left(\frac{\partial \rho}{\partial w_1}(0)\right)^2 \frac{\partial^2 \rho}{\partial w_2^2}.$$

In the shifted coordinates, where the boundary point w_0 is switched to zero, and considering condition (4.5.18) on the boundary of the domain D, the last equality goes to

$$\frac{\partial^2 \tilde{\rho}}{\partial z_1^2}(0) = 0.$$

Further, for convenience we will write $\rho(z)$ instead of the functions $\tilde{\rho}(z)$, defining the boundary of D. In other words,

$$\rho(z_1, z_2) = 0 \tag{4.5.20}$$

provided that

$$\begin{cases} \dfrac{\partial \rho}{\partial z_1}(0) = 0, \\ \dfrac{\partial^2 \rho}{\partial z_1^2}(0) = 0, \end{cases} \tag{4.5.21}$$

and further provided that the value of $\dfrac{\partial \rho}{\partial z_2}(0) \neq 0$ is purely imaginary.

The derivatives with respect to the complex variables can be expressed in terms of derivatives with respect to real variables as follows:

$$\frac{\partial\rho}{\partial z_1} = \frac{1}{2}\left(\frac{\partial\rho}{\partial x_1} - i\frac{\partial\rho}{\partial x_2}\right), \qquad \frac{\partial\rho}{\partial z_2} = \frac{1}{2}\left(\frac{\partial\rho}{\partial x_3} - i\frac{\partial\rho}{\partial x_4}\right).$$

Thus from these relations and the set of conditions from (4.5.21), it follows that

$$\frac{\partial\rho}{\partial x_1}(0) = 0, \qquad \frac{\partial\rho}{\partial x_2}(0) = 0, \qquad \frac{\partial\rho}{\partial x_3}(0) = 0. \tag{4.5.22}$$

Next, we write the second condition in system (4.5.21) in terms of real variables. We have

$$\frac{\partial\rho}{\partial x_1} = \left(\frac{\partial}{\partial z_1} + \frac{\partial}{\partial \bar{z}_1}\right)\rho, \qquad \frac{\partial\rho}{\partial x_2} = i\left(\frac{\partial}{\partial z_1} - \frac{\partial}{\partial \bar{z}_1}\right)\rho,$$

$$\frac{\partial^2\rho}{\partial x_1^2} = \left(\frac{\partial}{\partial z_1} + \frac{\partial}{\partial \bar{z}_1}\right)\left(\frac{\partial\rho}{\partial z_1} + \frac{\partial\rho}{\partial \bar{z}_1}\right) =$$

$$= \frac{\partial^2\rho}{\partial z_1^2} + \frac{\partial^2\rho}{\partial \bar{z}_1 \partial z_1} + \frac{\partial^2\rho}{\partial z_1 \partial \bar{z}_1} + \frac{\partial^2\rho}{\partial \bar{z}_1^2} = \frac{\partial^2\rho}{\partial z_1^2} + \frac{\partial^2\rho}{\partial \bar{z}_1^2} + 2\frac{\partial^2\rho}{\partial z_1 \partial \bar{z}_1},$$

$$\frac{\partial^2\rho}{\partial x_2^2} = i\left(\frac{\partial}{\partial z_1} - \frac{\partial}{\partial \bar{z}_1}\right)i\left(\frac{\partial\rho}{\partial z_1} - \frac{\partial\rho}{\partial \bar{z}_1}\right) =$$

$$= -\left(\frac{\partial^2\rho}{\partial z_1^2} - \frac{\partial^2\rho}{\partial \bar{z}_1 \partial z_1} - \frac{\partial^2\rho}{\partial z_1 \partial \bar{z}_1} + \frac{\partial^2\rho}{\partial \bar{z}_1^2}\right) = 2\frac{\partial^2\rho}{\partial z_1 \partial \bar{z}_1} - \frac{\partial^2\rho}{\partial z_1^2} - \frac{\partial^2\rho}{\partial \bar{z}_1^2},$$

$$\frac{\partial^2\rho}{\partial x_1 \partial x_2} = \left(\frac{\partial}{\partial z_1} + \frac{\partial}{\partial \bar{z}_1}\right)i\left(\frac{\partial\rho}{\partial z_1} - \frac{\partial\rho}{\partial \bar{z}_1}\right) =$$

$$= i\left(\frac{\partial^2\rho}{\partial z_1^2} - \frac{\partial^2\rho}{\partial \bar{z}_1 \partial z_1} + \frac{\partial^2\rho}{\partial z_1 \partial \bar{z}_1} - \frac{\partial^2\rho}{\partial \bar{z}_1^2}\right) = i\left(\frac{\partial^2\rho}{\partial z_1^2} - \frac{\partial^2\rho}{\partial \bar{z}_1^2}\right).$$

Thus, by the second condition in system (4.5.21) and taking into account the real-analyticity of the function ρ, from the above calculations it follows that conditions for the function $\rho(x_1, x_2, x_3, x_4)$ will have the form

$$\frac{\partial^2\rho}{\partial x_1^2}(0) = \frac{\partial^2\rho}{\partial x_2^2}(0), \qquad \frac{\partial^2\rho}{\partial x_1 \partial x_2}(0) = 0. \tag{4.5.23}$$

Due to the transition to real coordinates the function defining the boundary of D takes the form

$$\rho(x_1, x_2, x_3, x_4) = 0.$$

Since the gradient of $\rho(x_1, x_2, x_3, x_4)$ is nonzero, then by virtue of relations (4.5.22) we obtain $\dfrac{\partial \rho}{\partial x_4}(0) \neq 0$. Then, by the standard implicit function theorem in a neighborhood of the boundary point 0, the boundary defining function takes the form

$$x_4 = \varphi(x_1, x_2, x_3), \qquad\qquad (4.5.24)$$

where

$$\frac{\partial \varphi}{\partial x_k} = -\frac{\dfrac{\partial \rho}{\partial x_k}\big(x_1, x_2, x_3, \varphi(x_1, x_2, x_3)\big)}{\dfrac{\partial \rho}{\partial x_4}\big(x_1, x_2, x_3, \varphi(x_1, x_2, x_3)\big)}, \qquad k = 1, 2, 3.$$

So the function φ satisfies the conditions $\varphi(0) = 0$, $\dfrac{\partial \varphi}{\partial x_k}(0) = 0$, $k = 1, 2, 3$.

Next, using relations (4.5.22) and (4.5.23), we find conditions on the function $\varphi(x_1, x_2, x_3)$. For this we consider the derivatives $\dfrac{\partial^2 \varphi}{\partial x_k \partial x_j}$, $j = 1, 2, 3$. We obtain

$$\frac{\partial}{\partial x_j}\frac{\partial \rho}{\partial x_k}\big(x_1, x_2, x_3, \varphi(x_1, x_2, x_3)\big) = \frac{\partial^2 \rho}{\partial x_k \partial x_j} + \frac{\partial^2 \rho}{\partial x_k \partial x_4}\frac{\partial \varphi}{\partial x_j} =$$

$$= \frac{\partial^2 \rho}{\partial x_k \partial x_j} - \frac{\dfrac{\partial^2 \rho}{\partial x_k \partial x_4}\dfrac{\partial \rho}{\partial x_j}}{\dfrac{\partial \rho}{\partial x_4}},$$

$$\frac{\partial}{\partial x_j}\frac{\partial \rho}{\partial x_4}\big(x_1, x_2, x_3, \varphi(x_1, x_2, x_3)\big) = \frac{\partial^2 \rho}{\partial x_4 \partial x_j} + \frac{\partial^2 \rho}{\partial x_4 \partial x_4}\frac{\partial \varphi}{\partial x_j} =$$

$$= \frac{\partial^2 \rho}{\partial x_4 \partial x_j} - \frac{\dfrac{\partial^2 \rho}{\partial x_4^2}\dfrac{\partial \rho}{\partial x_j}}{\dfrac{\partial \rho}{\partial x_4}}.$$

Thus

$$\frac{\partial^2 \varphi}{\partial x_k \partial x_j} = -\frac{\left(\dfrac{\partial^2 \rho}{\partial x_k \partial x_j}\dfrac{\partial \rho}{\partial x_4} - \dfrac{\partial^2 \rho}{\partial x_k \partial x_4}\dfrac{\partial \rho}{\partial x_j}\right)\dfrac{\partial \rho}{\partial x_4} - \left(\dfrac{\partial^2 \rho}{\partial x_4 \partial x_j}\dfrac{\partial \rho}{\partial x_4} - \dfrac{\partial^2 \rho}{\partial x_4^2}\dfrac{\partial \rho}{\partial x_j}\right)\dfrac{\partial \rho}{\partial x_k}}{\left(\dfrac{\partial \rho}{\partial x_4}\right)^3},$$

wherein

$$\frac{\partial^2 \varphi}{\partial x_k^2} = -\frac{\left(\frac{\partial^2 \rho}{\partial x_k^2}\frac{\partial \rho}{\partial x_4} - \frac{\partial^2 \rho}{\partial x_k \partial x_4}\frac{\partial \rho}{\partial x_k}\right)\frac{\partial \rho}{\partial x_4} - \left(\frac{\partial^2 \rho}{\partial x_4 \partial x_k}\frac{\partial \rho}{\partial x_4} - \frac{\partial^2 \rho}{\partial x_4^2}\frac{\partial \rho}{\partial x_k}\right)\frac{\partial \rho}{\partial x_k}}{\left(\frac{\partial \rho}{\partial x_4}\right)^3}.$$

Taking into account conditions (4.5.22) and (4.5.23), it is easy to see that

$$\frac{\partial^2 \varphi}{\partial x_1 \partial x_2}(0) = 0, \qquad \frac{\partial^2 \varphi}{\partial x_1^2}(0) = \frac{\partial^2 \varphi}{\partial x_2^2}(0).$$

$\square$

We continue proving the theorem. Later on we will consider a section $D_a(\tau)$ of the domain D

$$D_a(\tau) = \left(\frac{\tau}{1 + |a|^2}a, \frac{\tau}{1 + |a|^2}\right), \qquad \forall \tau \in \overline{\Delta}_a,$$

extending in the direction of the vector $(a, 1) \in \mathbb{C}^2$. The domain Δ_a of change of parameter τ is a domain in the complex plane with a real-analytic boundary (in a neighborhood of the boundary point 0).

Expanding the function $\varphi(x_1, x_2, x_3)$ in a Taylor series in the neighborhood of 0 in expression (4.5.24), from the conditions on φ we obtain

$$x_4 = T(x_1, x_2, x_3) + o\left(|x'|^2\right), \quad |x'| \to 0, \quad x' = (x_1, x_2, x_3), \qquad (4.5.25)$$

where $T(x_1, x_2, x_3) = c_{11}x_1^2 + c_{22}x_2^2 + c_{33}x_3^2 + c_{12}x_1x_2 + c_{13}x_1x_3 + c_{23}x_2x_3$ is a positive definite quadratic form (due to the strong convexity of ρ). Moreover, in view of conditions (4.5.19) on the function $\varphi(x_1, x_2, x_3)$ for coefficients of the form $T(x_1, x_2, x_3)$ we have the relations

$$c_{12} = 0, \quad c_{11} = c_{22}.$$

We select real and imaginary parts in the variables z_1, z_2 and write expressions for x_1, x_2, x_3, x_4. Let $\tau = u + iv, a = a_1 + ia_2$. Then

$$\frac{\tau}{1 + |a|^2}a = \frac{(ua_1 - va_2) + i(ua_2 + va_1)}{1 + |a|^2},$$

$$\frac{\tau}{1 + |a|^2} = \frac{u + iv}{1 + |a|^2}.$$

So

$$x_1 = \frac{ua_1 - va_2}{1 + |a|^2}, \quad x_2 = \frac{ua_2 + va_1}{1 + |a|^2}, \quad x_3 = \frac{u}{1 + |a|^2}, \quad x_4 = \frac{v}{1 + |a|^2}.$$

We write an expression for the quadratic form $T(x_1, x_2, x_3)$:

$$T(x_1, x_2, x_3) = c_{11}x_1^2 + c_{11}x_2^2 + c_{33}x_3^2 + c_{13}x_1x_3 + c_{23}x_2x_3$$

$$= \frac{1}{\left(1 + |a|^2\right)^2} \Big[c_{11} \left(u^2a_1^2 - 2uva_1a_2 + v^2a_2^2\right) + c_{11} \left(u^2a_2^2 + 2uva_1a_2 + v^2a_1^2\right)$$

$$+ c_{33}u^2 + c_{13} \left(u^2a_1 - uva_2\right) + c_{23} \left(u^2a_2 + uva_1\right) \Big]$$

$$= \frac{1}{\left(1 + |a|^2\right)^2} \Big[v^2 \left(c_{11}a_2^2 + c_{11}a_1^2\right) + v \left(-c_{13}ua_2 + c_{23}ua_1\right) +$$

$$+ \left(c_{11}u^2a_1^2 + c_{11}u^2a_2^2 + c_{33}u^2 + c_{13}u^2a_1 + c_{23}u^2a_2\right) \Big].$$

We substitute the values found for x_4 and $T(x_1, x_2, x_3)$ into Eq. (4.5.25) and reduce the similar terms. We get

$$v^2 \left(c_{11}a_2^2 + c_{11}a_1^2\right) + v \left(-c_{13}ua_2 + c_{23}ua_1 - 1 - |a|^2\right)$$

$$+ \left(c_{11}u^2a_1^2 + c_{11}u^2a_2^2 + c_{33}u^2 + c_{13}u^2a_1 + c_{23}u^2a_2\right) + o\left(|a|^2\right) = 0$$

as $|a| \to +\infty$. Choosing $|a|$ large enough, that is, replacing a by ta with $|a| = 1$, we obtain

$$v^2 \left(c_{11}a_2^2t^2 + c_{11}a_1^2t^2\right) + v \left(-c_{13}ua_2t + c_{23}ua_1t - 1 - |a|^2t^2\right)$$

$$+ \left(c_{11}u^2a_1^2t^2 + c_{11}u^2a_2^2t^2 + c_{33}u^2 + c_{13}u^2a_1t + c_{23}u^2a_2t\right) + o\left(|t|^2\right) = 0$$

as $t \to +\infty$. Thus, dividing by t^2 and passing over to the limit as $t \to +\infty$ in this expression, we obtain

$$v^2 \left(c_{11}a_2^2 + c_{11}a_1^2\right) - v|a|^2 + c_{11}u^2a_1^2 + c_{11}u^2a_2^2 = 0,$$

$$c_{11}v^2 - v + c_{11}u^2 = 0.$$

We write this equation in a complex form and obtain

$$c_{11}\left(v^2 - \frac{v}{c_{11}} + u^2\right) = 0,$$

$$\left|\tau - \frac{i}{2c_{11}}\right|^2 = \left(\frac{1}{2c_{11}}\right)^2. \tag{4.5.26}$$

Thus, we have shown that the domain Δ of change of the parameter τ in the limiting case when $|a| \to +\infty$, is a circle of radius $r_0 = \frac{1}{2c_{11}}$ and centered at $\tau_0 = \frac{i}{2c_{11}}$. The coefficient $c_{11} > 0$ due to the positive definiteness of the quadratic form $T(x_1, x_2, x_3)$. Relation (4.5.26) defines the boundary $\partial\Delta$.

It should be noted that the tangent to the boundary of the domain D drawn at the boundary point 0 is the line $\operatorname{Im} z_2 = 0$. It is easy to see that when $|a| \to +\infty$, the section $D_a(\tau)$ is close to the tangent to the boundary of D in the boundary point 0 since

$$\operatorname{Im} z_2 = \frac{v}{1 + |a|^2} \to 0, \quad \text{when } |a| \to +\infty.$$

Moreover, the section $D_a(\tau)$ lies in the neighborhood of $z_0 = 0$ when $|a| \to +\infty$. Namely, if $z \in D_a(\tau)$, then

$$|z - z_0|^2 = \frac{|\tau|^2 |a|^2}{(1 + |a|^2)^2} + \frac{|\tau|^2}{(1 + |a|^2)^2} = \frac{|\tau|^2}{1 + |a|^2} \to 0,$$

when $|a| \to +\infty$.

The function $\rho(z_1, z_2, \bar{z}_1, \bar{z}_2)$ being real-analytic, we solve Eq. (4.5.20) with respect to the variable $\bar{z}_2$. Since $\rho(z, \bar{z})$ is a real-analytic function, it can be expanded in a series in the neighborhood of $(0,0) \in \mathbb{C}^4 = \mathbb{C}^2 \times \mathbb{C}^2$. We pass over from the variables $\bar{z}$ to the variables ζ, i.e., we make the change

$$\bar{z}_1 = \zeta_1, \quad \bar{z}_2 = \zeta_2.$$

We obtain an analytic function $\hat{\rho}(z, \zeta)$ in z and ζ with the conditions

$$\begin{cases} \hat{\rho}(z, \zeta) = 0, \\ \zeta = \bar{z}. \end{cases}$$

Since the gradient of $\hat{\rho}(z_1, z_2, \zeta_1, \zeta_2)$ is nonzero, then the derivative with respect to one of the variables is different from zero, for example, the derivative $\dfrac{\partial \hat{\rho}}{\partial \zeta_2} \neq 0$. Then, applying the implicit function theorem for holomorphic functions, we can express the variable ζ_2 through other variables:

$$\begin{cases} \zeta_2 = \psi\,(z_1, z_2, \zeta_1)\,, \\ \bar{z}_1 = \zeta_1, \\ \bar{z}_2 = \zeta_2. \end{cases}$$

Then $f\,(z_1, z_2, \bar{z}_1, \bar{z}_2)\ =\ f\,(z_1, z_2, \bar{z}_1, \psi(z_1, z_2, \zeta_1))$ is a real-analytic function that can be expanded in a series of variables z_1, z_2, $\zeta_1\ =\ \bar{z}_1$, which converges in a neighborhood of the boundary point $(0, 0)$. Namely

$$f\,(z_1, \bar{z}_1, z_2) = \sum_{l=0}^{+\infty} \sum_{h+k+2m=l} b_{h,k,m} z_1^h \bar{z}_1^k z_2^m,$$

where we have re-denoted the element in the weight degree (giving a weight of 2 to z_2).

Choosing $|a|$ large enough, we consider moments of order N on sections $D_a(\tau)$:

$$G(a, N) = \int_{\partial \Delta_a} \tau^N f\Big(D_a(\tau)\Big)\, d\tau$$

$$= \int_{\partial \Delta_a} \tau^N \sum_{l=0}^{+\infty} \sum_{h+k+2m=l} b_{h,k,m} \left(\frac{\tau}{1+|a|^2}a\right)^h \left(\overline{\frac{\tau}{1+|a|^2}a}\right)^k \left(\frac{\tau}{1+|a|^2}\right)^m d\tau.$$

We prove that the coefficients $b_{h,k,m} = 0$ for $k > 0$. Let l_0 be the smallest weight degree with the property $b_{h,k,m} \neq 0$ for $k > 0$ and k_0 be the greatest degree in $\bar{z}_1$ for which this holds. By the hypothesis, we have $G(a, N) = 0$ for all N and a, in particular, for ta with $|a| = 1$ and $t \to +\infty$. Consider the limit

$$\lim_{t \to +\infty} G(ta, N) t^{l_0}$$

$$= \lim_{t \to +\infty} \int_{\partial \Delta_a} \tau^N \sum_{l=l_0}^{+\infty} \sum_{h+k+2m=l} b_{h,k,m} \left(\frac{\tau}{1+|ta|^2} ta\right)^h \left(\overline{\frac{\tau}{1+|ta|^2} ta}\right)^k \times$$

$$\times \left(\frac{\tau}{1+|ta|^2}\right)^m t^{l_0} d\tau$$

$$= \lim_{t \to +\infty} \int_{\partial \Delta_a} \tau^N \sum_{l=l_0}^{+\infty} \sum_{h+k+2m=l} b_{h,k,m} \tau^h \bar{\tau}^k \tau^m t^h t^k t^{l_0} a^h \bar{a}^k \left(\frac{1}{\frac{1}{t^2}+|a|^2}\right)^h \frac{1}{t^{2h}} \times$$

$$\times \left(\frac{1}{\frac{1}{t^2}+|a|^2}\right)^k \frac{1}{t^{2k}} \left(\frac{1}{\frac{1}{t^2}+|a|^2}\right)^m \frac{1}{t^{2m}} d\tau$$

$$
= \lim_{t\to+\infty} \sum_{l=l_0}^{+\infty} \sum_{h+k+2m=l} b_{h,k,m} \int_{\partial\Delta_a} \tau^N \tau^h \bar{\tau}^k \tau^m \, d\tau \; t^{h+k+l_0-(2h+2k+2m)} a^h \bar{a}^k \times
$$

$$
\times \left(\frac{1}{\frac{1}{t^2}+|a|^2} \right)^{h+k+m+m-m}
= \lim_{t\to+\infty} \sum_{l=l_0}^{+\infty} \sum_{h+k+2m=l} b_{h,k,m} \int_{\partial\Delta_a} \tau^N \tau^h \bar{\tau}^k \tau^m \, d\tau \; t^{l_0-l} a^h \bar{a}^k \times
$$

$$
\times \left(\frac{1}{t^2}+|a|^2 \right)^m \frac{1}{\left(\frac{1}{t^2}+|a|^2 \right)^l}
= \sum_{h+k+2m=l_0} b_{h,k,m} \int_{\partial\Delta} \tau^N \tau^h \bar{\tau}^k \tau^m \, d\tau \; \frac{a^h \bar{a}^k |a|^{2m}}{|a|^{2l_0}} = 0,
$$

where $\partial\Delta$ is defined by (4.5.26).

Let us calculate the value of the integral $\displaystyle\int_{\partial\Delta} \tau^N \tau^h \bar{\tau}^k \tau^m \, d\tau$. Expressing $\bar{\tau}$ as a fractional-linear function from (4.5.26), we obtain

$$
\left| \tau - \frac{i}{2c_{11}} \right|^2 = \left(\tau - \frac{i}{2c_{11}} \right) \bar{\tau} + \frac{i}{2c_{11}} \tau + \frac{1}{4c_{11}^2}.
$$

Then

$$
\left(\tau - \frac{i}{2c_{11}} \right) \bar{\tau} + \frac{i}{2c_{11}} \tau + \frac{1}{4c_{11}^2} = \frac{1}{4c_{11}^2},
$$

$$
\left(\tau - \frac{i}{2c_{11}} \right) \bar{\tau} = -\frac{i}{2c_{11}} \tau, \qquad \bar{\tau} = \frac{-\dfrac{i}{2c_{11}} \tau}{\tau - \dfrac{i}{2c_{11}}}.
$$

We substitute the value found for $\bar{\tau}$ into the integrand expression and obtain

$$
\int_{\partial\Delta} \tau^N \tau^h \bar{\tau}^k \tau^m \, d\tau = \left(-\frac{i}{2c_{11}} \right)^k \int_{\partial\Delta} \frac{\tau^{N+h+m} \tau^k}{\left(\tau - \dfrac{i}{2c_{11}} \right)^k} \, d\tau
$$

$$
= \left(-\frac{i}{2c_{11}} \right)^k \int_{\partial\Delta} \frac{\tau^{N+h+m+k}}{\left(\tau - \dfrac{i}{2c_{11}} \right)^k} \, d\tau.
$$

Now we compute the integral

$$
\int_{\partial\Delta} \frac{\tau^{N+h+m+k}}{\left(\tau - \dfrac{i}{2c_{11}}\right)^k} \, d\tau = 2\pi i \frac{1}{(k-1)!} \lim_{\tau\to\tau_0} \frac{d^{k-1}}{d\tau^{k-1}} \tau^{N+h+m+k}
$$

$$
= 2\pi i \frac{1}{(k-1)!} \lim_{\tau\to\tau_0} \frac{(N+h+m+k)!}{(N+h+m+1)!} \tau^{N+h+m+1}
$$

$$
= 2\pi i \frac{1}{(k-1)!} \frac{(N+h+m+k)!}{(N+h+m+1)!} \left(\frac{i}{2c_{11}}\right)^{N+h+m+1},
$$

where $\tau_0 = \dfrac{i}{2c_{11}}$. Thus

$$
\int_{\partial\Delta} \tau^N \tau^h \bar{\tau}^k \tau^m \, d\tau = \left(-\frac{i}{2c_{11}}\right)^k 2\pi i \frac{1}{(k-1)!} \frac{(N+h+m+k)!}{(N+h+m+1)!} \left(\frac{i}{2c_{11}}\right)^{N+h+m+1}.
$$

Finally, we conclude the proof of the theorem. Since

$$
\sum_{h+k+2m=l_0} b_{h,k,m} \int_{\partial\Delta} \tau^N \tau^h \bar{\tau}^k \tau^m \, d\tau \, \frac{a^h \bar{a}^k |a|^{2m}}{|a|^{2l_0}} = 0,
$$

then, substituting the value found for the integral into the expression, we obtain

$$
\sum_{h+k+2m=l_0} b_{h,k,m} (-1)^k \left(\frac{i}{2c_{11}}\right)^{N+h+m+k+1} 2\pi i \binom{N+h+m+k}{k-1} \frac{a^h \bar{a}^k |a|^{2m}}{|a|^{2l_0}} = 0.
$$

Choosing $N = k_0 - 1$, we obtain the following relation for the coefficients $b_{h,k,m}$

$$
\sum_{h+k_0+2m=l_0} (-1)^{k_0} \left(\frac{i}{2c_{11}}\right)^{2k_0+h+m} 2\pi i \binom{2k_0+h+m-1}{k_0-1} b_{h,k_0,m} a^{h+m} \bar{a}^{k_0+m} = 0.
$$

Substituting $a = e^{i\theta}$, we obtain a trigonometric polynomial (with respect to the variable θ)

$$
\sum_{h+k_0+2m=l_0} (-1)^{k_0} \left(\frac{i}{2c_{11}}\right)^{2k_0+h+m} 2\pi i \binom{2k_0+h+m-1}{k_0-1} b_{h,k_0,m} e^{i\theta(h-k_0)},
$$

identically equal to 0.

Thus, we have shown that the function f is holomorphic in a neighborhood of the boundary point 0. In view of the conditions of the theorem, the function f extends holomorphically to the intersection of D with each complex line passing through the boundary point 0. Consequently, by Hartogs' extension theorem [73] and subject to linear fractional transformation (where by the boundary point goes into infinity and the lines passing through the boundary point become parallel) the function f holomorphically continues to the whole domain $D \subset \mathbb{C}^2$. These arguments complete the proof. □

To conclude this section we consider examples of domains satisfying (4.5.18).

Example 4.5.1 Let $D = B^n$ be a ball of radius R centered at the origin, i.e.,

$$D = \{\zeta : |\zeta| < R\}.$$

Example 4.5.2 Let $\zeta_j = \dfrac{L_j(w)}{L(w)}, j = 1, \ldots, n$, where $L_j(w)$, $L(w)$ are the linear functions. Then the image of the ball B^n under this mapping (unless it is degenerate) is a domain, for which condition (4.5.18) is satisfied.

Example 4.5.3 Let the function ρ defining the boundary of D have the form

$$\rho(w_1, \ldots, w_n) = |w_1|^2 + \ldots + |w_n|^2 - R^2 + \sum_j |L_j(w)|^2,$$

where $L_j(w)$ are the linear functions. Then the domain $D = \{w : \rho(w) < 0\}$ satisfies condition (4.5.18).

Example 4.5.4 Let the function $\rho(w, \bar{w})$ be linearly dependent on w and arbitrarily dependent on $\bar{w}$. Then the domain $D = \{w : \rho(w) < 0\}$ satisfies (4.5.18).

4.5.4 Computation of Moment Integrals

We continue proving Theorem 4.5.1 in the two-dimensional case. Recall that the tangent to the boundary of D, drawn at the boundary point 0, is the line $\mathrm{Im}\, z_2 = 0$. It is easy to see that when $|a| \to +\infty$, the section $D_a(\tau)$ is close to the tangent to the boundary of D at the boundary point 0 as

$$\mathrm{Im}\, z_2 = \frac{v}{1 + |a|^2} \to 0$$

for $|a| \to +\infty$. Moreover, the section $D_a(\tau)$ lies in a sufficiently small neighborhood of $z_0 = 0$ for $|a| \to +\infty$. Namely, if $z \in D_a(\tau)$ then

$$|z - z_0|^2 = \left| \frac{\tau}{1 + |a|^2} a \right|^2 + \left| \frac{\tau}{1 + |a|^2} \right|^2 = \frac{|\tau|^2}{1 + |a|^2} \to 0$$

as $|a| \to +\infty$.

Using the real-analyticity of the function $\rho(z_1, z_2, \bar{z}_1, \bar{z}_2)$, as we did in the preceding subsection, we will solve the equation $\rho(z_1, z_2, \bar{z}_1, \bar{z}_2) = 0$ with respect to the variable $\bar{z}_2$. Since $\rho(z, \bar{z})$ is a real-analytic function, then it can be expanded in a Taylor series in the neighborhood of $(0, 0) \in \mathbb{C}^4 = \mathbb{C}^2 \times \mathbb{C}^2$. We pass over from the coordinates $\bar{z}$ to the variables ζ, i.e., make the change

$$\bar{z}_1 = \zeta_1, \quad \bar{z}_2 = \zeta_2.$$

We obtain an analytic function $\hat{\rho}(z, \zeta)$ of z and ζ with the conditions

$$\begin{cases} \hat{\rho}(z, \zeta) = 0, \\ \zeta = \bar{z}. \end{cases}$$

Since the gradient of the function $\hat{\rho}(z_1, z_2, \zeta_1, \zeta_2)$ is nonzero, so the derivative with respect to one of the variables is different from zero, for example, the derivative $\dfrac{\partial \hat{\rho}}{\partial \zeta_2} \neq 0$. Then, applying the implicit function theorem for holomorphic functions, we can express the variable ζ_2 through other variables:

$$\begin{cases} \zeta_2 = \psi(z_1, z_2, \zeta_1), \\ \bar{z}_1 = \zeta_1, \\ \bar{z}_2 = \zeta_2. \end{cases}$$

Then $f(z_1, z_2, \bar{z}_1, \bar{z}_2) = f(z_1, z_2, \zeta_1, \psi(z_1, z_2, \zeta_1))$ is a real-analytic function which can be expanded in a series of variables $z_1, z_2, \zeta_1 = \bar{z}_1$ converging in a neighborhood of the boundary point $(0, 0)$. Namely

$$f(z_1, \bar{z}_1, z_2) = \sum_{l=0}^{+\infty} \sum_{h+k+2m=l} b_{h,k,m} z_1^h \bar{z}_1^k z_2^m$$

on ∂D, where we have re-denoted the summation index giving a weight of 2 to z_2.

Choosing $|a|$ large enough, we consider moments $G(a, N)$ on sections $D_a(\tau)$:

$$G(a, N) = \int_{\partial \Delta_a} \tau^N f(D_a(\tau)) \, d\tau$$

$$= \int_{\partial \Delta_a} \tau^N \sum_{l=0}^{+\infty} \sum_{h+k+2m=l} b_{h,k,m} \left(\frac{\tau}{1+|a|^2} a \right)^h \left(\overline{\frac{\tau}{1+|a|^2} a} \right)^k \left(\frac{\tau}{1+|a|^2} \right)^m d\tau.$$

Let us prove that the coefficients $b_{h,k,m} = 0$ for $k > 0$. Let l_0 be the smallest weight degree with the property $b_{h,k,m} \neq 0$ for $k > 0$ and k_0 be the greatest degree in $\bar{z}_1$ for which this holds. Then from the condition of Theorem 4.5.1 $G(a, N) = 0$ for all N and a, in particular, for ta with $|a| = 1$ and $t \to +\infty$. Consider the limit

$$\lim_{t \to +\infty} G(ta, N) \, t^{l_0}$$

$$= \lim_{t \to +\infty} \int_{\partial \Delta_a} \tau^N \sum_{l=l_0}^{+\infty} \sum_{h+k+2m=l} b_{h,k,m} \left(\frac{\tau}{1 + |ta|^2} ta \right)^h \left(\frac{\tau}{1 + |ta|^2} ta \right)^k \times$$

$$\times \left(\frac{\tau}{1 + |ta|^2} \right)^m t^{l_0} d\tau = \lim_{t \to +\infty} \int_{\partial \Delta_a} \tau^N \sum_{l=l_0}^{+\infty} \sum_{h+k+2m=l} b_{h,k,m} \tau^h \bar{\tau}^k \tau^m t^h t^k t^{l_0} a^h \bar{a}^k \times$$

$$\times \left(\frac{1}{\frac{1}{t^2} + |a|^2} \right)^h \frac{1}{t^{2h}} \left(\frac{1}{\frac{1}{t^2} + |a|^2} \right)^k \frac{1}{t^{2k}} \left(\frac{1}{\frac{1}{t^2} + |a|^2} \right)^m \frac{1}{t^{2m}} d\tau$$

$$= \lim_{t \to +\infty} \sum_{l=l_0}^{+\infty} \sum_{h+k+2m=l} b_{h,k,m} t^{h+k+l_0-(2h+2k+2m)} a^h \bar{a}^k \left(\frac{1}{\frac{1}{t^2} + |a|^2} \right)^{l-m} \times$$

$$\times \int_{\partial \Delta_a} \tau^N \tau^h \bar{\tau}^k \tau^m \, d\tau = \lim_{t \to +\infty} \sum_{l=l_0}^{+\infty} \sum_{h+k+2m=l} b_{h,k,m} t^{l_0 - l} a^h \bar{a}^k \left(\frac{1}{t^2} + |a|^2 \right)^m \times$$

$$\times \frac{1}{\left(\frac{1}{t^2} + |a|^2 \right)^l} \int_{\partial \Delta_a} \tau^N \tau^h \bar{\tau}^k \tau^m \, d\tau = \sum_{h+k+2m=l_0} b_{h,k,m} a^h \bar{a}^k \int_{\partial \Delta} \tau^N \tau^h \bar{\tau}^k \tau^m \, d\tau = 0,$$

where $\partial \Delta$ is defined by (4.5.3).

We recall that the expressions for λ_1 and λ_2 are defined by (4.5.6), the angle α is found from relation (4.5.8) and the coefficients b'_1, and b'_2 from (4.5.10).

Proposition 4.5.2 *The value of the integral equals*

$$\int_{\partial \Delta} \tau^N \tau^h \bar{\tau}^k \tau^m \, d\tau$$

$$= e^{i\alpha(N+h+m+1-k)} \, 2\pi i \left(\frac{1}{2} \right)^{N+h+m+k+1} (-1)^{N+h+m+k} \left(\frac{1}{\sqrt{\lambda_1}} + \frac{1}{\sqrt{\lambda_2}} \right)^{N+h+m+1} \times$$

$$\times \left(\frac{1}{\sqrt{\lambda_1}} - \frac{1}{\sqrt{\lambda_2}} \right)^k \left(\frac{b'_1}{\sqrt{\lambda_1}} + i \frac{b'_2}{\sqrt{\lambda_2}} \right)^{N+h+m+k+1} \times$$

$$
\times \sum_{\substack{\alpha_1+\alpha_2+\alpha_3+\alpha_4 \\ =N+h+m+k+1}} \binom{k}{\alpha_3}\binom{k}{\alpha_4} \left(\frac{\sqrt{\lambda_2}-\sqrt{\lambda_1}}{\sqrt{\lambda_2}+\sqrt{\lambda_1}}\right)^{\alpha_1-\alpha_4} \left(\frac{\dfrac{b_1'}{\sqrt{\lambda_1}}-i\dfrac{b_2'}{\sqrt{\lambda_2}}}{\dfrac{b_1'}{\sqrt{\lambda_1}}+i\dfrac{b_2'}{\sqrt{\lambda_2}}}\right)^{\alpha_1+\alpha_4} \times
$$

$$
\times \left[-\binom{N+h+m}{\alpha_1-1}\binom{N+h+m}{\alpha_2-1}+\binom{N+h+m}{\alpha_1}\binom{N+h+m}{\alpha_2}\right].
$$

Proof Conformal mapping [57] of the exterior of an ellipse

$$
\frac{u''^2}{A^2}+\frac{v''^2}{B^2}=1
$$

on the exterior of a unit circle is performed by the function

$$
\omega = \frac{\tilde{\tau}+\sqrt{\tilde{\tau}^2-c^2}}{A+B},
$$

where $c=\sqrt{A^2-B^2}$. Then

$$
\tilde{\tau}=\frac{A+B}{2}\omega+\frac{A-B}{2}\frac{1}{\omega}. \tag{4.5.27}
$$

Thus, taking into account coordinate transformation (4.5.17) and representation (4.5.27) for $\tilde{\tau}$ we have the required mapping of the exterior of the ellipse (4.5.3) on the exterior of the unit circle by the formula

$$
\tau = \left(-\left[\frac{b_1'}{\lambda_1}+i\frac{b_2'}{\lambda_2}\right]+\frac{A+B}{2}\omega+\frac{A-B}{2}\frac{1}{\omega}\right)e^{i\alpha},
$$

where the semi-axes A and B are defined by (4.5.16). Then

$$
\bar{\tau} = \left(-\left[\frac{b_1'}{\lambda_1}-i\frac{b_2'}{\lambda_2}\right]+\frac{A+B}{2}\bar{\omega}+\frac{A-B}{2}\frac{1}{\bar{\omega}}\right)e^{-i\alpha},
$$

$$
d\tau = \left(\frac{A+B}{2}-\frac{A-B}{2}\frac{1}{\omega^2}\right)e^{i\alpha}\,d\omega.
$$

Let us find the value of the integral $\int_{\partial\Delta} \tau^N \tau^h \bar{\tau}^k \tau^m \, d\tau$ by substituting the resulting expressions for τ, $\bar{\tau}$ and the differential $d\tau$. We get

$$
I = \int_{\partial\Delta} \tau^{N+h+m} \bar{\tau}^k \, d\tau = e^{i\alpha(N+h+m+1-k)} \int_{|\omega|=1} \left(-\left[\frac{b_1'}{\lambda_1} + i\frac{b_2'}{\lambda_2} \right] \right.
$$

$$
+ \frac{A+B}{2}\omega + \frac{A-B}{2}\frac{1}{\omega} \Bigg)^{N+h+m} \left(-\left[\frac{b_1'}{\lambda_1} - i\frac{b_2'}{\lambda_2} \right] + \frac{A+B}{2}\frac{1}{\omega} + \frac{A-B}{2}\omega \right)^k \times
$$

$$
\times \left(\frac{A+B}{2} - \frac{A-B}{2}\frac{1}{\omega^2} \right) d\omega.
$$

We make the substitution $\omega = \dfrac{1}{\zeta}$ under the integral sign while $d\omega = -\dfrac{1}{\zeta^2}\,d\zeta$, then

$$
I = e^{i\alpha(N+h+m+1-k)} \int_{|\zeta|=1} \left(-\left[\frac{b_1'}{\lambda_1} + i\frac{b_2'}{\lambda_2} \right] + \frac{A+B}{2}\frac{1}{\zeta} + \frac{A-B}{2}\zeta \right)^{N+h+m} \times
$$

$$
\times \left(-\left[\frac{b_1'}{\lambda_1} - i\frac{b_2'}{\lambda_2} \right] + \frac{A+B}{2}\zeta + \frac{A-B}{2}\frac{1}{\zeta} \right)^k \left(\frac{A+B}{2} - \frac{A-B}{2}\zeta^2 \right)\left(-\frac{1}{\zeta^2} \right) d\zeta
$$

$$
= e^{i\alpha(N+h+m+1-k)} \int_{|\zeta|=1} \left(-\left[\frac{b_1'}{\lambda_1} + i\frac{b_2'}{\lambda_2} \right] + \frac{A+B}{2}\frac{1}{\zeta} + \frac{A-B}{2}\zeta \right)^{N+h+m} \times
$$

$$
\times \left(-\left[\frac{b_1'}{\lambda_1} - i\frac{b_2'}{\lambda_2} \right] + \frac{A+B}{2}\zeta + \frac{A-B}{2}\frac{1}{\zeta} \right)^k \left(\frac{A-B}{2}\zeta^2 - \frac{A+B}{2} \right)\frac{1}{\zeta^2} d\zeta
$$

$$
= e^{i\alpha(N+h+m+1-k)} \int_{|\zeta|=1} \Omega(\zeta)\, d\zeta,
$$

where

$$
\Omega(\zeta) = \frac{\left(\dfrac{A-B}{2}\zeta^2 - \left[\dfrac{b_1'}{\lambda_1} + i\dfrac{b_2'}{\lambda_2} \right]\zeta + \dfrac{A+B}{2} \right)^{N+h+m}}{\zeta^{N+h+m+k+2}} \times
$$

$$
\times \left(\frac{A+B}{2}\zeta^2 - \left[\frac{b_1'}{\lambda_1} - i\frac{b_2'}{\lambda_2} \right]\zeta + \frac{A-B}{2} \right)^k \left(\frac{A-B}{2}\zeta^2 - \frac{A+B}{2} \right).
$$

We now factorize the quadratic trinomial of the variable ζ, under the integral sign. First we find the value of the expression $A^2 - B^2$. Using relation (4.5.16) we obtain

$$A^2 - B^2 = \frac{b_1'^2 \lambda_2 + b_2'^2 \lambda_1}{\lambda_1^2 \lambda_2} - \frac{b_1'^2 \lambda_2 + b_2'^2 \lambda_1}{\lambda_1 \lambda_2^2} = \frac{b_1'^2}{\lambda_1^2} + \frac{b_2'^2}{\lambda_1 \lambda_2} - \frac{b_1'^2}{\lambda_1 \lambda_2} - \frac{b_2'^2}{\lambda_2^2}.$$

We find the roots of the quadratic equation

$$\frac{A-B}{2}\zeta^2 - \left[\frac{b_1'}{\lambda_1} + i\frac{b_2'}{\lambda_2}\right]\zeta + \frac{A+B}{2} = 0. \qquad (4.5.28)$$

Its discriminant is

$$\left(\frac{b_1'}{\lambda_1} + i\frac{b_2'}{\lambda_2}\right)^2 - 4\frac{A^2-B^2}{4} = \frac{b_1'^2}{\lambda_1^2} - \frac{b_2'^2}{\lambda_2^2} + 2i\frac{b_1'b_2'}{\lambda_1\lambda_2} - \left(\frac{b_1'^2}{\lambda_1^2} + \frac{b_2'^2}{\lambda_1\lambda_2} - \frac{b_1'^2}{\lambda_1\lambda_2} - \frac{b_2'^2}{\lambda_2^2}\right)$$

$$= 2i\frac{b_1'b_2'}{\lambda_1\lambda_2} - \frac{b_2'^2}{\lambda_1\lambda_2} + \frac{b_1'^2}{\lambda_1\lambda_2} = \frac{1}{\lambda_1\lambda_2}\left(b_1'^2 + 2ib_1'b_2' - b_2'^2\right) = \frac{1}{\lambda_1\lambda_2}\left(b_1' + ib_2'\right)^2.$$

Then the roots of quadratic equation (4.5.28) are defined by

$$\zeta_{1,2} = \frac{\dfrac{b_1'}{\lambda_1} + i\dfrac{b_2'}{\lambda_2} \pm \dfrac{1}{\sqrt{\lambda_1\lambda_2}}\left(b_1' + ib_2'\right)}{A-B}.$$

Thus

$$\zeta_1 = \frac{\dfrac{b_1'}{\lambda_1} + i\dfrac{b_2'}{\lambda_2} + \dfrac{1}{\sqrt{\lambda_1\lambda_2}}\left(b_1' + ib_2'\right)}{A-B} = \frac{\dfrac{b_1'}{\lambda_1} + \dfrac{b_1'}{\sqrt{\lambda_1\lambda_2}} + i\left(\dfrac{b_2'}{\lambda_2} + \dfrac{b_2'}{\sqrt{\lambda_1\lambda_2}}\right)}{A-B}$$

$$= \frac{\left(\dfrac{1}{\sqrt{\lambda_1}} + \dfrac{1}{\sqrt{\lambda_2}}\right)\left(\dfrac{b_1'}{\sqrt{\lambda_1}} + i\dfrac{b_2'}{\sqrt{\lambda_2}}\right)}{A-B},$$

$$\zeta_2 = \frac{\dfrac{b_1'}{\lambda_1} + i\dfrac{b_2'}{\lambda_2} - \dfrac{1}{\sqrt{\lambda_1\lambda_2}}\left(b_1' + ib_2'\right)}{A-B} = \frac{\dfrac{b_1'}{\lambda_1} - \dfrac{b_1'}{\sqrt{\lambda_1\lambda_2}} + i\left(\dfrac{b_2'}{\lambda_2} - \dfrac{b_2'}{\sqrt{\lambda_1\lambda_2}}\right)}{A-B}$$

$$= \frac{\left(\dfrac{1}{\sqrt{\lambda_1}} - \dfrac{1}{\sqrt{\lambda_2}}\right)\left(\dfrac{b_1'}{\sqrt{\lambda_1}} - i\dfrac{b_2'}{\sqrt{\lambda_2}}\right)}{A-B}.$$

We also consider the quadratic equation

$$\frac{A+B}{2}\zeta^2 - \left[\frac{b_1'}{\lambda_1} - i\frac{b_2'}{\lambda_2}\right]\zeta + \frac{A-B}{2} = 0. \tag{4.5.29}$$

Its discriminant is

$$\left(\frac{b_1'}{\lambda_1} - i\frac{b_2'}{\lambda_2}\right)^2 - 4\frac{A^2-B^2}{4} = \frac{b_1'^2}{\lambda_1^2} - \frac{b_2'^2}{\lambda_2^2} - 2i\frac{b_1'b_2'}{\lambda_1\lambda_2} - \left(\frac{b_1'^2}{\lambda_1^2} + \frac{b_2'^2}{\lambda_1\lambda_2} - \frac{b_1'^2}{\lambda_1\lambda_2} - \frac{b_2'^2}{\lambda_2^2}\right)$$

$$= -2i\frac{b_1'b_2'}{\lambda_1\lambda_2} - \frac{b_2'^2}{\lambda_1\lambda_2} + \frac{b_1'^2}{\lambda_1\lambda_2} = \frac{1}{\lambda_1\lambda_2}\left(b_1'^2 - 2ib_1'b_2' - b_2'^2\right) = \frac{1}{\lambda_1\lambda_2}\left(b_1' - ib_2'\right)^2.$$

Then the roots of quadratic equation (4.5.29) are defined by

$$\zeta_{3,4} = \frac{\dfrac{b_1'}{\lambda_1} - i\dfrac{b_2'}{\lambda_2} \pm \dfrac{1}{\sqrt{\lambda_1\lambda_2}}\left(b_1' - ib_2'\right)}{A+B}.$$

Therefore,

$$\zeta_3 = \frac{\dfrac{b_1'}{\lambda_1} - i\dfrac{b_2'}{\lambda_2} + \dfrac{1}{\sqrt{\lambda_1\lambda_2}}\left(b_1' - ib_2'\right)}{A+B} = \frac{\dfrac{b_1'}{\lambda_1} + \dfrac{b_1'}{\sqrt{\lambda_1\lambda_2}} - i\left(\dfrac{b_2'}{\lambda_2} + \dfrac{b_2'}{\sqrt{\lambda_1\lambda_2}}\right)}{A+B}$$

$$= \frac{\left(\dfrac{1}{\sqrt{\lambda_1}} + \dfrac{1}{\sqrt{\lambda_2}}\right)\left(\dfrac{b_1'}{\sqrt{\lambda_1}} - i\dfrac{b_2'}{\sqrt{\lambda_2}}\right)}{A+B},$$

$$\zeta_4 = \frac{\dfrac{b_1'}{\lambda_1} - i\dfrac{b_2'}{\lambda_2} - \dfrac{1}{\sqrt{\lambda_1\lambda_2}}\left(b_1' - ib_2'\right)}{A+B} = \frac{\dfrac{b_1'}{\lambda_1} - \dfrac{b_1'}{\sqrt{\lambda_1\lambda_2}} - i\left(\dfrac{b_2'}{\lambda_2} - \dfrac{b_2'}{\sqrt{\lambda_1\lambda_2}}\right)}{A+B}$$

$$= \frac{\left(\dfrac{1}{\sqrt{\lambda_1}} - \dfrac{1}{\sqrt{\lambda_2}}\right)\left(\dfrac{b_1'}{\sqrt{\lambda_1}} + i\dfrac{b_2'}{\sqrt{\lambda_2}}\right)}{A+B}.$$

Thus, we have the following factorization for the square trinomials of the variable ζ under the integral sign:

$$\frac{A-B}{2}\zeta^2 - \left[\frac{b_1'}{\lambda_1} + i\frac{b_2'}{\lambda_2}\right]\zeta + \frac{A+B}{2} = \frac{A-B}{2}(\zeta - \zeta_1)(\zeta - \zeta_2),$$

$$\frac{A+B}{2}\zeta^2 - \left[\frac{b_1'}{\lambda_1} - i\frac{b_2'}{\lambda_2}\right]\zeta + \frac{A-B}{2} = \frac{A+B}{2}(\zeta - \zeta_3)(\zeta - \zeta_4).$$

Then the original integral I can be written as

$$I = e^{i\alpha(N+h+m+1-k)} \left(\frac{A-B}{2}\right)^{N+h+m} \left(\frac{A+B}{2}\right)^k \int_{|\zeta|=1} \frac{f_1 f_2 f_3 f_4 f_5}{\zeta^{N+h+m+k+2}} d\zeta,$$

where the functions

$$f_1 = (\zeta - \zeta_1)^{N+h+m}, \qquad f_2 = (\zeta - \zeta_2)^{N+h+m}, \qquad f_3 = (\zeta - \zeta_3)^k,$$

$$f_4 = (\zeta - \zeta_4)^k, \qquad f_5 = \frac{A-B}{2}\zeta^2 - \frac{A+B}{2}.$$

The point $\zeta = 0$ is a pole of order $(N+h+m+k+2)$. We will need the value of the derivative of the function

$$(f_1 f_2 f_3 f_4 f_5)^{(N+h+m+k+1)}$$

$$= \sum_{\substack{\alpha_1+\alpha_2+\alpha_3+\alpha_4+\alpha_5= \\ =N+h+m+k+1}} f_1^{(\alpha_1)} f_2^{(\alpha_2)} f_3^{(\alpha_3)} f_4^{(\alpha_4)} f_5^{(\alpha_5)} \frac{(N+h+m+k+1)!}{\alpha_1!\alpha_2!\alpha_3!\alpha_4!\alpha_5!}.$$

Expressions for the function derivatives are as follows:

$$f_1^{(\alpha_1)} = \frac{(N+h+m)!}{(N+h+m-\alpha_1)!}(\zeta - \zeta_1)^{N+h+m-\alpha_1},$$

$$f_2^{(\alpha_2)} = \frac{(N+h+m)!}{(N+h+m-\alpha_2)!}(\zeta - \zeta_2)^{N+h+m-\alpha_2},$$

$$f_3^{(\alpha_3)} = \frac{k!}{(k-\alpha_3)!}(\zeta - \zeta_3)^{k-\alpha_3}, \qquad\qquad f_4^{(\alpha_4)} = \frac{k!}{(k-\alpha_4)!}(\zeta - \zeta_4)^{k-\alpha_4},$$

$$f_5^{(2)} = 2\frac{A-B}{2}, \qquad\qquad f_5^{(0)} = \frac{A-B}{2}\zeta^2 - \frac{A+B}{2}.$$

The order of the derivative is subject to natural limitations:

$$\alpha_1 \le N+h+m, \quad \alpha_2 \le N+h+m, \quad \alpha_3 \le k, \quad \alpha_4 \le k.$$

Thus

$$\frac{(f_1 f_2 f_3 f_4 f_5)^{(N+h+m+k+1)}}{(N+h+m+k+1)!}(0) = \sum_{\substack{\alpha_1+\alpha_2+\alpha_3+\alpha_4+2= \\ =N+h+m+k+1}} \binom{N+h+m}{\alpha_1}\binom{N+h+m}{\alpha_2} \times$$

$$\times \binom{k}{\alpha_3}\binom{k}{\alpha_4}(-\zeta_1)^{N+h+m-\alpha_1}(-\zeta_2)^{N+h+m-\alpha_2}(-\zeta_3)^{k-\alpha_3}(-\zeta_4)^{k-\alpha_4}\left(\frac{A-B}{2}\right)$$

$$+ \sum_{\substack{\alpha_1+\alpha_2+\alpha_3+\alpha_4=\\=N+h+m+k+1}} \binom{N+h+m}{\alpha_1}\binom{N+h+m}{\alpha_2}\binom{k}{\alpha_3}\binom{k}{\alpha_4}(-\zeta_1)^{N+h+m-\alpha_1} \times$$

$$\times (-\zeta_2)^{N+h+m-\alpha_2}(-\zeta_3)^{k-\alpha_3}(-\zeta_4)^{k-\alpha_4}\left(-\frac{A+B}{2}\right).$$

Then, the original integral I can be represented as

$$I = e^{i\alpha(N+h+m+1-k)}\, 2\pi i\, (\sigma_1 + \sigma_2),$$

where σ_1 corresponds to the part of the sum in the expression for the product derivative

$$\frac{(f_1 f_2 f_3 f_4 f_5)^{(N+h+m+k+1)}}{(N+h+m+k+1)!}(0),$$

in which the parameter $\alpha_5 = 2$ and σ_2 corresponds to the sum with the parameter $\alpha_5 = 0$, i.e.,

$$\sigma_1 = \left(\frac{A-B}{2}\right)^{N+h+m}\left(\frac{A+B}{2}\right)^k \sum_{\substack{\alpha_1+\alpha_2+\alpha_3+\alpha_4+2=\\=N+h+m+k+1}} \binom{N+h+m}{\alpha_1}\binom{N+h+m}{\alpha_2} \times$$

$$\times \binom{k}{\alpha_3}\binom{k}{\alpha_4}(-\zeta_1)^{N+h+m-\alpha_1}(-\zeta_2)^{N+h+m-\alpha_2}(-\zeta_3)^{k-\alpha_3}(-\zeta_4)^{k-\alpha_4}\left(\frac{A-B}{2}\right),$$

$$\sigma_2 = \left(\frac{A-B}{2}\right)^{N+h+m}\left(\frac{A+B}{2}\right)^k \sum_{\substack{\alpha_1+\alpha_2+\alpha_3+\alpha_4=\\=N+h+m+k+1}} \binom{N+h+m}{\alpha_1}\binom{N+h+m}{\alpha_2} \times$$

$$\times \binom{k}{\alpha_3}\binom{k}{\alpha_4}(-\zeta_1)^{N+h+m-\alpha_1}(-\zeta_2)^{N+h+m-\alpha_2}(-\zeta_3)^{k-\alpha_3}(-\zeta_4)^{k-\alpha_4}\left(\frac{A+B}{2}\right).$$

In the expression for σ_1 we introduce $\alpha_1 + 1 = \alpha_1'$, $\alpha_2 + 1 = \alpha_2'$, and obtain

$$\sigma_1 = \left(\frac{A-B}{2}\right)^{N+h+m}\left(\frac{A+B}{2}\right)^k \sum_{\substack{\alpha_1'+\alpha_2'+\alpha_3+\alpha_4=\\=N+h+m+k+1}} \binom{N+h+m}{\alpha_1'-1}\binom{N+h+m}{\alpha_2'-1}\binom{k}{\alpha_3} \times$$

$$\times \binom{k}{\alpha_4}(-\zeta_1)^{N+h+m-\alpha_1'+1}(-\zeta_2)^{N+h+m-\alpha_2'+1}(-\zeta_3)^{k-\alpha_3}(-\zeta_4)^{k-\alpha_4}\left(\frac{A-B}{2}\right).$$

We add σ_1 and σ_2, and get

$$
\sigma = \sigma_1 + \sigma_2 = \left(\frac{A-B}{2}\right)^{N+h+m} \left(\frac{A+B}{2}\right)^{k} \sum_{\substack{\alpha_1+\alpha_2+\alpha_3+\alpha_4= \\ =N+h+m+k+1}} \binom{k}{\alpha_3}\binom{k}{\alpha_4} \times
$$

$$
\times (-\zeta_1)^{N+h+m-\alpha_1} (-\zeta_2)^{N+h+m-\alpha_2} (-\zeta_3)^{k-\alpha_3} (-\zeta_4)^{k-\alpha_4} \times
$$

$$
\times \left[\binom{N+h+m}{\alpha_1-1}\binom{N+h+m}{\alpha_2-1} \zeta_1\zeta_2 \left(\frac{A-B}{2}\right)\right.
$$

$$
\left. + \binom{N+h+m}{\alpha_1}\binom{N+h+m}{\alpha_2} \left(-\frac{A+B}{2}\right)\right].
$$

Next we calculate the degree of (-1):

$$
N+h+m-\alpha_1 + N+h+m-\alpha_2 + k-\alpha_3 + k-\alpha_4 = 2(N+h+m+k)
$$

$$
-(\alpha_1+\alpha_2+\alpha_3+\alpha_4) = 2(N+h+m+k)-(N+h+m+k+1) = N+h+m+k-1,
$$

then

$$
\sigma = \left(\frac{A-B}{2}\right)^{N+h+m}\left(\frac{A+B}{2}\right)^{k}(-1)^{N+h+m+k-1}\zeta_1^{N+h+m}\zeta_2^{N+h+m}\zeta_3^{k}\zeta_4^{k}\times
$$

$$
\times \sum_{\substack{\alpha_1+\alpha_2+\alpha_3+\alpha_4= \\ =N+h+m+k+1}} \binom{k}{\alpha_3}\binom{k}{\alpha_4} \zeta_1^{-\alpha_1}\zeta_2^{-\alpha_2}\zeta_3^{-\alpha_3}\zeta_4^{-\alpha_4}\left[\binom{N+h+m}{\alpha_1-1}\binom{N+h+m}{\alpha_2-1}\right.\times
$$

$$
\left.\times \zeta_1\zeta_2\left(\frac{A-B}{2}\right) + \binom{N+h+m}{\alpha_1}\binom{N+h+m}{\alpha_2}\left(-\frac{A+B}{2}\right)\right].
$$

For the solutions ζ_1, ζ_2, and ζ_3, ζ_4 of the respective quadratic equations (4.5.28) and (4.5.29) the following relations

$$
\zeta_1\zeta_2 = \frac{A+B}{A-B}, \qquad \zeta_3\zeta_4 = \frac{A-B}{A+B}
$$

hold. Thus

$$
\sigma = \left(\frac{A-B}{2}\right)^{N+h+m}\left(\frac{A+B}{2}\right)^{k}(-1)^{N+h+m+k-1}\left(\frac{A+B}{A-B}\right)^{N+h+m}\left(\frac{A-B}{A+B}\right)^{k}\times
$$

$$
\times \sum_{\substack{\alpha_1+\alpha_2+\alpha_3+\alpha_4 \\ =N+h+m+k+1}} \binom{k}{\alpha_3}\binom{k}{\alpha_4} \zeta_1^{-\alpha_1}\zeta_2^{-\alpha_2}\zeta_3^{-\alpha_3}\zeta_4^{-\alpha_4}\left[\binom{N+h+m}{\alpha_1-1}\binom{N+h+m}{\alpha_2-1}\right.\times
$$

$$\times \left(\frac{A+B}{2}\right) + \binom{N+h+m}{\alpha_1}\binom{N+h+m}{\alpha_2}\left(-\frac{A+B}{2}\right)\Bigg]$$

$$= \left(\frac{1}{2}\right)^{N+h+m+k+1} (-1)^{N+h+m+k} (A+B)^{N+h+m+1} (A-B)^k \times$$

$$\times \sum_{\substack{\alpha_1+\alpha_2+\alpha_3+\alpha_4 \\ =N+h+m+k+1}} \binom{k}{\alpha_3}\binom{k}{\alpha_4} \zeta_1^{-\alpha_1} \zeta_2^{-\alpha_2} \zeta_3^{-\alpha_3} \zeta_4^{-\alpha_4} \times$$

$$\times \left[-\binom{N+h+m}{\alpha_1-1}\binom{N+h+m}{\alpha_2-1} + \binom{N+h+m}{\alpha_1}\binom{N+h+m}{\alpha_2} \right].$$

Denote $C = C(\alpha) = b_1'^2\lambda_2 + b_2'^2\lambda_1$. Then, using relations (4.5.16) and the introduced notation we find that

$$A = \frac{\sqrt{C}}{\lambda_1\sqrt{\lambda_2}}, \qquad B = \frac{\sqrt{C}}{\sqrt{\lambda_1}\lambda_2}.$$

So

$$A - B = \frac{\sqrt{C}}{\lambda_1\sqrt{\lambda_2}} - \frac{\sqrt{C}}{\sqrt{\lambda_1}\lambda_2} = \frac{\sqrt{C}}{\sqrt{\lambda_1}\sqrt{\lambda_2}}\left(\frac{1}{\sqrt{\lambda_1}} - \frac{1}{\sqrt{\lambda_2}}\right),$$

$$A + B = \frac{\sqrt{C}}{\lambda_1\sqrt{\lambda_2}} + \frac{\sqrt{C}}{\sqrt{\lambda_1}\lambda_2} = \frac{\sqrt{C}}{\sqrt{\lambda_1}\sqrt{\lambda_2}}\left(\frac{1}{\sqrt{\lambda_1}} + \frac{1}{\sqrt{\lambda_2}}\right).$$

Then the roots $\zeta_1, \zeta_2, \zeta_3, \zeta_4$ of quadratic equations (4.5.28) and (4.5.29) can be written as

$$\zeta_1 = \frac{\left(\frac{1}{\sqrt{\lambda_1}} + \frac{1}{\sqrt{\lambda_2}}\right)\left(\frac{b_1'}{\sqrt{\lambda_1}} + i\frac{b_2'}{\sqrt{\lambda_2}}\right)}{A-B}$$

$$= \frac{\sqrt{\lambda_1}\sqrt{\lambda_2}\left(\sqrt{\lambda_2}+\sqrt{\lambda_1}\right)}{\sqrt{C}\left(\sqrt{\lambda_2}-\sqrt{\lambda_1}\right)}\left(\frac{b_1'}{\sqrt{\lambda_1}} + i\frac{b_2'}{\sqrt{\lambda_2}}\right),$$

$$\zeta_2 = \frac{\left(\frac{1}{\sqrt{\lambda_1}} - \frac{1}{\sqrt{\lambda_2}}\right)\left(\frac{b_1'}{\sqrt{\lambda_1}} - i\frac{b_2'}{\sqrt{\lambda_2}}\right)}{A-B} = \frac{\sqrt{\lambda_1}\sqrt{\lambda_2}}{\sqrt{C}}\left(\frac{b_1'}{\sqrt{\lambda_1}} - i\frac{b_2'}{\sqrt{\lambda_2}}\right),$$

$$\zeta_3 = \frac{\left(\frac{1}{\sqrt{\lambda_1}} + \frac{1}{\sqrt{\lambda_2}}\right)\left(\frac{b_1'}{\sqrt{\lambda_1}} - i\frac{b_2'}{\sqrt{\lambda_2}}\right)}{A+B} = \frac{\sqrt{\lambda_1}\sqrt{\lambda_2}}{\sqrt{C}}\left(\frac{b_1'}{\sqrt{\lambda_1}} - i\frac{b_2'}{\sqrt{\lambda_2}}\right),$$

$$\zeta_4 = \frac{\left(\dfrac{1}{\sqrt{\lambda_1}} - \dfrac{1}{\sqrt{\lambda_2}}\right)\left(\dfrac{b_1'}{\sqrt{\lambda_1}} + i\dfrac{b_2'}{\sqrt{\lambda_2}}\right)}{A + B}$$

$$= \frac{\sqrt{\lambda_1}\sqrt{\lambda_2}\,(\sqrt{\lambda_2} - \sqrt{\lambda_1})}{\sqrt{C}\,(\sqrt{\lambda_2} + \sqrt{\lambda_1})}\left(\frac{b_1'}{\sqrt{\lambda_1}} + i\frac{b_2'}{\sqrt{\lambda_2}}\right).$$

Now we find the value of the expression $\zeta_1^{-\alpha_1}\zeta_2^{-\alpha_2}\zeta_3^{-\alpha_3}\zeta_4^{-\alpha_4}$, under the sign of the sum in the expression for σ.

$$\zeta_1^{-\alpha_1}\zeta_2^{-\alpha_2}\zeta_3^{-\alpha_3}\zeta_4^{-\alpha_4} = \left(\frac{\sqrt{\lambda_1}\sqrt{\lambda_2}}{\sqrt{C}}\right)^{-\alpha_1}\left(\frac{\sqrt{\lambda_2} + \sqrt{\lambda_1}}{\sqrt{\lambda_2} - \sqrt{\lambda_1}}\right)^{-\alpha_1} \times$$

$$\times \left(\frac{b_1'}{\sqrt{\lambda_1}} + i\frac{b_2'}{\sqrt{\lambda_2}}\right)^{-\alpha_1}\left(\frac{\sqrt{\lambda_1}\sqrt{\lambda_2}}{\sqrt{C}}\right)^{-\alpha_2}\left(\frac{b_1'}{\sqrt{\lambda_1}} - i\frac{b_2'}{\sqrt{\lambda_2}}\right)^{-\alpha_2}\left(\frac{\sqrt{\lambda_1}\sqrt{\lambda_2}}{\sqrt{C}}\right)^{-\alpha_3} \times$$

$$\times \left(\frac{b_1'}{\sqrt{\lambda_1}} - i\frac{b_2'}{\sqrt{\lambda_2}}\right)^{-\alpha_3}\left(\frac{\sqrt{\lambda_1}\sqrt{\lambda_2}}{\sqrt{C}}\right)^{-\alpha_4}\left(\frac{\sqrt{\lambda_2} - \sqrt{\lambda_1}}{\sqrt{\lambda_2} + \sqrt{\lambda_1}}\right)^{-\alpha_4}\left(\frac{b_1'}{\sqrt{\lambda_1}} + i\frac{b_2'}{\sqrt{\lambda_2}}\right)^{-\alpha_4}$$

$$= \left(\frac{\sqrt{C}}{\sqrt{\lambda_1}\sqrt{\lambda_2}}\right)^{\alpha_1 + \alpha_2 + \alpha_3 + \alpha_4}\left(\frac{\sqrt{\lambda_2} - \sqrt{\lambda_1}}{\sqrt{\lambda_2} + \sqrt{\lambda_1}}\right)^{\alpha_1 - \alpha_4}\left(\frac{b_1'}{\sqrt{\lambda_1}} - i\frac{b_2'}{\sqrt{\lambda_2}}\right)^{-\alpha_2 - \alpha_3} \times$$

$$\times \left(\frac{b_1'}{\sqrt{\lambda_1}} + i\frac{b_2'}{\sqrt{\lambda_2}}\right)^{-\alpha_1 - \alpha_4} = \left(\frac{\sqrt{C}}{\sqrt{\lambda_1}\sqrt{\lambda_2}}\right)^{N+h+m+k+1}\left(\frac{\sqrt{\lambda_2} - \sqrt{\lambda_1}}{\sqrt{\lambda_2} + \sqrt{\lambda_1}}\right)^{\alpha_1 - \alpha_4} \times$$

$$\times \left(\frac{b_1'}{\sqrt{\lambda_1}} - i\frac{b_2'}{\sqrt{\lambda_2}}\right)^{\alpha_1 + \alpha_4}\left(\frac{b_1'}{\sqrt{\lambda_1}} - i\frac{b_2'}{\sqrt{\lambda_2}}\right)^{-(N+h+m+k+1)}\left(\frac{b_1'}{\sqrt{\lambda_1}} + i\frac{b_2'}{\sqrt{\lambda_2}}\right)^{-\alpha_1 - \alpha_4}$$

$$= \left(\frac{\sqrt{C}}{\sqrt{\lambda_1}\sqrt{\lambda_2}}\right)^{N+h+m+k+1}\left(\frac{b_1'}{\sqrt{\lambda_1}} - i\frac{b_2'}{\sqrt{\lambda_2}}\right)^{-(N+h+m+k+1)}\left(\frac{\sqrt{\lambda_2} - \sqrt{\lambda_1}}{\sqrt{\lambda_2} + \sqrt{\lambda_1}}\right)^{\alpha_1 - \alpha_4} \times$$

$$\times \left(\frac{\dfrac{b_1'}{\sqrt{\lambda_1}} - i\dfrac{b_2'}{\sqrt{\lambda_2}}}{\dfrac{b_1'}{\sqrt{\lambda_1}} + i\dfrac{b_2'}{\sqrt{\lambda_2}}}\right)^{\alpha_1 + \alpha_4}.$$

Thus

$$\sigma = \left(\frac{1}{2}\right)^{N+h+m+k+1}(-1)^{N+h+m+k}(A + B)^{N+h+m+1}(A - B)^k \times$$

$$\times \left(\frac{\sqrt{C}}{\sqrt{\lambda_1}\sqrt{\lambda_2}}\right)^{N+h+m+k+1}\left(\frac{b_1'}{\sqrt{\lambda_1}} - i\frac{b_2'}{\sqrt{\lambda_2}}\right)^{-(N+h+m+k+1)} \times$$

$$\times \sum_{\substack{\alpha_1+\alpha_2+\alpha_3+\alpha_4= \\ =N+h+m+k+1}} \binom{k}{\alpha_3}\binom{k}{\alpha_4} \left(\frac{\sqrt{\lambda_2}-\sqrt{\lambda_1}}{\sqrt{\lambda_2}+\sqrt{\lambda_1}}\right)^{\alpha_1-\alpha_4} \left(\frac{\dfrac{b_1'}{\sqrt{\lambda_1}}-i\dfrac{b_2'}{\sqrt{\lambda_2}}}{\dfrac{b_1'}{\sqrt{\lambda_1}}+i\dfrac{b_2'}{\sqrt{\lambda_2}}}\right)^{\alpha_1+\alpha_4} \times$$

$$\times \left[-\binom{N+h+m}{\alpha_1-1}\binom{N+h+m}{\alpha_2-1}+\binom{N+h+m}{\alpha_1}\binom{N+h+m}{\alpha_2}\right].$$

We transform the factor standing in front of the sum and obtain

$$(A+B)^{N+h+m+1}\,(A-B)^k \left(\frac{\sqrt{C}}{\sqrt{\lambda_1}\sqrt{\lambda_2}}\right)^{N+h+m+k+1} \left(\frac{b_1'}{\sqrt{\lambda_1}}-i\frac{b_2'}{\sqrt{\lambda_2}}\right)^{-(N+h+m+k+1)}$$

$$=\left(\frac{\sqrt{C}}{\sqrt{\lambda_1}\sqrt{\lambda_2}}\right)^{N+h+m+1} \left(\frac{1}{\sqrt{\lambda_1}}+\frac{1}{\sqrt{\lambda_2}}\right)^{N+h+m+1} \left(\frac{\sqrt{C}}{\sqrt{\lambda_1}\sqrt{\lambda_2}}\right)^{k} \left(\frac{1}{\sqrt{\lambda_1}}-\frac{1}{\sqrt{\lambda_2}}\right)^{k} \times$$

$$\times \left(\frac{\sqrt{C}}{\sqrt{\lambda_1}\sqrt{\lambda_2}}\right)^{N+h+m+k+1} \left(\frac{b_1'}{\sqrt{\lambda_1}}-i\frac{b_2'}{\sqrt{\lambda_2}}\right)^{-(N+h+m+k+1)}$$

$$=\left(\frac{1}{\sqrt{\lambda_1}}+\frac{1}{\sqrt{\lambda_2}}\right)^{N+h+m+1} \left(\frac{1}{\sqrt{\lambda_1}}-\frac{1}{\sqrt{\lambda_2}}\right)^{k} \left(\frac{C}{\lambda_1\lambda_2}\right)^{N+h+m+k+1} \times$$

$$\times \left(\frac{b_1'}{\sqrt{\lambda_1}}-i\frac{b_2'}{\sqrt{\lambda_2}}\right)^{-(N+h+m+k+1)} =\left(\frac{1}{\sqrt{\lambda_1}}+\frac{1}{\sqrt{\lambda_2}}\right)^{N+h+m+1} \left(\frac{1}{\sqrt{\lambda_1}}-\frac{1}{\sqrt{\lambda_2}}\right)^{k} \times$$

$$\times \left(\frac{b_1'^2}{\lambda_1}+\frac{b_2'^2}{\lambda_2}\right)^{N+h+m+k+1} \left(\frac{b_1'}{\sqrt{\lambda_1}}-i\frac{b_2'}{\sqrt{\lambda_2}}\right)^{-(N+h+m+k+1)}$$

$$=\left(\frac{1}{\sqrt{\lambda_1}}+\frac{1}{\sqrt{\lambda_2}}\right)^{N+h+m+1} \left(\frac{1}{\sqrt{\lambda_1}}-\frac{1}{\sqrt{\lambda_2}}\right)^{k} \left(\frac{b_1'}{\sqrt{\lambda_1}}+i\frac{b_2'}{\sqrt{\lambda_2}}\right)^{N+h+m+k+1}.$$

Finally, after all of these transformations and calculations have been done, we have

$$\sigma=\left(\frac{1}{2}\right)^{N+h+m+k+1} (-1)^{N+h+m+k} \left(\frac{1}{\sqrt{\lambda_1}}+\frac{1}{\sqrt{\lambda_2}}\right)^{N+h+m+1} \left(\frac{1}{\sqrt{\lambda_1}}-\frac{1}{\sqrt{\lambda_2}}\right)^{k} \times$$

$$\times \left(\frac{b_1'}{\sqrt{\lambda_1}}+i\frac{b_2'}{\sqrt{\lambda_2}}\right)^{N+h+m+k+1} \sum_{\substack{\alpha_1+\alpha_2+\alpha_3+\alpha_4= \\ =N+h+m+k+1}} \binom{k}{\alpha_3}\binom{k}{\alpha_4} \left(\frac{\sqrt{\lambda_2}-\sqrt{\lambda_1}}{\sqrt{\lambda_2}+\sqrt{\lambda_1}}\right)^{\alpha_1-\alpha_4} \times$$

$$\times \left(\frac{\dfrac{b'_1}{\sqrt{\lambda_1}} - i\dfrac{b'_2}{\sqrt{\lambda_2}}}{\dfrac{b'_1}{\sqrt{\lambda_1}} + i\dfrac{b'_2}{\sqrt{\lambda_2}}} \right)^{\alpha_1 + \alpha_4} \left[-\binom{N+h+m}{\alpha_1 - 1}\binom{N+h+m}{\alpha_2 - 1} \right.$$

$$\left. + \binom{N+h+m}{\alpha_1}\binom{N+h+m}{\alpha_2} \right]. \qquad (4.5.30)$$

These arguments complete the proof. □

Lemma 4.5.2 *Expression* (4.5.30) *is not identically zero with respect to the variable equal to* $\dfrac{\dfrac{b'_1}{\sqrt{\lambda_1}} - i\dfrac{b'_2}{\sqrt{\lambda_2}}}{\dfrac{b'_1}{\sqrt{\lambda_1}} + i\dfrac{b'_2}{\sqrt{\lambda_2}}}$.

Proof We will consider the sum in the expression for σ, to be a polynomial of the variable equal to the fraction

$$\frac{\dfrac{b'_1}{\sqrt{\lambda_1}} - i\dfrac{b'_2}{\sqrt{\lambda_2}}}{\dfrac{b'_1}{\sqrt{\lambda_1}} + i\dfrac{b'_2}{\sqrt{\lambda_2}}}.$$

Thus, to show that $\sigma \not\equiv 0$, it suffices to show that the sum has at least one term that is nonzero. We show that the coefficient of power $\alpha_1 + \alpha_4 = 1$ is nonzero. In this case, the sets of indices have the following options:

$$\alpha_1 = 1, \ \alpha_4 = 0 \quad \text{or} \quad \alpha_1 = 0, \ \alpha_4 = 1.$$

Then the expression for the coefficient of a power of $\alpha_1 + \alpha_4 = 1$ will have the form

$$\sum_{\alpha_2 + \alpha_3 = N+h+m+k} \binom{k}{\alpha_3}\left(\frac{\sqrt{\lambda_2} - \sqrt{\lambda_1}}{\sqrt{\lambda_2} + \sqrt{\lambda_1}} \right)^1 \left[-\binom{N+h+m}{\alpha_2 - 1} + \binom{N+h+m}{1} \times \right.$$

$$\left. \times \binom{N+h+m}{\alpha_2} \right] + \sum_{\alpha_2 + \alpha_3 = N+h+m+k} \binom{k}{\alpha_3}\binom{k}{1}\left(\frac{\sqrt{\lambda_2} - \sqrt{\lambda_1}}{\sqrt{\lambda_2} + \sqrt{\lambda_1}} \right)^{-1} \left[\binom{N+h+m}{\alpha_2} \right]$$

$$= \sum_{\alpha_2 + \alpha_3 = N+h+m+k} \binom{k}{\alpha_3}\left(\frac{\sqrt{\lambda_2} - \sqrt{\lambda_1}}{\sqrt{\lambda_2} + \sqrt{\lambda_1}} \right)^1 \left[-\binom{N+h+m}{\alpha_2 - 1} \right]$$

$$
+ \sum_{\alpha_2+\alpha_3=N+h+m+k} \binom{k}{\alpha_3} \left(\frac{\sqrt{\lambda_2}-\sqrt{\lambda_1}}{\sqrt{\lambda_2}+\sqrt{\lambda_1}} \right)^1 \left[(N+h+m)\binom{N+h+m}{\alpha_2} \right]
$$

$$
+ \sum_{\alpha_2+\alpha_3=N+h+m+k} \binom{k}{\alpha_3} k \left(\frac{\sqrt{\lambda_2}-\sqrt{\lambda_1}}{\sqrt{\lambda_2}+\sqrt{\lambda_1}} \right)^{-1} \left[\binom{N+h+m}{\alpha_2} \right]
$$

$$
= \left(\frac{\sqrt{\lambda_2}-\sqrt{\lambda_1}}{\sqrt{\lambda_2}+\sqrt{\lambda_1}} \right)^1 [-(N+h+m)-k] + \left(\frac{\sqrt{\lambda_2}-\sqrt{\lambda_1}}{\sqrt{\lambda_2}+\sqrt{\lambda_1}} \right)^1 (N+h+m)
$$

$$
+ k \left(\frac{\sqrt{\lambda_2}-\sqrt{\lambda_1}}{\sqrt{\lambda_2}+\sqrt{\lambda_1}} \right)^{-1} = -k \left(\frac{\sqrt{\lambda_2}-\sqrt{\lambda_1}}{\sqrt{\lambda_2}+\sqrt{\lambda_1}} \right)^1 + k \left(\frac{\sqrt{\lambda_2}-\sqrt{\lambda_1}}{\sqrt{\lambda_2}+\sqrt{\lambda_1}} \right)^{-1}
$$

$$
= k \frac{4\sqrt{\lambda_2}\sqrt{\lambda_1}}{\lambda_2-\lambda_1} \neq 0.
$$

Thus, we have shown that the coefficient of a power of $\alpha_1+\alpha_4=1$ is different from zero in the sum in the expression for σ. It follows that $\sigma \neq 0$. These arguments complete the proof of the lemma.   □

We continue proving the theorem. Since

$$
\sum_{h+k+2m=l_0} b_{h,k,m} a^h \bar{a}^k \int_{\partial \Delta} \tau^N \tau^h \bar{\tau}^k \tau^m \, d\tau = 0,
$$

then, substituting the found value for the integral into the expression and choosing $N = k_0 - 1$, we obtain

$$
\sum_{h+k_0+2m=l_0} b_{h,k_0,m} e^{i\alpha(h+m)} 2\pi i \left(\frac{1}{2} \right)^{2k_0+h+m} (-1)^{2k_0+h+m-1} \left(\frac{1}{\sqrt{\lambda_1}} + \frac{1}{\sqrt{\lambda_2}} \right)^{k_0+h+m} \times
$$

$$
\times \left(\frac{1}{\sqrt{\lambda_1}} - \frac{1}{\sqrt{\lambda_2}} \right)^{k_0} \left(\frac{b_1'}{\sqrt{\lambda_1}} + i\frac{b_2'}{\sqrt{\lambda_2}} \right)^{2k_0+h+m} \sum_{\substack{\alpha_1+\alpha_2+\alpha_3+\alpha_4= \\ =2k_0+h+m}} \binom{k_0}{\alpha_3}\binom{k_0}{\alpha_4} \times
$$

$$
\times \left(\frac{\sqrt{\lambda_2}-\sqrt{\lambda_1}}{\sqrt{\lambda_2}+\sqrt{\lambda_1}} \right)^{\alpha_1-\alpha_4} \left(\frac{\frac{b_1'}{\sqrt{\lambda_1}} - i\frac{b_2'}{\sqrt{\lambda_2}}}{\frac{b_1'}{\sqrt{\lambda_1}} + i\frac{b_2'}{\sqrt{\lambda_2}}} \right)^{\alpha_1+\alpha_4} \left[-\binom{k_0-1+h+m}{\alpha_1-1} \times \right.
$$

$$
\left. \times \binom{k_0-1+h+m}{\alpha_2-1} + \binom{k_0-1+h+m}{\alpha_1}\binom{k_0-1+h+m}{\alpha_2} \right] a^{h-k_0} = 0.
$$

We denote $\psi = e^{i\alpha}$, $\xi = \dfrac{b_1'}{\sqrt{\lambda_1}} + i\dfrac{b_2'}{\sqrt{\lambda_2}}$ and accordingly $\bar{\xi} = \dfrac{b_1'}{\sqrt{\lambda_1}} - i\dfrac{b_2'}{\sqrt{\lambda_2}}$.

Lemma 4.5.3 *The following relations hold true*

$$\psi = i\left[\xi\left(\sqrt{\lambda_2} - \sqrt{\lambda_1}\right) - \bar{\xi}\left(\sqrt{\lambda_2} + \sqrt{\lambda_1}\right)\right], \qquad a^4 = \frac{1}{\psi^4}\frac{c_{11} - c_{22} + c_{12}i}{c_{11} - c_{22} - c_{12}i}.$$

Proof We first show the validity of the first relation. In accordance with the notation, we have

$$\frac{b_1'}{\sqrt{\lambda_1}} = \frac{\xi + \bar{\xi}}{2}, \qquad \frac{b_2'}{\sqrt{\lambda_2}} = \frac{\xi - \bar{\xi}}{2i}.$$

Then

$$b_1' = \frac{\sqrt{\lambda_1}}{2}\left(\xi + \bar{\xi}\right), \qquad b_2' = \frac{\sqrt{\lambda_2}}{2i}\left(\xi - \bar{\xi}\right).$$

On the other hand, according to relation (4.5.10), we have

$$b_1' = -\frac{1}{2}\sin\alpha, \qquad b_2' = -\frac{1}{2}\cos\alpha.$$

Thus, equating both sides of the expressions for b_1' and b_2' respectively, yields

$$-\frac{1}{2}\sin\alpha = \frac{\sqrt{\lambda_1}}{2}\left(\xi + \bar{\xi}\right), \qquad -\frac{1}{2}\cos\alpha = \frac{\sqrt{\lambda_2}}{2i}\left(\xi - \bar{\xi}\right),$$

or

$$\sin\alpha = -\sqrt{\lambda_1}\left(\xi + \bar{\xi}\right), \qquad \cos\alpha = -\frac{\sqrt{\lambda_2}}{i}\left(\xi - \bar{\xi}\right) = i\sqrt{\lambda_2}\left(\xi - \bar{\xi}\right).$$

So we obtain

$$\psi = e^{i\alpha} = \cos\alpha + i\sin\alpha = i\sqrt{\lambda_2}\left(\xi - \bar{\xi}\right) - i\sqrt{\lambda_1}\left(\xi + \bar{\xi}\right)$$
$$= i\left[\xi\left(\sqrt{\lambda_2} - \sqrt{\lambda_1}\right) - \bar{\xi}\left(\sqrt{\lambda_2} + \sqrt{\lambda_1}\right)\right]$$

and thus we have proved the first relation.

To prove the second relation, we write the left- and right-hand sides of expression (4.5.8), using the previously introduced notation. First we transform the left-hand side of expression (4.5.8), and obtain

$$\frac{\cos 2\alpha}{\sin 2\alpha} = \frac{1 - 2\sin^2 \alpha}{2\sin \alpha \cos \alpha} = \frac{1 - 2\left(\dfrac{\psi - \bar\psi}{2i}\right)^2}{2\left(\dfrac{\psi - \bar\psi}{2i}\right)\left(\dfrac{\psi + \bar\psi}{2}\right)} = \frac{\psi^4 + 1}{\psi^4 - 1}i.$$

Next we transform the right-hand side of expression (4.5.8), and obtain

$$\frac{b_{11} - b_{22}}{2b_{12}} = \frac{c_{11}a_1^2 + c_{22}a_2^2 + c_{12}a_1a_2 - c_{11}a_2^2 - c_{22}a_1^2 + c_{12}a_1a_2}{2\left[-c_{11}a_1a_2 + c_{22}a_1a_2 + \dfrac{c_{12}}{2}a_1^2 - \dfrac{c_{12}}{2}a_2^2\right]}$$

$$= \frac{c_{11}\left(a_1^2 - a_2^2\right) + c_{22}\left(a_2^2 - a_1^2\right) + 2c_{12}a_1a_2}{2\left[(c_{22} - c_{11})a_1a_2 + \dfrac{c_{12}}{2}\left(a_1^2 - a_2^2\right)\right]} = \frac{\left(a_1^2 - a_2^2\right)(c_{11} - c_{22}) + 2c_{12}a_1a_2}{2\left[(c_{22} - c_{11})a_1a_2 + \dfrac{c_{12}}{2}\left(a_1^2 - a_2^2\right)\right]}$$

$$= \frac{\left(1 - 2a_2^2\right)(c_{11} - c_{22}) + 2c_{12}a_1a_2}{2\left[(c_{22} - c_{11})a_1a_2 + \dfrac{c_{12}}{2}\left(1 - 2a_2^2\right)\right]}.$$

Since $a = a_1 + ia_2, \bar a = a_1 - ia_2, |a| = 1$, then

$$a_1 = \frac{a + \bar a}{2} = \frac{a^2 + 1}{2a}, \qquad a_2 = \frac{a - \bar a}{2i} = \frac{a^2 - 1}{2ia},$$

hence

$$a_1 a_2 = \frac{\left(a^2 + 1\right)\left(a^2 - 1\right)}{2a} = \frac{1 - a^4}{4a^2}i,$$

$$1 - 2a_2^2 = 1 + \frac{2\left(a^2 - 1\right)^2}{4a^2} = \frac{4a^2 + 2\left(a^4 - 2a^2 + 1\right)}{4a^2} = \frac{a^4 + 1}{2a^2}.$$

Substitution of these expressions results in expression

$$\frac{b_{11} - b_{22}}{2b_{12}} = \frac{\dfrac{\left(a^4 + 1\right)}{2a^2}(c_{11} - c_{22}) + 2c_{12}\dfrac{\left(1 - a^4\right)}{4a^2}i}{2\left[(c_{22} - c_{11})\dfrac{\left(1 - a^4\right)}{4a^2}i + \dfrac{c_{12}}{2}\dfrac{\left(a^4 + 1\right)}{2a^2}\right]}$$

$$= \frac{\dfrac{\left(a^4 + 1\right)(c_{11} - c_{22})}{2a^2} + \dfrac{c_{12}\left(1 - a^4\right)i}{2a^2}}{\dfrac{(c_{22} - c_{11})\left(1 - a^4\right)i}{2a^2} + \dfrac{c_{12}\left(a^4 + 1\right)}{2a^2}} = \frac{\left(a^4 + 1\right)(c_{11} - c_{22}) + c_{12}\left(1 - a^4\right)i}{(c_{22} - c_{11})\left(1 - a^4\right)i + c_{12}\left(a^4 + 1\right)}.$$

Thus, equating the converted left- and right-hand sides of expression (4.5.8) we obtain the relation

$$\frac{\psi^4 + 1}{\psi^4 - 1} i = \frac{(a^4 + 1)(c_{11} - c_{22}) + c_{12}(1 - a^4) i}{(c_{22} - c_{11})(1 - a^4) i + c_{12}(a^4 + 1)}.$$

We solve this relation with respect to a^4, and obtain

$$\frac{\psi^4 + 1}{\psi^4 - 1} i = \frac{a^4 (c_{11} - c_{22} - c_{12}i) + (c_{11} - c_{22} + c_{12}i)}{a^4 (c_{11}i - c_{22}i + c_{12}) + (c_{22}i - c_{11}i + c_{12})},$$

$$a^4 (c_{11} - c_{22} - c_{12}i)(\psi^4 - 1) + (c_{11} - c_{22} + c_{12}i)(\psi^4 - 1)$$
$$= a^4 (c_{22} - c_{11} + c_{12}i)(\psi^4 + 1) + (c_{11} - c_{22} + c_{12}i)(\psi^4 + 1),$$

$$a^4 (c_{11} - c_{22} - c_{12}i)(\psi^4 - 1) - a^4 (c_{22} - c_{11} + c_{12}i)(\psi^4 + 1)$$
$$= (c_{11} - c_{22} + c_{12}i)(\psi^4 + 1) - (c_{11} - c_{22} + c_{12}i)(\psi^4 - 1),$$

$$a^4 (c_{11} - c_{22} - c_{12}i)(\psi^4 - 1) + a^4 (c_{11} - c_{22} - c_{12}i)(\psi^4 + 1) = 2(c_{11} - c_{22} + c_{12}i),$$

$$a^4 (c_{11} - c_{22} - c_{12}i) 2\psi^4 = 2(c_{11} - c_{22} + c_{12}i),$$

$$a^4 = \frac{1}{\psi^4} \frac{c_{11} - c_{22} + c_{12}i}{c_{11} - c_{22} - c_{12}i},$$

and thereby our second relation has been proved. These arguments complete the proof.   □

4.5.5   Transformation of Moment Conditions

We continue proving the theorem. Denoting

$$x = \frac{\sqrt{\lambda_2} - \sqrt{\lambda_1}}{\sqrt{\lambda_2} + \sqrt{\lambda_1}}, \quad 0 < x < 1, \tag{4.5.31}$$

we obtain the following expression for the coefficients $b_{h,k_0,m}$

$$\sum_{h+k_0+2m=l_0} b_{h,k_0,m} \left(i \left[\xi \left(\sqrt{\lambda_2} - \sqrt{\lambda_1} \right) - \bar{\xi} \left(\sqrt{\lambda_2} + \sqrt{\lambda_1} \right) \right] \right)^{h+m} 2\pi i \left(\frac{1}{2} \right)^{2k_0+h+m} \times$$

$$\times (-1)^{2k_0+h+m-1} \left(\frac{1}{\sqrt{\lambda_1}} + \frac{1}{\sqrt{\lambda_2}} \right)^{k_0+h+m} \left(\frac{1}{\sqrt{\lambda_1}} - \frac{1}{\sqrt{\lambda_2}} \right)^{k_0} \xi^{2k_0+h+m} a^{h-k_0} \times$$

$$\times \sum_{\substack{\alpha_1+\alpha_2+\alpha_3+\alpha_4 \\ =2k_0+h+m}} \binom{k_0}{\alpha_3}\binom{k_0}{\alpha_4} x^{\alpha_1-\alpha_4} \left(\frac{\bar{\xi}}{\xi}\right)^{\alpha_1+\alpha_4} \left[-\binom{k_0-1+h+m}{\alpha_1-1}\binom{k_0-1+h+m}{\alpha_2-1}\right.$$

$$\left.+\binom{k_0-1+h+m}{\alpha_1}\binom{k_0-1+h+m}{\alpha_2}\right]=0.$$

The set $\{\eta=\bar{\xi}\}$ is a generic manifold in $\mathbb{C}^2$, which is a set of uniqueness for holomorphic functions [67], that is this equation is valid for any ξ and η. Therefore, the equality

$$\sum_{h+k_0+2m=l_0} b_{h,k_0,m} \left(i\left[\xi\left(\sqrt{\lambda_2}-\sqrt{\lambda_1}\right)-\eta\left(\sqrt{\lambda_2}+\sqrt{\lambda_1}\right)\right]\right)^{h+m} 2\pi i x$$

$$\times\left(\frac{1}{2}\right)^{2k_0+h+m} (-1)^{2k_0+h+m-1}\left(\frac{1}{\sqrt{\lambda_1}}+\frac{1}{\sqrt{\lambda_2}}\right)^{k_0+h+m}\left(\frac{1}{\sqrt{\lambda_1}}-\frac{1}{\sqrt{\lambda_2}}\right)^{k_0}\times$$

$$\times \xi^{2k_0+h+m} a^{h-k_0}\sum_{\substack{\alpha_1+\alpha_2+\alpha_3+\alpha_4= \\ =2k_0+h+m}}\binom{k_0}{\alpha_3}\binom{k_0}{\alpha_4} x^{\alpha_1-\alpha_4}\left(\frac{\eta}{\xi}\right)^{\alpha_1+\alpha_4}\left[-\binom{k_0-1+h+m}{\alpha_1-1}\right.\times$$

$$\left.\times\binom{k_0-1+h+m}{\alpha_2-1}+\binom{k_0-1+h+m}{\alpha_1}\binom{k_0-1+h+m}{\alpha_2}\right]=0$$

holds. Moreover, we have the following limitations

$$\alpha_1 \le k_0+h+m, \quad \alpha_2 \le k_0+h+m, \quad \alpha_3 \le k_0, \quad \alpha_4 \le k_0$$

for the summation indices. In the latter relation we reduce similar monomials in the inner sum, introducing new variables s_1 and s_2 to denote the powers of η and ξ, i.e.,

$$s_1 = \alpha_1 + \alpha_4, \quad s_2 = 2k_0 + h + m - (\alpha_1 + \alpha_4).$$

We obtain

$$\sum_{s_1,s_2} b_{h,k_0,m}\, \eta^{s_1}\xi^{s_2}\left(i\left[\xi\left(\sqrt{\lambda_2}-\sqrt{\lambda_1}\right)-\eta\left(\sqrt{\lambda_2}+\sqrt{\lambda_1}\right)\right]\right)^{s_1+s_2-2k_0}\times$$

$$\times 2\pi i\left(\frac{1}{2}\right)^{s_1+s_2}(-1)^{s_1+s_2-1}\left(\frac{1}{\sqrt{\lambda_1}}+\frac{1}{\sqrt{\lambda_2}}\right)^{s_1+s_2-k_0}\left(\frac{1}{\sqrt{\lambda_1}}-\frac{1}{\sqrt{\lambda_2}}\right)^{k_0}\times$$

$$\times\, a^{2(s_1+s_2)-l_0-4k_0} \sum_{\substack{\alpha_1+\alpha_2+\alpha_3+\alpha_4= \\ =s_1+s_2}} \binom{k_0}{\alpha_3}\binom{k_0}{\alpha_4} x^{\alpha_1-\alpha_4}\left[-\binom{s_1+s_2-k_0-1}{\alpha_1-1}\right.\times$$

$$\left.\times \binom{s_1+s_2-k_0-1}{\alpha_2-1}+\binom{s_1+s_2-k_0-1}{\alpha_1}\binom{s_1+s_2-k_0-1}{\alpha_2}\right]=0,$$

where

$$h = 2(s_1+s_2)-l_0-3k_0, \qquad m = l_0+k_0-(s_1+s_2).$$

Indeed, since on the one hand $h+k_0+2m=l_0$, and, on the other hand, according to the above notation $h+m=s_1+s_2-2k_0$, then $s_1+s_2-k_0+m=l_0$ and now the required expressions for h and m are easy to obtain.

Next we write the resulting relation of homogeneous monomials of degree $r = s_1+s_2$. We obtain

$$\sum_{r=2k_0}^{l_0+k_0} \sum_{s_1+s_2=r} b_{h,k_0,m}\,\eta^{s_1}\xi^{s_2}\left(i\left[\xi\left(\sqrt{\lambda_2}-\sqrt{\lambda_1}\right)-\eta\left(\sqrt{\lambda_2}+\sqrt{\lambda_1}\right)\right]\right)^{s_1+s_2-2k_0}\times$$

$$\times\, 2\pi i\left(\frac{1}{2}\right)^{s_1+s_2}(-1)^{s_1+s_2-1}\left(\frac{1}{\sqrt{\lambda_1}}+\frac{1}{\sqrt{\lambda_2}}\right)^{s_1+s_2-k_0}\left(\frac{1}{\sqrt{\lambda_1}}-\frac{1}{\sqrt{\lambda_2}}\right)^{k_0}\times$$

$$\times\, a^{2(s_1+s_2)-l_0-4k_0} \sum_{\substack{\alpha_1+\alpha_2+\alpha_3+\alpha_4= \\ =s_1+s_2}} \binom{k_0}{\alpha_3}\binom{k_0}{\alpha_4} x^{\alpha_1-\alpha_4}\left[-\binom{s_1+s_2-k_0-1}{\alpha_1-1}\right.\times$$

$$\left.\times \binom{s_1+s_2-k_0-1}{\alpha_2-1}+\binom{s_1+s_2-k_0-1}{\alpha_1}\binom{s_1+s_2-k_0-1}{\alpha_2}\right]=0,$$

where

$$h = 2r-l_0-3k_0, \qquad m = l_0+k_0-r, \qquad h+m = r-2k_0, \qquad (4.5.32)$$

and we have the following limitations

$$\alpha_1 \le r-k_0, \quad \alpha_2 \le r-k_0, \quad \alpha_3 \le k_0, \quad \alpha_4 \le k_0 \qquad (4.5.33)$$

for the summation indices. Then each homogeneous component is zero, i.e.,

$$\sum_{s_1+s_2=r} b_{h,k_0,m}\,\eta^{s_1}\xi^{s_2}\left(i\left[\xi\left(\sqrt{\lambda_2}-\sqrt{\lambda_1}\right)-\eta\left(\sqrt{\lambda_2}+\sqrt{\lambda_1}\right)\right]\right)^{s_1+s_2-2k_0}\times$$

$$\times\, 2\pi i\left(\frac{1}{2}\right)^{s_1+s_2}(-1)^{s_1+s_2-1}\left(\frac{1}{\sqrt{\lambda_1}}+\frac{1}{\sqrt{\lambda_2}}\right)^{s_1+s_2-k_0}\left(\frac{1}{\sqrt{\lambda_1}}-\frac{1}{\sqrt{\lambda_2}}\right)^{k_0}\times$$

$$\times\, a^{2(s_1+s_2)-l_0-4k_0} \sum_{\substack{\alpha_1+\alpha_2+\alpha_3+\alpha_4=\\=s_1+s_2}} \binom{k_0}{\alpha_3}\binom{k_0}{\alpha_4} x^{\alpha_1-\alpha_4}\left[-\binom{s_1+s_2-k_0-1}{\alpha_1-1}\times\right.$$

$$\left.\times\binom{s_1+s_2-k_0-1}{\alpha_2-1} + \binom{s_1+s_2-k_0-1}{\alpha_1}\binom{s_1+s_2-k_0-1}{\alpha_2}\right] = 0.$$

Or, factoring out the terms independent of the summation indices from the summation sign, we get

$$\left(i\left[\xi\left(\sqrt{\lambda_2}-\sqrt{\lambda_1}\right)-\eta\left(\sqrt{\lambda_2}+\sqrt{\lambda_1}\right)\right]\right)^{r-2k_0} 2\pi i\left(\frac{1}{2}\right)^r(-1)^{r-1}\times$$

$$\times\left(\frac{1}{\sqrt{\lambda_1}}+\frac{1}{\sqrt{\lambda_2}}\right)^{r-k_0}\left(\frac{1}{\sqrt{\lambda_1}}-\frac{1}{\sqrt{\lambda_2}}\right)^{k_0} a^{2r-l_0-4k_0}\sum_{s_1+s_2=r} b_{h,k_0,m}\,\eta^{s_1}\xi^{s_2}\times$$

$$\times\sum_{\alpha_1+\alpha_2+\alpha_3+\alpha_4=r}\binom{k_0}{\alpha_3}\binom{k_0}{\alpha_4}x^{\alpha_1-\alpha_4}\left[-\binom{r-k_0-1}{\alpha_1-1}\binom{r-k_0-1}{\alpha_2-1}\right.$$

$$\left.+\binom{r-k_0-1}{\alpha_1}\binom{r-k_0-1}{\alpha_2}\right] = 0.$$

Using the fact that the terms outside the summation sign are nonzero, we obtain the relation

$$\sum_{s_1+s_2=r} b_{h,k_0,m}\,\eta^{s_1}\xi^{s_2}\sum_{\alpha_1+\alpha_2+\alpha_3+\alpha_4=r}\binom{k_0}{\alpha_3}\binom{k_0}{\alpha_4}x^{\alpha_1-\alpha_4}\left[-\binom{r-k_0-1}{\alpha_1-1}\times\right.$$

$$\left.\times\binom{r-k_0-1}{\alpha_2-1}+\binom{r-k_0-1}{\alpha_1}\binom{r-k_0-1}{\alpha_2}\right] = 0.$$

If a polynomial of two variables is identically zero, then all its coefficients are zeros. Thus we obtain the following assertion.

Proposition 4.5.3 *The following equality holds*

$$b_{h,k_0,m}\sum_{\alpha_1+\alpha_2+\alpha_3+\alpha_4=r}\binom{k_0}{\alpha_3}\binom{k_0}{\alpha_4}x^{\alpha_1-\alpha_4}\left[-\binom{r-k_0-1}{\alpha_1-1}\binom{r-k_0-1}{\alpha_2-1}\right.$$

$$\left.+\binom{r-k_0-1}{\alpha_1}\binom{r-k_0-1}{\alpha_2}\right] = 0,$$

where the variable x is defined by (4.5.31), relations (4.5.32) are satisfied for h and m, and the summation indices $\alpha_1,\alpha_2,\alpha_3,\alpha_4$ satisfy limitations (4.5.33).

4.5.6 Completion of the Proof of Theorem 4.5.1 in the Two-Dimensional Case

Denote

$$
g(x) = \sum_{\alpha_1+\alpha_2+\alpha_3+\alpha_4=r} \binom{k_0}{\alpha_3}\binom{k_0}{\alpha_4} x^{\alpha_1-\alpha_4+k_0} \left[-\binom{r-k_0-1}{\alpha_1-1}\binom{r-k_0-1}{\alpha_2-1} \right.
$$
$$
\left. + \binom{r-k_0-1}{\alpha_1}\binom{r-k_0-1}{\alpha_2} \right].
$$

Write the difference of products of binomial coefficients in the expression for $g(x)$. We get

$$
-\binom{r-k_0-1}{\alpha_1-1}\binom{r-k_0-1}{\alpha_2-1} + \binom{r-k_0-1}{\alpha_1}\binom{r-k_0-1}{\alpha_2}
$$
$$
= \frac{(r-k_0-1)!\,(r-k_0-1)!}{\alpha_1!\,\alpha_2!\,(r-k_0-\alpha_1)!\,(r-k_0-\alpha_2)!} \left[-\alpha_1\alpha_2 + (r-k_0-\alpha_1)(r-k_0-\alpha_2)\right]
$$
$$
= \binom{r-k_0}{\alpha_1}\binom{r-k_0}{\alpha_2}\left[1 - \frac{\alpha_1+\alpha_2}{r-k_0}\right].
$$

Then $g(x)$ can be written as

$$
g(x) = \sum_{\alpha_1+\alpha_2+\alpha_3+\alpha_4=r} \binom{k_0}{\alpha_3}\binom{k_0}{\alpha_4}\binom{r-k_0}{\alpha_1}\binom{r-k_0}{\alpha_2}\left[1 - \frac{\alpha_1+\alpha_2}{r-k_0}\right] x^{\alpha_1-\alpha_4+k_0}.
$$

Introduce $p = \alpha_1 - \alpha_4 + k_0,\ 0 \leqslant p \leqslant r$. Then

$$
g(x) = \sum_{2\alpha_1+\alpha_2+\alpha_3-p=r-k_0} \binom{k_0}{\alpha_3}\binom{k_0}{\alpha_1+k_0-p}\binom{r-k_0}{\alpha_1}\binom{r-k_0}{\alpha_2}\left[1 - \frac{\alpha_1+\alpha_2}{r-k_0}\right] x^p.
$$

Thus, we can write $g(x)$ as $g(x) = \sum_{p=0}^{r} c_p x^p$, where the coefficients c_p have the form

$$
c_p = \sum_{2\alpha_1+\alpha_2+\alpha_3=r+p-k_0} \binom{k_0}{\alpha_3}\binom{k_0}{\alpha_1+k_0-p}\binom{r-k_0}{\alpha_1}\binom{r-k_0}{\alpha_2}\left[1 - \frac{\alpha_1+\alpha_2}{r-k_0}\right].
$$

Proposition 4.5.4 *Coefficients c_p and symmetric thereto coefficients c_{r-p} of the polynomial $g(x)$ relate as*

$$c_p + c_{r-p} = 0, \qquad (4.5.34)$$

i.e., unity is the root of the polynomial $g(x)$.

Proof Consider the coefficient

$$c_{r-p} = \sum_{\alpha_1 = r-p-k_0}^{r-k_0} \sum_{\substack{\alpha_2+\alpha_3= \\ =r+r-p-k_0-2\alpha_1}} \binom{k_0}{\alpha_3}\binom{k_0}{\alpha_1+k_0-r+p}\binom{r-k_0}{\alpha_1}\binom{r-k_0}{\alpha_2} \times$$

$$\times \left[1 - \frac{\alpha_1+\alpha_2}{r-k_0}\right].$$

Making the change of variables

$$\alpha_1' = \alpha_1 - r + p + k_0, \qquad \alpha_1 = \alpha_1' + r - p - k_0,$$

we arrive at the following expression for the coefficient

$$c_{r-p} = \sum_{\alpha_1'=0}^{p} \sum_{\substack{\alpha_2+\alpha_3=r+r-p-k_0- \\ -2\alpha_1'-2r+2p+2k_0}} \binom{k_0}{\alpha_3}\binom{k_0}{\alpha_1'}\binom{r-k_0}{\alpha_1'+r-p-k_0}\binom{r-k_0}{\alpha_2} \times$$

$$\times \left[1 - \frac{\alpha_1'+r-p-k_0+\alpha_2}{r-k_0}\right]$$

$$= -\sum_{\alpha_1'=0}^{p} \sum_{\substack{\alpha_2+\alpha_3= \\ =p+k_0-2\alpha_1'}} \binom{k_0}{\alpha_3}\binom{k_0}{\alpha_1'}\binom{r-k_0}{p-\alpha_1'}\binom{r-k_0}{\alpha_2} \left[\frac{\alpha_1'+\alpha_2-p}{r-k_0}\right]$$

$$= -\sum_{\alpha_1'=0}^{p}\sum_{\alpha_3=0}^{k_0} \binom{k_0}{\alpha_3}\binom{k_0}{\alpha_1'}\binom{r-k_0}{p-\alpha_1'}\binom{r-k_0}{p+k_0-2\alpha_1'-\alpha_3} \times$$

$$\times \left[\frac{\alpha_1'+p+k_0-2\alpha_1'-\alpha_3-p}{r-k_0}\right]$$

$$= -\sum_{\alpha_1'=0}^{p}\sum_{\alpha_3=0}^{k_0} \binom{k_0}{\alpha_3}\binom{k_0}{\alpha_1'}\binom{r-k_0}{p-\alpha_1'}\binom{r-k_0}{p+k_0-2\alpha_1'-\alpha_3} \left[\frac{k_0-\alpha_1'-\alpha_3}{r-k_0}\right].$$

On the other hand

$$
c_p = \sum_{\alpha_1=0}^{p} \sum_{\substack{\alpha_2+\alpha_3= \\ =r+p-k_0-2\alpha_1}} \binom{k_0}{\alpha_3}\binom{k_0}{\alpha_1+k_0-p}\binom{r-k_0}{\alpha_1}\binom{r-k_0}{\alpha_2}\left[1-\frac{\alpha_1+\alpha_2}{r-k_0}\right]
$$

$$
= \sum_{\alpha_1=0}^{p} \sum_{\alpha_3=0}^{k_0} \binom{k_0}{\alpha_3}\binom{k_0}{\alpha_1+k_0-p}\binom{r-k_0}{\alpha_1}\binom{r-k_0}{2\alpha_1+\alpha_3-p}\left[\frac{\alpha_1+\alpha_3-p}{r-k_0}\right]
$$

$$
= \sum_{\alpha_1=0}^{p} \sum_{\alpha_3=0}^{k_0} \binom{k_0}{\alpha_3}\binom{k_0}{p-\alpha_1}\binom{r-k_0}{\alpha_1}\binom{r-k_0}{2\alpha_1+\alpha_3-p}\left[\frac{\alpha_1+\alpha_3-p}{r-k_0}\right].
$$

Making the change of variables

$$
\alpha_1' = p - \alpha_1, \qquad \alpha_1 = p - \alpha_1',
$$

we arrive at the following expression for the coefficient

$$
c_p = \sum_{\alpha_1'=0}^{p} \sum_{\alpha_3=0}^{k_0} \binom{k_0}{\alpha_3}\binom{k_0}{\alpha_1'}\binom{r-k_0}{p-\alpha_1'}\binom{r-k_0}{p-2\alpha_1'+\alpha_3}\left[\frac{\alpha_3-\alpha_1'}{r-k_0}\right].
$$

Making the change of variables

$$
\alpha_3' = k_0 - \alpha_3, \qquad \alpha_3 = k_0 - \alpha_3',
$$

we arrive at the final expression for the coefficient

$$
c_p = \sum_{\alpha_1'=0}^{p} \sum_{\alpha_3'=0}^{k_0} \binom{k_0}{\alpha_3'}\binom{k_0}{\alpha_1'}\binom{r-k_0}{p-\alpha_1'}\binom{r-k_0}{p-2\alpha_1'+k_0-\alpha_3'}\left[\frac{k_0-\alpha_3'-\alpha_1'}{r-k_0}\right].
$$

Thus, relation (4.5.34) has been proved for the coefficients of the polynomial $g(x)$.
$\square$

Proposition 4.5.5 *The polynomial $g(x)$ has a unique positive root $x = 1$.*

Proof According to Descartes' theorem, the number of positive roots of a polynomial, each counted as many times as its multiplicity is equal to the number of sign changes in the system of coefficients of the polynomial (leaving out zero coefficients), or less than this number by an even number. Thus, for the positive roots of $g(x)$ to be unique, it is necessary to show the presence of exactly one sign change in the system of coefficients.

Consider for example the coefficients c_0, c_1, c_r.

$$c_0 = \sum_{\alpha_2+\alpha_3=r-k_0} \binom{k_0}{\alpha_3}\binom{r-k_0}{\alpha_2}\left[1 - \frac{\alpha_2}{r-k_0}\right] > 0,$$

$$c_1 = \sum_{\alpha_2+\alpha_3=r+1-k_0} \binom{k_0}{\alpha_3} k_0 \binom{r-k_0}{\alpha_2}\left[1 - \frac{\alpha_2}{r-k_0}\right] +$$

$$+ \sum_{\alpha_2+\alpha_3=r-1-k_0} \binom{k_0}{\alpha_3}(r-k_0)\binom{r-k_0}{\alpha_2}\left[1 - \frac{1+\alpha_2}{r-k_0}\right] > 0,$$

$$c_r = \sum_{\substack{2(r-k_0)+\alpha_2+\alpha_3= \\ r+r-k_0}} \binom{k_0}{\alpha_3}\binom{k_0}{r-k_0+k_0-r}\binom{r-k_0}{r-k_0}\binom{r-k_0}{\alpha_2}\left[1 - \frac{r-k_0+\alpha_2}{r-k_0}\right]$$

$$= \sum_{\alpha_2+\alpha_3=k_0} \binom{k_0}{\alpha_3}\binom{r-k_0}{\alpha_2}\left[-\frac{\alpha_2}{r-k_0}\right] = -\sum_{\alpha_2+\alpha_3=k_0} \binom{k_0}{\alpha_3}\binom{r-k_0-1}{\alpha_2-1} < 0.$$

Let us show that the coefficients $c_p \geq 0$, when $0 \leq p \leq r/2 \leq r - k_0$. Given relation (4.5.34), this will mean that there is exactly one sign change in the system of coefficients of the polynomial $g(x)$. Making the change $\beta = \alpha_1 + \alpha_2$, $\alpha_2 = \beta - \alpha_1$, we obtain the following expression for the coefficient

$$c_p = \sum_{\substack{\alpha_1+\beta+\alpha_3 \\ r+p-k_0}} \binom{k_0}{\alpha_3}\binom{k_0}{\alpha_1+k_0-p}\binom{r-k_0}{\alpha_1}\binom{r-k_0}{\beta-\alpha_1}\left[1 - \frac{\beta}{r-k_0}\right]$$

$$= \sum_{\beta=0}^{r-k_0} \sum_{\substack{\alpha_1+\alpha_3= \\ r+p-k_0-\beta}} \binom{k_0}{\alpha_3}\binom{k_0}{\alpha_1+k_0-p}\binom{r-k_0}{\alpha_1}\binom{r-k_0}{\beta-\alpha_1}\left[\frac{r-k_0-\beta}{r-k_0}\right]$$

$$- \sum_{\beta=r-k_0}^{r+p-k_0} \sum_{\substack{\alpha_1+\alpha_3= \\ r+p-k_0-\beta}} \binom{k_0}{\alpha_3}\binom{k_0}{\alpha_1+k_0-p}\binom{r-k_0}{\alpha_1}\binom{r-k_0}{\beta-\alpha_1}\left[\frac{\beta-(r-k_0)}{r-k_0}\right].$$

In the second sum we make the substitution $\beta' = \beta - (r - k_0)$, which yields

$$\sum_{\beta=r-k_0}^{r+p-k_0} \sum_{\substack{\alpha_1+\alpha_3=\\r+p-k_0-\beta}} \binom{k_0}{\alpha_3}\binom{k_0}{\alpha_1+k_0-p}\binom{r-k_0}{\alpha_1}\binom{r-k_0}{\beta-\alpha_1}\left[\frac{\beta-(r-k_0)}{r-k_0}\right]$$

$$= \sum_{\alpha_1=0}^{p}\sum_{\beta'=0}^{p-\alpha_1}\binom{k_0}{p-\beta'-\alpha_1}\binom{k_0}{\alpha_1+k_0-p}\binom{r-k_0}{\alpha_1}\binom{r-k_0}{\alpha_1-\beta'}\left[\frac{\beta'}{r-k_0}\right].$$

In the first sum of the original expression for c_p we arrange the β-terms in reverse order. Making the substitution $\beta' = r - k_0 - \beta$, we obtain

$$\sum_{\beta=0}^{r-k_0} \sum_{\alpha_1+\alpha_3=r+p-k_0-\beta} \binom{k_0}{\alpha_3}\binom{k_0}{\alpha_1+k_0-p}\binom{r-k_0}{\alpha_1}\binom{r-k_0}{\beta-\alpha_1}\left[\frac{r-k_0-\beta}{r-k_0}\right]$$

$$= \sum_{\alpha_1=0}^{p}\sum_{\beta'=0}^{p-\alpha_1}\binom{k_0}{p+\beta'-\alpha_1}\binom{k_0}{\alpha_1+k_0-p}\binom{r-k_0}{\alpha_1}\binom{r-k_0}{\alpha_1+\beta'}\left[\frac{\beta'}{r-k_0}\right]$$

$$+ \sum_{\alpha_1=0}^{p}\sum_{\beta'=p-\alpha_1+1}^{r-k_0-\alpha_1}\binom{k_0}{p+\beta'-\alpha_1}\binom{k_0}{\alpha_1+k_0-p}\binom{r-k_0}{\alpha_1}\binom{r-k_0}{\alpha_1+\beta'}\left[\frac{\beta'}{r-k_0}\right].$$

Thus

$$c_p = \sum_{\alpha_1=0}^{p}\sum_{\beta'=p-\alpha_1+1}^{r-k_0-\alpha_1}\binom{k_0}{p+\beta'-\alpha_1}\binom{k_0}{\alpha_1+k_0-p}\binom{r-k_0}{\alpha_1}\binom{r-k_0}{\alpha_1+\beta'}\left[\frac{\beta'}{r-k_0}\right]$$

$$+ \sum_{\alpha_1=0}^{p}\sum_{\beta'=0}^{p-\alpha_1}\binom{k_0}{p+\beta'-\alpha_1}\binom{k_0}{\alpha_1+k_0-p}\binom{r-k_0}{\alpha_1}\binom{r-k_0}{\alpha_1+\beta'}\left[\frac{\beta'}{r-k_0}\right]$$

$$- \sum_{\alpha_1=0}^{p}\sum_{\beta'=0}^{p-\alpha_1}\binom{k_0}{p-\beta'-\alpha_1}\binom{k_0}{\alpha_1+k_0-p}\binom{r-k_0}{\alpha_1}\binom{r-k_0}{\alpha_1-\beta'}\left[\frac{\beta'}{r-k_0}\right]$$

$$= \sum_{\alpha_1=0}^{p}\sum_{\beta'=p-\alpha_1+1}^{r-k_0-\alpha_1}\binom{k_0}{p+\beta'-\alpha_1}\binom{k_0}{\alpha_1+k_0-p}\binom{r-k_0}{\alpha_1}\binom{r-k_0}{\alpha_1+\beta'}\left[\frac{\beta'}{r-k_0}\right]$$

$$+ \sum_{\alpha_1=0}^{p}\sum_{\beta'=0}^{p-\alpha_1}\binom{k_0}{\alpha_1+k_0-p}\binom{r-k_0}{\alpha_1}\left[\frac{\beta'}{r-k_0}\right]\left[\binom{k_0}{p+\beta'-\alpha_1}\binom{r-k_0}{\alpha_1+\beta'}\right.$$

$$\left. - \binom{k_0}{p-\beta'-\alpha_1}\binom{r-k_0}{\alpha_1-\beta'}\right].$$

We write down the difference of products of binomial coefficients, and obtain

$$
\binom{k_0}{p + \beta' - \alpha_1}\binom{r - k_0}{\alpha_1 + \beta'} - \binom{k_0}{p - \beta' - \alpha_1}\binom{r - k_0}{\alpha_1 - \beta'}
$$

$$
= \frac{k_0!}{(p + \beta' - \alpha_1)! \, (k_0 - p + \beta' + \alpha_1)!} \cdot \frac{(r - k_0)!}{(\alpha_1 + \beta')! \, (r - k_0 - \alpha_1 + \beta')!} \times
$$

$$
\times \big[(k_0 - p - \beta' + \alpha_1 + 1) \cdots (k_0 - p + \beta' + \alpha_1) \, (r - k_0 - \alpha_1 - \beta' + 1) \cdots \times
$$

$$
\times (r - k_0 - \alpha_1 + \beta') - (p - \beta' - \alpha_1 + 1) \cdots
$$

$$
\cdots (p + \beta' - \alpha_1) \, (\alpha_1 - \beta' + 1) \cdots (\alpha_1 + \beta') \big].
$$

Next we show that for given values α_1 and β' and for $0 \le p \le r/2 \le r - k_0$ the following relation holds

$$
(k_0 - p - \beta' + \alpha_1 + 1) \cdots (k_0 - p + \beta' + \alpha_1) \, (r - k_0 - \alpha_1 - \beta' + 1) \cdots \times
$$

$$
\times (r - k_0 - \alpha_1 + \beta') - (p - \beta' - \alpha_1 + 1) \cdots (p + \beta' - \alpha_1) \times
$$

$$
\times (\alpha_1 - \beta' + 1) \cdots (\alpha_1 + \beta') \ge 0. \qquad (4.5.35)
$$

Note that for $0 \le p \le r/2 \le r - k_0$ the following inequality holds

$$
(p - \beta' - \alpha_1 + 1) \cdots (p + \beta' - \alpha_1) \le (r - k_0 - \alpha_1 - \beta' + 1) \cdots (r - k_0 - \alpha_1 + \beta').
$$

Moreover, if $0 \le p \le k_0 \le r/2 \le r - k_0$, then

$$
(\alpha_1 - \beta' + 1) \cdots (\alpha_1 + \beta') \le (k_0 - p - \beta' + \alpha_1 + 1) \cdots (k_0 - p + \beta' + \alpha_1).
$$

Thus, the desired inequality (4.5.35) holds for $p \le k_0 \le r/2 \le r - k_0$.

Now we consider the case when $k_0 \le p \le r/2 \le r - k_0$. In this case, to prove inequality (4.5.35), it sufficient to prove the following relations:

$$
(k_0 - p - \beta' + \alpha_1 + 1) \, (r - k_0 - \alpha_1 - \beta' + 1) \ge (p - \beta' - \alpha_1 + 1) \, (\alpha_1 - \beta' + 1)
$$

$$
\vdots
$$

$$
(k_0 - p + \beta' + \alpha_1) \, (r - k_0 - \alpha_1 + \beta') \ge (p + \beta' - \alpha_1) \, (\alpha_1 + \beta').
$$

We will prove the first and last relations in this group. The interim relations are met because of their monotony in β'. Let us prove the first relation:

$$
(k_0 - p - \beta' + \alpha_1 + 1) \, (r - k_0 - \alpha_1 - \beta' + 1)
$$

$$
\ge (p - \beta' - \alpha_1 + 1) \, (\alpha_1 - \beta' + 1). \qquad (4.5.36)
$$

After expansion we obtain

$$(k_0 - p)(r - k_0) - \alpha_1 k_0 - \beta' k_0 + \alpha_1 p + \beta' p + k_0 - p - \beta'(r - k_0) + \beta' \alpha_1 + \beta'^2$$
$$- \beta' + \alpha_1 (r - k_0) - \alpha_1^2 - \alpha_1 \beta' + \alpha_1 + r - k_0 - \alpha_1 - \beta' + 1$$
$$\geq \alpha_1 p - \beta' p + p - \beta' \alpha_1 + \beta'^2 - \beta' - \alpha_1^2 + \alpha_1 \beta' - \alpha_1 + \alpha_1 - \beta' + 1.$$

By combining similar terms we obtain

$$(k_0 - p)(r - k_0) - \alpha_1 k_0 - \beta' k_0 + \beta' p + k_0 - p - \beta'(r - k_0) + \alpha_1 (r - k_0) + r - k_0$$
$$\geq -\beta' p + p,$$
$$-(r - k_0)(k_0 - p - \beta' + \alpha_1) + k_0 (\alpha_1 + \beta') - 2\beta' p + 2p - r \leq 0.$$

Note that $k_0 - p - \beta' + \alpha_1 \geq 0$, since this is a binomial coefficient. For given values α_1 and β' we consider the positive part $k_0 (\alpha_1 + \beta') - 2\beta' p$ of this inequality:

$$k_0 (\alpha_1 + \beta') - 2\beta' p = k_0 \alpha_1 + k_0 \beta' - \beta' p - \beta' p$$
$$= k_0 \alpha_1 - \beta' p + (k_0 - p) \beta' = k_0 \alpha_1 - \beta' p - (p - k_0) \beta'.$$

For the positive part to have maximum value, we need to take $\alpha_1 = p - 1$, $\beta' = 1$, and obtain

$$-(r - k_0)(k_0 - 2) + (k_0 - 2) p + 2p - r \leq 0.$$

This inequality holds because $p \leq r/2 \leq r - k_0$. Thus, we have proved the validity of (4.5.36).

Let us prove the last relation in the group, namely show that

$$(k_0 - p + \beta' + \alpha_1)(r - k_0 - \alpha_1 + \beta') \geq (p + \beta' - \alpha_1)(\alpha_1 + \beta'). \qquad (4.5.37)$$

After expansion we obtain

$$(k_0 - p)(r - k_0) - \alpha_1 k_0 + \alpha_1 p + \beta' k_0 - \beta' p + \beta'(r - k_0) - \alpha_1 \beta' + \beta'^2$$
$$+ \alpha_1 (r - k_0) - \alpha_1^2 + \alpha_1 \beta' \geq p\alpha_1 + p\beta' + \beta' \alpha_1 + \beta'^2 - \alpha_1^2 - \alpha_1 \beta'.$$

Summation of similar terms yields

$$(k_0 - p)(r - k_0) - \alpha_1 k_0 + \beta' k_0 - \beta' p + \beta'(r - k_0) + \alpha_1 (r - k_0) - p\beta' \geq 0,$$
$$-(r - k_0)(k_0 - p + \beta' + \alpha_1) + k_0 (\alpha_1 - \beta') + 2p\beta' \leq 0.$$

Note that $k_0 - p + \beta' + \alpha_1 \geq 0, \alpha_1 - \beta' \geq 0$, since these are the binomial coefficients. For given values α_1 and β', we consider the positive part $k_0 (\alpha_1 - \beta') + 2p\beta'$ of this inequality:

$$k_0 (\alpha_1 - \beta') + 2p\beta' = k_0\alpha_1 + p\beta' + p\beta' - k_0\beta' = k_0\alpha_1 + p\beta' + (p - k_0) \beta'.$$

For the positive part to have maximum value we need to take $\alpha_1 = p/2, \beta' = p/2$, then

$$- (r - k_0) (k_0 - p + p/2 + p/2) + k_0 (p/2 - p/2) + 2p \frac{p}{2} \leq 0,$$

$$- (r - k_0) k_0 + k_0^2 \leq 0.$$

The last inequality holds because $r \geq 2k_0$.

Thus, we have proved the validity of (4.5.37). From the above arguments and calculations, we see that the coefficients $c_p \geq 0$, when $0 \leq p \leq r/2$. These arguments complete the proof.   □

Applying the above proposition, from Proposition 4.5.3 we get that $b_{h,k_0,m} = 0$ for $h + k_0 + 2m = l_0$. Thus, when $k \geq 1$ we have $b_{h,k,m} = 0$ for any weight degree l.

We have shown that the function f is holomorphic in a neighborhood of the boundary point 0. From the conditions of the theorem, the function f extends holomorphically to the intersection of D with each complex line passing through the boundary point 0. Consequently, by Hartogs' extension theorem [73] the function f will continue holomorphically to the whole domain $D \subset \mathbb{C}^2$. These arguments complete the proof of Theorem 4.5.1 in the two-dimensional case.

□

4.6   On a Boundary Analogue of Hartogs' Theorem in a Ball

In Sect. 4.1 we proved that a family of complex lines intersecting the germ of a generic manifold γ, is sufficient for holomorphic extension. In Sect. 4.3 we considered a family of complex lines passing through the germ of a complex hypersurface, the germ of a generic manifold in a complex hypersurface, and the germ of a real-analytic manifold of real dimension $(n - 1)$. In particular, in $\mathbb{C}^2$ this can be any real-analytic curve. Various other families are given in [5, 6, 12, 25]. We emphasize here the papers [6, 25], which show that a family of complex lines passing through a finite number of points arranged in some way is sufficient for holomorphic extension. However this is only asserted for real-analytic or infinitely differentiable functions defined on the boundary of a ball. So, Agranovskii and Globevnik showed that, in $\mathbb{C}^2$, for real-analytic functions defined on the boundary of a ball just two points lying in the closure of the ball are enough.

4.6.1 Main Results

Let $B = \{z \in \mathbb{C}^n : |z| < 1\}$ be a unit ball in $\mathbb{C}^n$ centered at the origin and let $S = \partial B$ be the boundary of the ball. We recall we say that the function $f \in \mathscr{C}(S)$ has the one-dimensional holomorphic extension property along the family $\mathfrak{L}_\Gamma$, if it has the one-dimensional holomorphic extension property along any complex line $l_{z,b} \in \mathfrak{L}_\Gamma$, where $l_{z,b}$ is a complex line of the form (3.2.1).

Recall we will also say that the set $\mathfrak{L}_\Gamma$ is sufficient for holomorphic extension, if the function $f \in \mathscr{C}(S)$ has the one-dimensional holomorphic extension property along all complex lines in the family $\mathfrak{L}_\Gamma$, and the function f holomorphically extends to B (i.e., f is a CR-function on S). In [13, 24, 49, 53] it is shown that for a class of continuous functions given on the boundary of a ball, a family of complex lines passing through finite points in the ball will be a sufficient family. Baracco was the first to prove this result, which was earlier explicitly conjectured by Agranovskii in [6]. Globevnik [24] suggested an alternative proof, even for the case when the vertices lie outside the ball. Those results were obtained by completely different methods.

Theorem A *Suppose* $n = 2$ *and the function* $f(\zeta) \in \mathscr{C}(S)$ *has the one-dimensional holomorphic extension property along the family* $\mathfrak{L}_{\{a,c,d\}}$, *and the points* a, c, $d \in B$ *do not lie on one complex line in* $\mathbb{C}^2$, *then* $f(\zeta)$ *extends holomorphically into* B.

We denote by $\mathscr{A}$ a set of points $a_k \in B \subset \mathbb{C}^n$, $k = 1, \ldots, n+1$, lying outside the complex hyperplane $\mathbb{C}^n$.

Theorem B *Let a function* $f(\zeta) \in \mathscr{C}(S)$ *have the one-dimensional holomorphic extension property along the family* $\mathfrak{L}_\mathscr{A}$, *then* $f(\zeta)$ *extends holomorphically into* B.

4.6.2 The Example

Now we give an example based on Globevnik's example, which shows that for continuous functions on the boundary S of the family $\mathfrak{L}_\mathscr{A}$, where $\mathscr{A}$ is a set of points $a_k \in B \subset \mathbb{C}^n$, $k = 1, \ldots, n$ is not enough for holomorphic extension.

Consider part of a complex hyperplane

$$\Gamma = \{(z', w) \in B : w = 0\},$$

in the ball $B = \{(z', w) \in \mathbb{C}^n : |z'|^2 + |w|^2 < 1\}$, where $z' = (z_1, \ldots, z_{n-1})$, $w \in \mathbb{C}$ and $|z'|^2 = |z_1|^2 + \cdots + |z_{n-1}|^2$. Then the function $f = \dfrac{w^{k+2}}{\overline{w}}$ $(k \in \mathbb{Z}, \ k \geq 0)$ has the one-dimensional property of holomorphic extension from ∂B along the complex line of the family $\mathfrak{L}_\Gamma$, which is smooth on ∂B, but does not extend holomorphically to B.

Consider complex lines intersecting Γ:

$$l_{a'} = \{(z', w) \in \mathbb{C}^n : z' = a' + b't, \; w = ct, \; t \in \mathbb{C}\}. \tag{4.6.1}$$

These lines pass through the point $(a', 0) \in \Gamma$. When $|a'| < 1$ the point $(a', 0) \in B$, while for $|a'| > 1$ the point $(a', 0) \notin \bar{B}$. Without loss of generality, we suppose that $|b'|^2 + |c|^2 = 1$. The intersection $l_{a'} \cap \partial B$ forms a circle

$$|t|^2 + \langle a', \bar{b}' \rangle \bar{t} + \langle \bar{a}', b' \rangle t = 1 - |a'|^2, \quad \text{or} \quad |t + a'\bar{b}'|^2 = 1 - |c|^2 |a'|^2, \tag{4.6.2}$$

where $\langle a', b' \rangle = a_1 b_1 + \cdots + a_{n-1} b_{n-1}$.

Indeed, since for complex lines of the form (4.6.1) on ∂B

$$\bar{t} = \frac{1 - |a'|^2 - \langle \bar{a}', b' \rangle t}{t + \langle a', \bar{b}' \rangle},$$

the function f on ∂B becomes

$$f = \frac{(t + \langle a', \bar{b}' \rangle)}{1 - |a'|^2 - \langle \bar{a}', b' \rangle t} \, (ct)^{k+2}.$$

The denominator of the fraction is equal to 0 at $t_0 = \dfrac{1 - |a'|^2}{\langle \bar{a}', b' \rangle}$. Substituting this point into expression (4.6.2), we obtain

$$\frac{(1 - |a'|^2)^2}{|\langle a', b \rangle|^2} + 1 - |a'|^2 > 0, \quad \text{if} \quad |a'| < 1.$$

Therefore the point of the line $l_{a'}$ corresponding to t_0 lies outside the ball B. So the function f holomorphically extends to $l_{a'} \cap B$. Consider the finite set $\mathscr{A} = \{a_1, \ldots, a_{n-1}, 0\} \in B$, then there exists a complex hyperplane containing $\mathscr{A}$. We can suppose, that this is the hyperplane Γ.

4.6.3 Complexification of the Poisson Kernel

Consider the invariant Poisson kernel [71, p. 48]

$$P(z, \zeta) = c_n \frac{\left(1 - |z|^2\right)^n}{\left|1 - \langle z, \bar{\zeta} \rangle\right|^{2n}} = c_n \frac{\left(1 - \langle z, \bar{z} \rangle\right)^n}{\left(1 - \langle z, \bar{\zeta} \rangle\right)^n \left(1 - \langle \zeta, \bar{z} \rangle\right)^n},$$

where $c_n = \dfrac{(n-1)!}{2\pi^n}$, $\langle z, \zeta \rangle = z_1 \zeta_1 + \cdots + z_n \zeta_n$.

If the function $f(z)$ is $\mathcal{M}$-harmonic in B and continuous on $\overline{B}$, then the integral representation

$$F(z) = \int_S f(\zeta) P(z, \zeta) \, d\sigma(\zeta), \qquad (4.6.3)$$

holds, where

$$d\sigma(\zeta) = \frac{2}{i^n} \sum_{k=1}^{n} (-1)^{k-1} \bar{\zeta}_k \, d\bar{\zeta}[k] \wedge d\zeta \Big|_S \qquad (4.6.4)$$

is the Lebesgue measure on S, $d\zeta = d\zeta_1 \wedge \cdots \wedge d\zeta_n$, $d\bar{\zeta}[k] = d\bar{\zeta}_1 \wedge \cdots \wedge d\bar{\zeta}_{k-1} \wedge d\bar{\zeta}_{k+1} \wedge \cdots \wedge d\bar{\zeta}_n$. The function $F(z)$ extends on $\overline{B}$ as a continuous function and the boundary values of the function $F(z)$ coincides with $f(\zeta)$, i.e., $F(z)\big|_S = f(\zeta)$. Recall that an $\mathcal{M}$-invariant harmonic function satisfies the Laplace equation [71, pp. 55–56]

$$\tilde{\Delta} F(z) = 0,$$

where

$$\tilde{\Delta} F(z) = 4\big(1 - |z|^2\big) \sum_{j,k=1}^{n} (\delta_{jk} - z_j \bar{z}_k) \frac{\partial^2 F(z)}{\partial z_j \partial \bar{z}_k}$$

and δ_{jk} is the Kronecker symbol. The holomorphic functions in the ball B are $\mathcal{M}$-harmonic. Therefore, formula (4.6.3) is an integral representation for holomorphic functions. Consider a complex line of the form

$$l_{z,b} = \{\zeta \in \mathbb{C}^n : \zeta = z + bt, \ t \in \mathbb{C}\}, \qquad (4.6.5)$$

where $z \in \mathbb{C}^n$, $b \in \mathbb{CP}^{n-1}$.

Consider a complexification of the Poisson kernel of the form

$$Q(z, w, \zeta) = c_n \frac{\big(1 - \langle z, w \rangle\big)^n}{\big(1 - \langle z, \bar{\zeta} \rangle\big)^n \big(1 - \langle \zeta, w \rangle\big)^n}. \qquad (4.6.6)$$

It is obvious that $P(z, \zeta) = Q(z, \bar{z}, \zeta)$. Introduce the function

$$\Phi(z, w) = \int_S f(\zeta) Q(z, w, \zeta) \, d\sigma(\zeta).$$

This function is holomorphic in the variables (z, w) in $B \times B \subset \mathbb{C}^{2n}$, because if $\zeta \in S$ and $z, w \in B$, then the denominator in kernel (4.6.6) does not vanish. Note that $\Phi(z, \bar{z}) = F(z)$, and the derivatives

$$\frac{\partial^{\alpha+\beta}\Phi}{\partial z^{\alpha}\,\partial w^{\beta}}\bigg|_{w=\bar{z}} = \frac{\partial^{\alpha+\beta}F}{\partial z^{\alpha}\,\partial\bar{z}^{\beta}}, \tag{4.6.7}$$

where

$$\frac{\partial^{\alpha+\beta}\Phi}{\partial z^{\alpha}\,\partial w^{\beta}} = \frac{\partial^{\|\alpha\|+\|\beta\|}\Phi}{\partial z_1^{\alpha_1}\cdots\partial z_n^{\alpha_n}\,\partial w_1^{\beta_1}\cdots\partial w_n^{\beta_n}}$$

and $\alpha = (\alpha_1, \ldots, \alpha_n)$, $\beta = (\beta_1, \ldots, \beta_n)$ are multi-indices, $\|\alpha\| = \alpha_1 + \cdots + \alpha_n$, $\|\beta\| = \beta_1 + \cdots + \beta_n$.

Proposition 4.6.1 *Let the function $f(\zeta) \in \mathscr{C}(S)$ have the one-dimensional holomorphic extension property along the family $\mathfrak{L}_{\{0\}}$, then the integral*

$$\Phi(z, w) = \int_S f(\zeta) Q(z, w, \zeta)\, d\sigma(\zeta)$$

has the properties $\Phi(0, w) = \mathrm{const}$ and the derivatives $\dfrac{\partial^{\alpha}\Phi(0, w)}{\partial z^{\alpha}}$ are polynomials in w of degree not higher than $\|\alpha\|$.

Proof Consider the derivative

$$\frac{\partial^{\alpha+\beta}}{\partial z^{\alpha}\,\partial w^{\beta}}\left(\frac{1}{(1 - \langle z, \bar{\zeta}\rangle)^n (1 - \langle \zeta, w\rangle)^n}\right)$$

$$= \frac{C_{\alpha,\beta}\,\bar{\zeta}^{\alpha}\,\zeta^{\beta}}{(1 - \langle z, \bar{\zeta}\rangle)^{n+\|\alpha\|}(1 - \langle \zeta, w\rangle)^{n+\|\beta\|}}. \tag{4.6.8}$$

Calculate the derivative of $Q(z, w, \zeta)$

$$\frac{\partial^{\alpha+\beta}Q(z, w, \zeta)}{\partial z^{\alpha}\,\partial w^{\beta}} = c_n \frac{\partial^{\alpha+\beta}}{\partial z^{\alpha}\,\partial w^{\beta}}\left[(1 - \langle z, w\rangle)^n \frac{1}{(1 - \langle z, \bar{\zeta}\rangle)^n (1 - \langle \zeta, w\rangle)^n}\right].$$

It is clear that

$$\frac{\partial^{\alpha}}{\partial z^{\alpha}}(fg) = \sum_{0 \le \gamma \le \alpha} b_{\gamma} \frac{\partial^{\gamma}f}{\partial z^{\gamma}} \frac{\partial^{\alpha-\gamma}g}{\partial z^{\alpha-\gamma}},$$

where $\gamma \leq \alpha$ means that $\gamma_1 \leq \alpha_1, \ldots, \gamma_n \leq \alpha_n$, and b_γ is some constant. Therefore

$$
\frac{\partial^\alpha Q(z, w, \zeta)}{\partial z^\alpha}
$$

$$
= c_n \sum_{0 \leq \gamma' \leq \alpha} b_{\gamma'} \frac{\partial^{\alpha - \gamma'}}{\partial z^{\alpha - \gamma'}} \left(\frac{1}{(1 - \langle z, \bar{\zeta} \rangle)^n (1 - \langle \zeta, w \rangle)^n} \right) \frac{\partial^{\gamma'}}{\partial z^{\gamma'}} (1 - \langle z, w \rangle)^n.
$$

The derivatives

$$
\frac{\partial^{\gamma'}}{\partial z^{\gamma'}} (1 - \langle z, w \rangle)^n = (-1)^{\|\gamma'\|} n(n - 1) \cdots (n - \|\gamma'\|)(1 - \langle z, w \rangle)^{n - \|\gamma'\|} w^{\gamma'}.
$$

Then

$$
\frac{\partial^{\gamma'}}{\partial z^{\gamma'}} (1 - \langle z, w \rangle)^n \bigg|_{z=0} = (-1)^{\|\gamma'\|} n(n - 1) \cdots (n - \|\gamma'\|) w^{\gamma'}. \qquad (4.6.9)
$$

Therefore formulas (4.6.8) and (4.6.9) imply

$$
\frac{\partial^\alpha Q(0, w, \zeta)}{\partial z^\alpha} = c_n \sum_{0 \leq \gamma' \leq \alpha} C_{\gamma'} \frac{\bar{\zeta}^{\alpha - \gamma'} w^{\gamma'}}{(1 - \langle \zeta, w \rangle)^n}.
$$

From here we obtain, that the derivative $\dfrac{\partial^{\alpha + \beta} Q(0, w, \zeta)}{\partial z^\alpha \partial w^\beta}$ is a sum of terms of the

form $\dfrac{C_{\alpha, \beta, \gamma, \delta} \bar{\zeta}^{\alpha - \gamma'} \zeta^{\beta - \gamma''} w^\delta}{(1 - \langle \zeta, w \rangle)^{n + \|\beta\| - \|\gamma''\|}}$ for $\|\delta\| \leq n$ and $\gamma'' \leq \gamma'$. Therefore the derivative

$\dfrac{\partial^{\alpha + \beta} \Phi(0, w)}{\partial z^\alpha \partial w^\beta}$ is a linear combination of integrals with the coefficients depending on
w

$$
\int_S f(\zeta) \frac{\bar{\zeta}^{\alpha - \gamma'} \zeta^{\beta - \gamma''}}{(1 - \langle \zeta, w \rangle)^{n + \|\beta\| - \|\gamma''\|}} \, d\sigma(\zeta). \qquad (4.6.10)
$$

The form $d\sigma(\zeta)$ in the variables b and t, where $\zeta = bt$, $t \in \mathbb{C}$ will change as follows [49]

$$
d\sigma(bt) = \frac{2}{i^n} |t|^{2n - 2} \bar{t} \, dt \wedge \left(\sum_{k=1}^n (-1)^{k-1} b_k \, db[k] \right) \wedge \left(\sum_{k=1}^n (-1)^{k-1} \bar{b}_k \, d\bar{b}[k] \right).
$$

Since on the sphere S the equality $1 = |\zeta| = |bt|$ holds, we have $|t| = \dfrac{1}{|b|}$ and $\bar{t} = \dfrac{1}{t|b|^2}$. Then

$$d\sigma(bt) = \frac{2}{i^n} \frac{1}{t|b|^{2n}} dt \wedge \left(\sum_{k=1}^{n} (-1)^{k-1} b_k \, db[k] \right) \wedge \left(\sum_{k=1}^{n} (-1)^{k-1} \bar{b}_k \, d\bar{b}[k] \right)$$

$$= \lambda(b) \wedge \frac{dt}{t}.$$

By Fubini's theorem we obtain

$$\int_S f(\zeta) \frac{\bar{\zeta}^{\alpha - \gamma'} \zeta^{\beta - \gamma''}}{(1 - \langle \zeta, w \rangle)^{n + \|\beta\| - \|\gamma''\|}} \, d\sigma(\zeta)$$

$$= c_n \int_{\mathbb{CP}^{n-1}} \lambda(b) \int_{S \cap l_{0,b}} f(bt) \frac{\bar{t}^{\|\alpha\| - \|\gamma'\|} t^{\|\beta\| - \|\gamma''\|}}{t(1 - t\langle b, w \rangle)^{n + \|\beta\| - \|\gamma''\|}} \, dt$$

$$= c_n \int_{\mathbb{CP}^{n-1}} \lambda(b) \int_{S \cap l_{0,b}} f(bt) \frac{t^{\|\beta\| - \|\gamma''\|}}{t^{\|\alpha\| - \|\gamma'\| + 1}(1 - t\langle b, w \rangle)^{n + \|\beta\| - \|\gamma''\|}} \, dt$$

$$= c_n \int_{\mathbb{CP}^{n-1}} \lambda(b) \int_{S \cap l_{0,b}} f(bt) \frac{t^{\|\beta\| - \|\alpha\| + \|\gamma'\| - \|\gamma''\| - 1}}{(1 - t\langle b, w \rangle)^{n + \|\beta\| - \|\gamma''\|}} \, dt = 0,$$

if $\|\beta\| > \|\alpha\|$ (then $\|\beta\| - \|\gamma''\| > \|\alpha\| - \|\gamma'\|$), and the function $\dfrac{1}{1 - t\langle b, w \rangle}$ is holomorphic in the closed disk $\left\{ t \in \mathbb{C} : |t| \leq \dfrac{1}{|b|} \right\}$, i.e., in $\bar{B} \cap l_{0,b}$. In fact, the singular point $t_0 = \dfrac{1}{\langle b, w \rangle}$ does not lie in the disk $\bar{B} \cap l_{0,b}$, since $|\langle b, w \rangle| \leq |b||w|$ and therefore $|t_0| > \dfrac{1}{|b|}$ if $|w| < 1$. Hence by virtue of (4.6.10)

$$\frac{\partial^{\alpha + \beta} \Phi(0, w)}{\partial z^\alpha \partial w^\beta} = 0 \tag{4.6.11}$$

for $\|\beta\| > \|\alpha\|$.

Therefore by the Taylor formula for the function $\Phi(z, w)$ at the point $(0, 0)$ we get that $\Phi(0, w) = \text{const}$ and the derivatives $\dfrac{\partial^\alpha \Phi(0, w)}{\partial z^\alpha}$ are polynomials in w of degree not higher than $\|\alpha\|$. $\square$

Corollary 4.6.1 *Under the hypotheses of Proposition 4.6.1 the equality*

$$\frac{\partial^{\alpha+\beta} F(0,0)}{\partial z^{\alpha} \partial \bar{z}^{\beta}} = 0$$

holds for $\|\beta\| > \|\alpha\|$.

Proof Substituting $w = \bar{z}$ into Eq. (4.6.11) and using Eq. (4.6.7), we obtain the desired result. $\qquad\square$

4.6.4 The Application of Automorphisms of a Ball

Recall that the of a ball $\varphi_a(u)$ transforming the point a into 0, and vice versa, has the form

$$z = \varphi_a(u) = \frac{a - \dfrac{\langle u, \bar{a} \rangle}{\langle a, \bar{a} \rangle} a - \sqrt{1 - |a|^2}\left(u - \dfrac{\langle u, \bar{a} \rangle}{\langle a, \bar{a} \rangle} a\right)}{1 - \langle u, \bar{a} \rangle}$$

and $\varphi_a(u)$ is an involution, i.e., $\varphi_a^{-1} = \varphi_a$ [71, p. 34]. Note that $\big(\varphi_a(u), \varphi_{\bar{a}}(v)\big)$ is the automorphism of $B \times B$, taking the point $(a, \bar{a})$ to $(0, 0)$ and vice versa.

Lemma 4.6.1 *The automorphism $\varphi_c(u)$ of a ball B transforms the complex line $l_{c,d}$, passing through points $c, d \in B$ to a complex line $l_{0,a}$, passing through points 0 and $a = \varphi_c(d) \in B$.*

Proof Let us show that the automorphism of a ball preserves complex lines. In fact, let $u = c + (d - c)t$. We compute $\langle u, c \rangle = |c|^2 + t(\langle d, \bar{c} \rangle - |c|^2)$, then

$$\varphi_c\big(c + (d - c)t\big) = t \, \frac{c\big(|c|^2 - \langle d, \bar{c} \rangle\big) - \sqrt{1 - |c|^2}\big(d|c|^2 - \langle d, \bar{c} \rangle c\big)}{|c|^2\big(1 - |c|^2 - t(\langle d, \bar{c} \rangle - |c|^2)\big)},$$

and the singular point of this function does not lie in $\bar{B}$. For $t = 0$ we get $\varphi_c\big(c + (d - c)0\big) = \varphi_c(c) = 0$, and for $t = 1$ we get

$$a = \varphi_c\big(c + (d - c)1\big) = \varphi_c(d) = \frac{c\big(|c|^2 - \langle d, \bar{c} \rangle\big) - \sqrt{1 - |c|^2}\big(d|c|^2 - \langle d, \bar{c} \rangle c\big)}{|c|^2\big(1 - \langle d, \bar{c} \rangle\big)},$$

and for other points t it takes the form

$$\varphi_c\big(c + (d - c)t\big) = \frac{tg}{e_1 - e_2 t},$$

where g is some vector of $\mathbb{C}^n$. We put $z_1 = \dfrac{tg_1}{e_1 - e_2 t}$, then $t = \dfrac{e_1 z_1}{g_1 + e_2 z_1}$, and $z_j = \dfrac{tg_j}{e_1 - e_2 t} = \dfrac{e_1 z_1 g_1}{e_1 g_1}$. Therefore $\varphi_c(c + (d - c)t)$ defines a complex line passing through the points 0 and a. $\square$

As shown in [71, p. 52, Remark] the following equality

$$d\sigma(\varphi_a(\eta)) = P(a, \eta)\, d\sigma(\eta), \qquad \eta \in S$$

holds and using Eq. (4.6.6), we obtain

$$d\sigma(\varphi_a(\eta)) = P(a, \eta)\, d\sigma(\eta) = Q(a, \bar{a}, \eta)\, d\sigma(\eta), \qquad \eta \in S.$$

By Theorem 2.2.2 [71, p. 34] the automorphism $\varphi_a(u)$ is a homeomorphism of the ball $\bar{B}$ to itself and homeomorphisms $S \to S$. Also by Theorem 3.3.5 [71, p. 50], the equality

$$P\big(\varphi_a(u), \varphi_a(\eta)\big) = \frac{P(u, \eta)}{P(a, \eta)}$$

holds. Therefore

$$Q\big(\varphi_a(u), \varphi_{\bar{a}}(\bar{u}), \varphi_a(\eta)\big) = \frac{Q(u, \bar{u}, \eta)}{Q(a, \bar{a}, \eta)}. \tag{4.6.12}$$

The manifold $v = \bar{u}$ is generic in $\mathbb{C}^{2n}$, and the functions from Eq. (4.6.12) are real-analytic. Hence

$$Q\big(\varphi_a(u), \varphi_{\bar{a}}(v), \varphi_a(\eta)\big) = \frac{Q(u, v, \eta)}{Q(a, \bar{a}, \eta)}.$$

Denote the function

$$\Phi_a(z, w) = \Phi\big(\varphi_a(u), \varphi_{\bar{a}}(v)\big) = \int_S f(\zeta) Q\big(\varphi_a(u), \varphi_{\bar{a}}(v), \zeta\big)\, d\sigma(\zeta).$$

Make the change $\zeta = \varphi_a(\eta)$ and denote $f(\varphi_a(\eta)) = f_a(\eta)$. We obtain

$$\Phi(z, w) = \int_S f(\varphi_a(\eta)) Q\big(\varphi_a(u), \varphi_{\bar{a}}(v), \varphi_a(\eta)\big)\, d\sigma(\varphi_a(\eta))$$

$$= \int_S f(\varphi_a(\eta)) \frac{Q(u, v, \eta) Q(a, \bar{a}, \eta)}{Q(a, \bar{a}, \eta)}\, d\sigma(\eta)$$

$$= \int_S f_a(\eta) Q(u, v, \eta)\, d\sigma(\eta) = \Phi_a(u, v). \tag{4.6.13}$$

Proposition 4.6.2 *Let the function* $f(\zeta) \in \mathscr{C}(S)$ *have the one-dimensional holomorphic extension property along the family* $\mathfrak{L}_{\{a\}}$, $a \in B$, *then* $\Phi(a, w) = \text{const}$ *and the derivatives* $\dfrac{\partial^\alpha \Phi(a, w)}{\partial z^\alpha}$ *are polynomials in* $\varphi_{\bar{a}}(w)$ *of degree not higher than* $\|\alpha\|$.

Proof With the help of the automorphism φ_a we translate the point a to 0. Then, by Proposition 4.6.1, $\Phi_a(0, v) = \text{const}$. Using Eq. (4.6.13) we obtain $\Phi(a, \varphi_{\bar{a}}(v)) = \text{const}$, i.e., $\Phi(a, w) = \text{const}$. Similarly, from Proposition 4.6.1 and equality (4.6.13) we get that the derivatives $\dfrac{\partial^\alpha \Phi(a, w)}{\partial z^\alpha}$ are polynomials in $\varphi_{\bar{a}}(w)$ of degree not higher than $\|\alpha\|$. $\square$

Represent the function $\Phi(z, w)$ as a sum of homogeneous polynomials in z and w. Expand $Q(z, w, \zeta)$ in a power series $\langle z, \bar{\zeta} \rangle$, $\langle \zeta, w \rangle$. Since

$$\frac{1}{(1 - \langle z, \bar{\zeta} \rangle)^n} = \sum_{k=0}^{\infty} C_{n+k-1}^k \langle z, \bar{\zeta} \rangle^k,$$

$$\frac{1}{(1 - \langle \zeta, w \rangle)^n} = \sum_{l=0}^{\infty} C_{n+l-1}^l \langle \zeta, w \rangle^k$$

(the series under consideration converges absolutely for $\zeta \in S$, $z, w \in B$, and uniformly on $S \times K$, where K is an arbitrary compact set in $B \times B$), therefore

$$Q(z, w, \zeta)$$
$$= c_n (1 - \langle z, w \rangle)^n \sum_{k=0}^{\infty} \sum_{l=0}^{\infty} C_{n+k-1}^k C_{n+l-1}^l \int_S f(\zeta) \langle z, \bar{\zeta} \rangle^k \langle \zeta, w \rangle^l d\sigma(\zeta). \qquad (4.6.14)$$

The integral $\displaystyle\int_S f(\zeta) \langle z, \bar{\zeta} \rangle^k \langle \zeta, w \rangle^l d\sigma(\zeta)$ is a homogeneous polynomial with the degree of homogeneity k on z and l on w. Multiplying the sum from equality (4.6.14) by the factor $(1 - \langle z, w \rangle)^n$ and regrouping the terms we find

$$\Phi(z, w) = \sum_{k,l=0}^{\infty} P_{k,l}(z, w), \qquad (4.6.15)$$

where $P_{k,l}(z, w)$ are the homogeneous holomorphic polynomials with the degree of homogeneity k in z and l in w, and the double series converges absolutely in $B \times B$ and uniformly on any compact set in $B \times B$.

Theorem 4.6.1 *Let the function* $f(\zeta) \in \mathscr{C}(S)$, *the point* $a \in B$ *and the function* $\Phi(z, w)$ *satisfy the conditions such that* $\Phi(0, w) = \text{const}$, *and* $\Phi(a, w) = \text{const}$, $\dfrac{\partial^\alpha \Phi(0, w)}{\partial z^\alpha}$ *be a polynomial on* w *of degree not higher than* $\|\alpha\|$, *then for any fixed* z, belonging to the complex line $l_{0,a} = \{z \in \mathbb{C}^n : z = at, |t| < 1\}$ it is true that $\Phi(z, w) = \text{const}$ on w, i.e., $\dfrac{\partial^\beta \Phi(z, w)}{\partial w^\beta} = 0$ at $\|\beta\| > 0$.

Proof We represent the function $\Phi(z, w)$ in the form (4.6.15):

$$\Phi(z, w) = \sum_{k,l=0}^{\infty} P_{k,l}(z, w).$$

By the hypothesis expansion (4.6.15) takes the form

$$\Phi(z, w) = \sum_{k \geq l \geq 0} P_{k,l}(z, w),$$

since the derivatives $\dfrac{\partial^{\alpha+\beta} \Phi(0, 0)}{\partial z^\alpha \partial w^\beta} = 0$ under $\|\beta\| > \|\alpha\|$.

Introduce the functions $\Phi_k(z, w) = \sum_{l=k}^{\infty} P_{k,l}(z, w)$, then

$$\Phi(z, w) = \sum_{k=0}^{\infty} \Phi_k(z, w). \tag{4.6.16}$$

Consider series (4.6.15) to be a double series converging absolutely in $B \times B$ and uniformly on compact subsets of $B \times B$, and series (4.6.16) is the same as in (4.6.15).

From the form of the series $\Phi_k(z, w)$ we get $\Phi_k(tz, w) = t^k \Phi_k(z, w)$ for each $t \in \mathbb{C}$. By the theorem's hypothesis

$$\Phi(0, w) = \Phi_0(0, w) = \sum_{l=0}^{\infty} P_{0,l}(0, w) = \text{const} \tag{4.6.17}$$

and

$$\Phi(a, w) = \sum_{k=0}^{\infty} \Phi_k(a, w) = \text{const}.$$

Consider

$$\Phi(at, w) = \sum_{k=0}^{\infty} t^k \Phi_k(a, w). \tag{4.6.18}$$

Calculate

$$\frac{d^m}{dt^m} \Phi(at, w) = m! \Phi_m(a, w) + \cdots + k(k-1) \cdots (k-m+1) t^{k-m} \Phi_k(a, w) + \cdots.$$

Calculate the same derivative as the derivative of a composite function

$$\frac{d^m}{dt^m}\Phi(at, w) = \sum_{\|\alpha\|=m} \frac{\partial^\alpha \Phi(at, w)}{\partial z^\alpha} a^\alpha.$$

Equating the derivatives, we obtain

$$\sum_{\|\alpha\|=m} \frac{\partial^\alpha \Phi(at, w)}{\partial z^\alpha} a^\alpha = \sum_{k=m}^{\infty} k(k-1)\cdots(k-m+1)t^{k-m}\Phi_k(a, w). \qquad (4.6.19)$$

Substituting $t = 0$ into Eq. (4.6.19) we obtain that $\dfrac{d^m}{dt^m}\Phi(0, w) = m!\Phi_m(a, w)$ is a polynomial of degree not higher than m in w, since the left-hand side of this equation is a polynomial of degree not higher than m in w by the hypothesis of the theorem. For $m = 0$ we get $\Phi(0, w) = \Phi_0(a, w) = \text{const}$ and from (4.6.17) we have $\Phi(0, w) = \Phi_0(a, w) = \Phi_0(0, w)$.

In Eq. (4.6.18) we substitute $t = 1$ and obtain

$$\Phi(a, w) = \sum_{k=0}^{\infty} \Phi_k(a, w) = \text{const}.$$

Since $\Phi_k(a, w) = \sum_{l=k}^{\infty} P_{k,l}(a, w)$ is a polynomial in w of degree not higher than k, then $\sum_{l=k}^{\infty} P_{k,l}(a, w) = P_{k,k}(a, w)$. Therefore

$$\text{const} = \Phi(a, w) = \sum_{k=0}^{\infty} \Phi_k(a, w) = \sum_{k=0}^{\infty} P_{k,k}(a, w).$$

Hence $P_{k,k}(a, w) = 0$ for $k > 0$. From here $\Phi_k(a, w) = 0$ for $k > 0$, so from (4.6.18) we get $\Phi(at, w) = \text{const}$ and $\dfrac{\partial^\beta \Phi(at, w)}{\partial w^\beta} = 0$ at $\|\beta\| > 0$. $\qquad \square$

Corollary 4.6.2 *Let the function $f(\zeta) \in \mathscr{C}(S)$ have the one-dimensional holomorphic extension property along the family $\mathfrak{L}_{\{0,a\}}$, then $\Phi(z, w) = \text{const}$ for points z belonging to the complex line $l_{0,a} \cap B$, i.e., $\dfrac{\partial^\beta \Phi(z, w)}{\partial w^\beta} = 0$ at $\|\beta\| > 0$.*

Proof follows from Proposition 4.6.1 and Theorem 4.6.1. $\qquad \square$

Corollary 4.6.3 *Under the hypotheses of Corollary 4.6.2 the equality*

$$\frac{\partial^\beta F(z)}{\partial \bar{z}^\beta} = 0$$

holds for all points $z \in l_{0,a} \cap B$ *and* $\|\beta\| > 0$.

Theorem 4.6.2 *Let the function* $f(\zeta) \in \mathscr{C}(S)$ *have the one-dimensional holomorphic extension property along the family* $\mathfrak{L}_{\{c,d\}}$ *and* $c, d \in B$, *then* $\Phi(c + (d - c)t, w) = \text{const on } w \text{ for } |t| < 1$, *i.e.,* $\dfrac{\partial^\beta \Phi(c + (d - c)t, w)}{\partial w^\beta} = 0 \text{ at } \|\beta\| > 0$.

Proof Suppose $c, d \in B$. Consider the automorphism $\varphi_c(z)$ mapping the point c to 0, and vice versa, i.e., $\varphi_c(c) = 0$ and $\varphi_c(0) = c$. Let the point d under this automorphism move to the point $a = \varphi_c(d)$. Denote $f_c(\zeta) = f(\varphi_c(\zeta))$ and

$$\Phi_c(z, w) = \int_S f_c(\zeta) Q(z, w, \zeta) \, d\sigma(\zeta).$$

From Eq. (4.6.13) we have $\Phi(z, w) = \Phi_c(\varphi_c(z), \varphi_{\bar{c}}(w))$. Proposition 4.6.1 implies that $\Phi_c(0, \varphi_{\bar{c}}(w)) = \text{const}$, i.e., $\Phi(c, w) = \text{const}$.

Let us show that an automorphism of the ball preserves complex lines. In fact, let $z = c + (d - c)t$. We compute $\langle z, c \rangle = |c|^2 + t(\langle d, \bar{c} \rangle - |c|^2)$, then

$$\varphi_c(c + (d - c)t) = t \frac{c(|c|^2 - \langle d, \bar{c} \rangle) - \sqrt{1 - |c|^2}(d|c|^2 - \langle d, \bar{c} \rangle c)}{|c|^2(1 - |c|^2 - t(\langle d, \bar{c} \rangle - |c|^2))}.$$

For $t = 0$ we get $\varphi_c(c + (d - c)0) = \varphi_c(c) = 0$, for $t = 1$ we get

$$a = \varphi_c(c + (d - c)1) = \varphi_c(d) = \frac{c(|c|^2 - \langle d, \bar{c} \rangle) - \sqrt{1 - |c|^2}(d|c|^2 - \langle d, \bar{c} \rangle c)}{|c|^2(1 - \langle d, \bar{c} \rangle)},$$

and for other points t it takes the form

$$\varphi_c(c + (d - c)t) = \frac{tg}{e_1 - e_2 t},$$

where g is the vector of $\mathbb{C}^n$. We take $z_1 = \dfrac{tg_1}{e_1 - e_2 t}$, then $t = \dfrac{e_1 z_1}{g_1 + e_2 z_1}$, and $z_j = \dfrac{tg_j}{e_1 - e_2 t} = \dfrac{e_1 z_1 g_1}{e_1 g_1}$. Therefore $\varphi_c(c + (d - c)t)$ defines a complex line passing through the points 0 and a.

By Lemma 4.6.1, the complex line $l_{c,d}$ transforms to a complex line $l_{0,a}$. Hence by Theorem 4.6.1 we have $\Phi_c(ct, w) = \text{const}$ and likewise, $\Phi(c + (d - c)t, w) = \text{const}$. $\qquad\square$

Corollary 4.6.4 *Under the hypotheses of Theorem 4.6.2 the equality*

$$\frac{\partial^\beta F(z)}{\partial \bar{z}^\beta}\bigg|_{z=c+(d-c)t} = 0$$

holds at $\|\beta\| > 0$.

4.6.5 Proof of the Main Results

Theorem 4.6.3 *Suppose* $n = 2$ *and the function* $f(\zeta) \in \mathscr{C}(S)$ *has the one-dimensional holomorphic extension property along the family* $\mathfrak{L}_{\{a,c,d\}}$, *and the points* $a, c, d \in B$ *do not lie on one complex line in* $\mathbb{C}^2$, *then* $\dfrac{\partial^\beta \Phi(z,w)}{\partial w^\beta} = 0$ *for any* $z \in B$ *and* $\|\beta\| > 0$, *i.e.,* $F(z)$ *is holomorphic in* B.

Proof We move the point d by automorphism φ to 0. The conditions for the points 0, $\varphi(a)$ and $\varphi(c)$ will remain the same. Therefore the points $\varphi(a)$ and $\varphi(c)$ again are denoted by a and c.

Let $\tilde{z}$ be an arbitrary point of the line $l_{a,c}$. Then by Theorem 4.6.2 we have $\dfrac{\partial^\beta \Phi(\tilde{z},w)}{\partial w^\beta} = 0$ at $\|\beta\| > 0$, and by Theorem 4.6.1 (Φ satisfies the conditions of Theorem 4.6.1 at zero), then $\dfrac{\partial^\beta \Phi(z,w)}{\partial w^\beta} = 0$ for all $z \in l_{0,\tilde{z}}$, i.e., for all points z in some open set in B. Substituting $w = \bar{z}$ into this equation and using equality (4.6.7), we get $\dfrac{\partial^\beta F(z)}{\partial \bar{z}^\beta} = 0$. Since the points 0, a, c do not lie on one complex line, we have $\dfrac{\partial^\beta F(z)}{\partial \bar{z}^\beta} = 0$ for all points z in some open set and consequently, at all points of the ball B due to the real-analyticity of the function $F(z)$. In particular, $\dfrac{\partial F(z)}{\partial \bar{z}_j} = 0$ for all $z \in B$ and $j = 1, \ldots, n$, therefore $f(\zeta)$ extends holomorphically into B. $\qquad\square$

Theorem 4.6.3 implies that in a ball $B \subset \mathbb{C}^2$ a sufficient set for a continuous function defined on the boundary of the ball is the set $\mathfrak{L}_{\{a,c,d\}}$, where a, c, d are arbitrary points of the ball not lying on one complex line.

Denote by $\mathscr{A}$ the set of points $a_k \in B \subset \mathbb{C}^n$, $k = 1, \ldots, n+1$ not lying on the complex hyperplane $\mathbb{C}^n$.

Theorem 4.6.4 *Suppose* $f(\zeta) \in \mathscr{C}(S)$ *has the one-dimensional holomorphic extension property along the family* $\mathfrak{L}_{\mathscr{A}}$, *then* $\dfrac{\partial^\beta \Phi(z,w)}{\partial w^\beta} = 0$ *for any* $z \in B$ *and* $\|\beta\| > 0$ *and* $f(\zeta)$ *extends holomorphically into* B.

Proof The proof is by induction on n, based on Theorem 4.6.3 ($n = 2$). Suppose the theorem is true for all dimensions $k < n$. Without loss of generality, we assume $a_{n+1} = 0$ when $k = n$,.

Consider a complex plane Γ passing through the points $a_1, \ldots, a_n$. By assumption, it has the dimension $n - 1$ and $0 \notin \Gamma$. The intersection $\Gamma \cap B$ is some ball in $\mathbb{C}^{n-1}$. The function $f|_{\Gamma \cap S}$ is continuous and has the property of holomorphic extension along the family $\mathfrak{L}_{\mathscr{A}_1}$, where $\mathscr{A}_1 = \{a_1, \ldots, a_n\}$. By the induction assumption $\dfrac{\partial^\beta \Phi(z', w)}{\partial w^\beta} = 0$ at $\|\beta\| > 0$ for all $z' \in \Gamma \cap B$. Connecting points $z' \in \Gamma$ with the point 0 by Theorem 4.6.1 we get $\dfrac{\partial^\beta \Phi(z, w)}{\partial w^\beta} = 0$ at $\|\beta\| > 0$ for some open set in B. Hence, as in Theorem 4.6.3, $F(z)$ is holomorphic in B. $\qquad\square$

Corollary 4.6.5 *Under the hypotheses of Theorem 4.6.4 the equality*

$$\frac{\partial^\beta F(z)}{\partial \bar{z}^\beta} = 0$$

holds for any $z \in B$ and $\|\beta\| > 0$ and $f(\zeta)$ extends holomorphically into B.

References

1. Agranovskii, M.L.: Invariant spaces and traces of holomorphic functions on the skeletons of classical domains. Sib. Math. J. **25**, 165–175 (1984)
2. Agranovskii, M.L.: Möbius spaces of functions on the Shilov boundaries of classical domains of tubular type. Sib. Math. J. **29**, 697–707 (1988)
3. Agranovskii, M.L., Semenov, A.M.: Boundary analogues of Hartog's theorem. Sib. Math. J. **32**, 137–139 (1991)
4. Agranovskii, M.L., Val'skii, R.E.: Maximality of invariant algebras of functions. Sib. Math. J. **12**, 1–7 (1971)
5. Agranovsky, M.: Propagation of boundary CR-foliations and Morera type theorems for manifolds with attached analytic discs. Adv. Math. **211**, 284–326 (2007)
6. Agranovsky, M.: Analogue of a theorem of Forelli for boundary values of holomorphic functions on the unit ball of $\mathbb{C}^n$. J. d'Analyse Math. **113**, 293–304 (2011)
7. Airapetyan, R.A., Khenkin, G.M.: Integral representations of differential forms on Cauchy-Riemann manifolds and the theory of CR-functions. Russ. Math. Surv. **39**, 41–118 (1984)
8. Aizenberg, L.A., Dautov, Sh.A.: Differential Forms Orthogonal to Holomorphic Functions or Forms, and Their Properties. AMS, Providence (1983)
9. Aizenberg, L.A., Yuzhakov, A.P.: Integral Representations and Residues in Multidimensional Complex Analysis. AMS, Providence (1983)
10. Aronov, A.M., Kytmanov, A.M.: Holomorphicity of functions representable by a Martinelli–Bochner integral. Funct. Anal. Appl. **9**, 254–255 (1975)
11. Baouendi, M.S., Ebenfelt, P., Rothschild, L.P.: Real Submanifolds in Complex Space and Their Mappings. Princeton University Press, Princeton (1998)
12. Baracco, L.: Holomorphic extension from the sphere to the ball. Math. Anal. Appl. **388**, 760–762 (2012)
13. Baracco, L.: Separate holomorphic extension along lines and holomorphic extension from the sphere to the ball. Am. J. Math. **135**, 493–497 (2013)
14. Baracco, L., Tumanov, A., Zampieri, G.: Extremal discs and the holomorphic extension from convex hypersurfaces. Ark. Mat. **45**, 1–13 (2007)
15. Beals, M., Fefferman, C., Grossman, R.: Strictly pseudoconvex domains in $\mathbb{C}^n$. Bull. Am. Math. Soc. **8**, 125–322 (1983)
16. Bochner, S.: Analytic and meromorphic continuation by means of Green's formula. Ann. Math. **44**, 652–673 (1943)
17. Carmona, J.J., Paramonov, P.V., Fedorovskii, K.Yu.: On uniform approximation by polyanalytic polynomials and the Dirichlet problem for bianalytic functions. Sb. Math. **193**, 1469–1492 (2002)

© Springer International Publishing Switzerland 2015
A.M. Kytmanov, S.G. Myslivets, *Multidimensional Integral Representations*,
DOI 10.1007/978-3-319-21659-1

18. Chirka, E.M.: Analytic representations of CR functions. Math. USSR Sb. **27**, 591–623 (1975)
19. Dautov, Sh.A., Kytmanov, A.M.: Granichnye znacheniya integrala tipa Martinalli-Bohnera (The boundary values of an integral of Martinelli–Bochner type. Properties of Holomorphic Functions of Several Complex Variables), pp. 49–54. Inst. Fyz. SO AN SSSR, Krasnoyarsk (1973) (in Russian)
20. Federer, H.: Geometric Measure Theory. Die Grundlehren der mathematischen. Springer, New York/Heidelberg/Berlin (1969)
21. Folland, G.B., Kohn, J.J.: The Neumann Problem for the Cauchy-Riemann Complex. Annals of Mathematics Studies, vol. 75. Princeton University Press/University of Tokyo Press, Princeton/Tokyo (1972)
22. Globevnik, J.: On holomorphic extension from spheres in $\mathbb{C}^2$. Proc. R. Soc. Edinb. Sect. A. **94**, 113–120 (1983)
23. Globevnik, J.: A family of lines for testing holomorphy in the ball of $\mathbb{C}^2$. Indiana Univ. Math. **36**, 639–644 (1987)
24. Globevnik, J.: Meromorphic extensions from small families of circles and holomorphic extensions from spheres. Trans. Am. Math. Soc. **364**, 5857–5880 (2012)
25. Globevnik, J.: Small families of complex lines for testing holomorphic extendibility. Am. J. Math. **134**, 1473–1490 (2012)
26. Globevnik, J., Stout, E.L.: Boundary Morera theorems for holomorphic functions of several complex variables. Duke Math. J. **64**, 571–615 (1991)
27. Govekar, D.: Morera conditions along real planes and a characterization of CR functions on boundaries of domains in $\mathbb{C}^n$. Math. Z. **216**, 195–207 (1994)
28. Griffiths, F., Harris, J.: Principles of Algebraic Geometry. Wiley, New York (1978)
29. Grinberg, E.: A boundary analogue of Morera's theorem in the unit ball of $\mathbb{C}^n$. Proc. Am. Math. Soc. **102**, 114–116 (1988)
30. Harvey, F.R., Lawson, H.B.: On boundaries of complex analytic varieties. Ann. Math. **102**, 223–290 (1975)
31. Hayman, W.K., Kennedy, P.B.: Subharmonic Functions, vol. 1. Academic Press, London (1976)
32. Hua, L.K.: Harmonic Analysis of Functions of Several Complex Variables in the Classical Domains (Translations of Mathematical Monographs). AMS, Providence (1963)
33. Khenkin, G.M.: The method of integral representations in complex analysis. Several Complex Variables I. Encyclopaedia of Mathematical Sciences, vol. 7, pp. 19–116. Springer, New York (1990)
34. Khenkin, G.M., Chirka, E.M.: Boundary properties of holomorphic functions of several complex variables. J. Sov. Math. **5**, 12–142 (1976)
35. Koppelman, W.: The Cauchy integral for differential forms. Bull. Am. Math. Soc. **73**, 554–556 (1967)
36. Koranyi, A.: The Poisson integral for generalized half-planes and bounded symmetric domaines. Ann. Math. **82**, 332–350 (1965)
37. Kosbergenov, S.: O granichnoi teoreme Morera dlya shara i polidiska (On the Boundary Morera Theorem for the Ball and the Polydisk). Complex Analysis and Mathematical Physics. Krasnoyarskii Gos. Universitet, Krasnoyarsk, pp. 59–65 (1998) (in Russian)
38. Kosbergenov, S., Kytmanov, A.M., Myslivets, S.G.: On a boundary Morera theorem for the classical domains. Sib. Math. J. **40**, 506–514 (1999)
39. Krantz, S.G.: Function Theory of Several Complex Variables, 2nd edn. Wadsworth & Brooks/Cole, Pacific Grove (1992)
40. Kuzovatov, V.I.: O nekotoryh semeistvah kompleksnyh pryamyh dostatochnyh dlya golomorfnogo prodolzheniya funktsii (On some families of complex lines sufficient for holomorphic extension of functions). Ufimskii Mat. Zh. **4**, 107–121 (2012) (in Russian)
41. Kuzovatov, V.I., Kytmanov, A.M.: On a boundary analogue of the Forelli theorem. Sib. Math. J. **54**, 841–856 (2013)
42. Kytmanov, A.M.: Holomorphic extension of integrable CR-functions from part of the boundary of the domain. Math. Notes **48**, 761–765 (1990)

43. Kytmanov, A.M.: Holomorphic extension of *CR* functions with singularities of hypersurface. Math. USSR Izv. **37**, 681–691 (1991)
44. Kytmanov, A.M.: On $\bar{\partial}$-Neumann problem for smooth functions and distributions. Math. USSR Sb. **70**, 79–92 (1991)
45. Kytmanov, A.M.: The Bochner–Martilnelli Integral and Its Applications. Birkhäuser Verlag, Basel (1995)
46. Kytmanov, A.M., Aizenberg, L.A.: O golomorfnosti nepreryfnyh funksii, predstavimyh integralom martinelli-Bohnera (The holomorphy of continuous functions that are representable by the Martinelli–Bochner integral). Izv. Akad. Nauk Armyan SSR. Ser. Mat. **13**, 158–169 (1978) (in Russian)
47. Kytmanov, A.M., Myslivets, S.G.: On a certain boundary analogue of Morera's theorem. Sib. Math. J. **36**, 1171–1174 (1995)
48. Kytmanov, A.M., Myslivets, S.G.: On holomorphy of functions representable by the logarithmic residue formula. Sib. Math. J. **38**, 302–311 (1997)
49. Kytmanov, A.M., Myslivets, S.G.: Higher-dimensional boundary analogues of the Morera theorem in problems of analytic continuation of functions. J. Math. Sci. **120**, 1842–1867 (2004)
50. Kytmanov, A.M., Myslivets, S.G.: On families of complex lines sufficient for holomorphic extension. Math. Notes **83**, 500–505 (2008)
51. Kytmanov, A.M., Myslivets, S.G.: Some families of complex lines sufficient for holomorphic extension of functions. Russ. Math. **55**, 60–66 (2011)
52. Kytmanov, A.M., Myslivets, S.G.: O semeistvah kompleksnyh pryamyh dostatochnyh dlya golomorfnogo prodolzheniya funktsii, zadannyh ns granitse oblasti (On the families of complex lines which are sufficient for holomorphic continuation of functions given on the boundary of the domain). J. Sib. Fed. Univ. Math. Phys. **5**, 213–222 (2012) (in Russian)
53. Kytmanov, A.M., Myslivets, S.G.: An analogue of the Hartogs theorem in a Ball of $\mathbb{C}^n$. Math. Nachr. **288**, 224–234 (2015)
54. Kytmanov, A.M., Yakimenko, M.Sh.: On holomorphic extension of hyperfunctions Sib. Math. J. **34**, 1101–1109 (1993)
55. Kytmanov, A.M., Myslivets, S.G., Kuzovatov, V.I.: Minimal dimension families of complex lines sufficient for holomorphic extension of functions. Sib. Math. J. **52**, 256–266 (2011)
56. Landkof, N.S.: Fondations of Modern Potential Theory. Springer, New York/Heidelberg/Berlin (1972)
57. Lavrent'ev, M.A., Shabat, B.V.: Metody teorii funktsii kompleksnogo peremennogo (Methods of the Functions Theory of Complex Variable). Lan', St.-Petersburg (2002) (in Russian)
58. Leray, J.: Fonction de variables complexe: sa représentation comme somme de puissances négatives de fonctions linéaires. Atti Accad. Naz. Lincei. Rend. Cl. Sci. Fis. Mat. Natur. **20**, 589–590 (1956)
59. Leray, J.: Le calcul differential et intrgral sur une variete analytique complexe (Probleme de Cauchy III). Bull. Soc. Math. Fr. **87**, 81–180 (1959)
60. Look, C.H., Zhong, T.D.: An extension of Privalof's theorem. Acta Math. Sin. **7**, 144–165 (1987)
61. Malgrange, B.: Ideals of Differentiable Functions. Tata Institute of Fundamental Research Studies in Mathematics, vol. 3. Oxford University Press, London (1967)
62. Martinelli, E.: Alcuni teoremi integrali per le funzioni analitiche di più variabili complesse. Mem. R. Accad. Ital. **9**, 269–283 (1938)
63. Martinelli, E.: Sopra una dimonstrazione de R.Fueter per un teorema di Hartogs. Comment. Math. Helv. **15**, 340–349 (1943)
64. Myslivets, S.G.: One boundary version of Morera's theorem. Sib. Math. J. **42**, 952–966 (2001)
65. Myslivets, S.G.: Boundary behavior of an integral of logarithmic residue type. Russ. Math. **46**, 43–48 (2002)
66. Nagel, A., Rudin, W.: Moebius-invariant function spaces on balls and spheres. Duke Math. J. **43**, 841–865 (1976)

67. Pinchuk, S.I.: Holomorphic maps in $\mathbb{C}^n$ and a problem of holomorphic equivalence. Several Complex Variables III. Encyclopaedia of Mathematical Sciences, vol. 9, pp. 173–199. Springer, New York (1989)
68. Prenov, B.B., Tarkhanov, N.N.: Martinelli–Bochner singular integral. Sib. Math. J. **33**, 355–359 (1992)
69. Range, R.M.: Holomorphic Functions and Integral Representations in Several Complex Variables. Springer, New York/Heidelberg/Berlin (1986)
70. Rossi, H., Vergne, M.: Equations de Cauchy–Riemann tangentiells, assosies a un domaine de Siegel. Ann. Ecole Norm. Super. **9**, 31–80 (1976)
71. Rudin, W.: Function Theory in the Unit Ball of $\mathbb{C}^n$. Springer, New York/Heidelberg/Berlin (1980)
72. Semenov, A.M.: Holomorphic extension from spheres in $\mathbb{C}^n$. Sib. Math. J. **30**, 445–450 (1989)
73. Shabat, B.V.: Introduction to Complex Analysis. Part 2: Functions of Several Complex Variables. AMS, Providence (1992)
74. Sobolev, S.L.: Introduction to the Theory of Cubatute Formulas. Gordon and Breach, Philadelphia (1992)
75. Sommen, F.: Martinelli–Bochner type formulae in complex Clifford analysis. Z. Anal. Anwend. **6**, 75–82 (1987)
76. Stein, E.M.: Boundary Behavior of Holomorphic Functions of Several Complex Variables. Princeton University Press, Princeton (1972)
77. Stein, E.M., Weiss, G.: Introduction to Fourier Analysis on Euclidian Spaces. Princeton University Press, Princeton (1975)
78. Stout, E.L.: The boundary values of holomorphic functions of several complex variables. Duke Math. J. **44**, 105–108 (1977)
79. Tumanov, A.E.: Extension of *CR* functions into a wedge from a manifold of finite type. Math. USSR Sb. **64**, 129–140 (1989)
80. Vasilevskii, N.L., Shapiro, M.V.: Some questions of hypercomplex analysis. In: Proceedings of the Conference on Complex Analysis and Applications (Varna 1987). Publishing House Bulgarian Academy of Sciences, Sofia, pp. 523–531 (1989)
81. Vladimirov, V.S.: Methods of the Theory of Functions of Many Complex Variables. MIT Press, Cambridge (1966)
82. Vladimirov, V.S.: Equations of Mathematical Physics. Mir, Moscow (1984)
83. Volkov, E.A.: Boundaries of subdomains, Hölder weight classes and the solutions of the Poisson equation in these classes. Proc. Steklov Inst. Math. **117**, 75–99 (1972)
84. Wells, R.O.: Differential Analysis on Complex Manifolds. Springer, New York/Heidelberg/Berlin (1980)

Index

© Springer International Publishing Switzerland 2015
A.M. Kytmanov, S.G. Myslivets, *Multidimensional Integral Representations*,
DOI 10.1007/978-3-319-21659-1

Printed in the United States
By Bookmasters